Stress Tolerance of Fungi

MYCOLOGY SERIES

Editor

Paul A. Lemke

Department of Botany and Microbiology
Auburn University
Auburn, Alabama

1. *Viruses and Plasmids in Fungi*, edited by Paul A. Lemke
2. *The Fungal Community: Its Organization and Role in the Ecosystem*, edited by Donald T. Wicklow and George C. Carroll
3. *Fungi Pathogenic for Humans and Animals* (in three parts), edited by Dexter H. Howard
4. *Fungal Differentiation: A Contemporary Synthesis*, edited by John E. Smith
5. *Secondary Metabolism and Differentiation in Fungi*, edited by Joan W. Bennett and Alex Ciegler
6. *Fungal Protoplasts*, edited by John F. Peberdy and Lajos Ferenczy
7. *Viruses of Fungi and Simple Eukaryotes*, edited by Yigal Koltin and Michael J. Leibowitz
8. *Molecular Industrial Mycology: Systems and Applications for Filamentous Fungi*, edited by Sally A. Leong and Randy M. Berka
9. *The Fungal Community: Its Organization and Role in the Ecosystem, Second Edition*, edited by George C. Carroll and Donald T. Wicklow
10. *Stress Tolerance of Fungi*, edited by D. H. Jennings

Additional Volumes in Preparation

Stress Tolerance of Fungi

edited by

D. H. Jennings

The University of Liverpool
Liverpool, England

Marcel Dekker, Inc. **New York • Basel • Hong Kong**

Library of Congress Cataloging-in-Publication Data

Stress tolerance of fungi / edited by D. H. Jennings.
p. cm. -- (Mycology series; 10)
Includes bibliographical references and index.
ISBN 0-8247-9061-8
1. Fungi--Effect of stress on. 2. Fungi--Ecophysiology.
I. Jennings, D. H. II. Series: Mycology series; v. 10.
QK604.2.E28S78 1993
589.2'045222--dc20 93-1187
CIP

The publisher offers discounts on this book when ordered in bulk quantities. For more information, write to Special Sales/Professional Marketing at the address below.

This book is printed on acid-free paper.

Marcel Dekker, Inc.
270 Madison Avenue, New York, New York 10016

Current printing (last digit):
10 9 8 7 6 5 4 3 2 1

PRINTED IN THE UNITED STATES OF AMERICA

Series Introduction

Mycology is the study of fungi, that vast assemblage of microorganisms which includes such things as molds, yeasts, and mushrooms. All of us in one way or another are influenced by fungi. Think of it for a moment—the good life without penicillin or a fine wine. Consider further the importance of fungi in the decomposition of wastes and the potential hazards of fungi as pathogens to plants and to humans. Yes, fungi are ubiquitous and important.

Mycologists study fungi either in nature or in the laboratory and at different experimental levels ranging from descriptive to molecular and from basic to applied. Since there are so many fungi and so many ways to study them, mycologists often find it difficult to communicate their results even to other mycologists, much less to other scientists or to society in general.

This series establishes a niche for publication of works dealing with all aspects of mycology. It is not intended to set the fungi apart, but rather to emphasize the study of fungi and of fungal processes as they relate to mankind and to science in general. Such a series of books is long overdue. It is broadly conceived as to scope, and should include textbooks and manuals as well as original and scholarly research works and monographs.

The scope of the series will be defined by, and hopefully will help define, progress in mycology.

Paul A. Lemke

Preface

This volume has been produced for those who wish to know more about how fungi respond to their environment—whether in the laboratory, the field, or in industry.

There is much information about the ecology of fungi. A great deal of that information is of a qualitative kind and deals with situations that are often not very precisely defined. Inevitably, the information conveys scant help in understanding how a fungus copes with the environmental pressures under consideration. This volume—the first specifically for fungi—focuses on clear situations, the so-called stressful ones, where one environmental variable is preeminent. Because there is this preeminence, these situations provide the best clues to the mechanisms that allow a fungus to adapt to environmental change. Essentially the book is about the physiology and biochemistry of fungi but with an environmental slant. However, in view of the proven power of molecular biology to probe how a fungus functions, there is much emphasis on the molecular aspects of the topics under consideration.

Beginning with an introductory chapter that puts the various topics in a general context, topics covered include: phenotypic plasticity; heat shock proteins; metal tolerance; membrane transport of toxic metals; oligotrophic growth; tolerance to freezing and thawing, to sulfur dioxide, and to ethanol; halophilic, xerophilic, and anaerobic fungi; drying and rewetting of lichens; and DNA repair.

Though the book consists of a series of relatively discrete topics written by acknowledged experts, it should not be difficult for fungal ecologists as well as plant pathologists and biotechnologists to see how some of the approaches can be extended to their own particular areas of research. Equally, physiologists, biochemists, and molecular biologists working with fungi will be able to see how the techniques that they routinely use can be applied to challenging problems of how a particular species responds to its environment.

This book provides important pointers as to how fungal ecology will develop in the future.

D. H. Jennings

Contents

Contributors

Lennart Adler Department of General and Marine Microbiology, University of Göteborg, Göteborg, Sweden

Anders Blomberg Department of General and Marine Microbiology, University of Göteborg, Göteborg, Sweden

Robert Brambl Department of Plant Biology and Plant Molecular Genetics Institute, University of Minnesota, St. Paul, Minnesota

M. X. Caddick Department of Genetics and Microbiology, The University of Liverpool, Liverpool, England

Ailsa D. Hocking Food Research Laboratory, Division of Food Science and Technology, Commonwealth Scientific and Industrial Research Organisation, North Ryde, New South Wales, Australia

D. H. Jennings Department of Genetics and Microbiology, The University of Liverpool, Liverpool, England

Naresh Magan Biotechnology Centre, Cranfield Institute of Technology, Cranfield, Bedford, England

Prashant Mishra Biochemical Engineering Research Centre, Indian Institute of Technology, New Delhi, India

Colin G. Orpin Division of Tropical Crops and Pastures, Commonwealth Scientific and Industrial Research Organisation, St. Lucia, Queensland, Australia

Fekret Osman Department of Genetics and Microbiology, The University of Liverpool, Liverpool, England

Nora Plesofsky-Vig Department of Genetics and Cell Biology and Department of Plant Biology, University of Minnesota, St. Paul, Minnesota

David H. S. Richardson Saint Mary's University, Halifax, Nova Scotia, Canada

I. S. Ross Department of Biological Sciences, University of Keele, Keele, Staffordshire, England

David Smith Genetic Resources Reference Collection, International Mycological Institute, Egham, Surrey, England

Peter Strike Department of Genetics and Microbiology, The University of Liverpool, Liverpool, England

A. Brian Tomsett Department of Genetics and Microbiology, The University of Liverpool, Liverpool, England

Milton Wainwright Department of Molecular Biology and Biotechnology, The University of Sheffield, Sheffield, England

Understanding Tolerance to Stress: Laboratory Culture Versus Environmental Actuality

D. H. Jennings

The University of Liverpool
Liverpool, England

This volume is about how fungi are able to live in what might be called stressful situations. To be honest, the term "stress" is not a good term to use (1). It is an anthropomorphism (along with words like adaptation and strategy) and emotive. The plant ecologist J. L. Harper (1), has said: " 'Stress' carries over easily from social and medical science to imply in biology 'what I don't think I would like if I was a (buttercup, kangaroo, flea, beetle, etc.).' " In another context, this time medical, G. Pickering (2) said: ". . . stress, again an old word, but now part of the popular modes of expression in jargonise, I am never sure what it means. The classical language of biology has two expressions—namely *stimulus* to describe change in environment and *response* to describe the resulting change in the organism." While all this is true, and I shall be presenting information for fungi later on which expands the point made by Pickering, the term stress has its value, if only as a useful first stage in the definition of more precise concepts, which in themselves need to generate a more objective view of the organism and at the same time be underpinned by hard experimental evidence.

I believe the term stress is useful now, because there is a good understanding by mycologists of what is under consideration. At one level, the term refers to those situations that restrict or prevent the growth and reproduction of the great variety of fungi. Those fungi that do live in stressful situations might, to use more general microbiological jargon, be called "extremophiles" (3). Extreme habitats are conventionally those in which there are high or low temperatures, low or high pH, low water activity, high salt concentrations, low nutrient availability, and, for almost fungi, low oxygen partial pressures. But at another level, speaking bluntly, a stressful environment is one that differs significantly from the normal culture conditions of an aqueous medium containing relatively high concentrations of nutrients, having a median pH and good aeration and maintained at laboratory (or possibly higher) temperature. A comparison of this sort can be considered analogous to that between the terrestrial green plant mesophyte and those plants that grow in saline (halophytes) or dry (xerophytes) conditions.

But to transfer these terms from botany to mycology would be dangerous. For the flowering plant, it is easy to define the environment. This is because the size of the organism means that the environment is essentially described by the bulk properties of air and soil and the spectral characteristics of the biologically active radiation. These properties and characteristics are easily determined. Thus, by and large, environments can be distinguished one from the other on the basis of bulk features. If the level of discrimination is broad, groups of plant growing in one environment (habitat) rather than another can be distinguished readily. Such groups provide candidates for physiological study. It is in such circumstances that the concept of the mesophyte is important, since plants in this group provide the point of comparison from which to assess how a more extreme habitat might require certain features in a flowering plant, if it is to grow there.

For fungi, discrimination on a similar basis is less easy. Because of their size, fungi live in environments that are more circumscribed, often considerably different from the bulk properties of the habitat in which a particular fungus resides. Further, again as a consequence of their size and also as a result of growth outward from a point source, fungi are able to change their immediate environment by excretion of protons, by sequestration of toxic metals, and in certain instances by the production of water either as a result of metabolic activity or via translocation (4). Thus characterizing a fungus in terms of its environment is not easy. The niche in which a particular fungus lives may have features masked by the bulk features of the substratum and may have been changed by the fungus itself. More correctly, the niche for a fungus is an amalgam of the local chemical and physical features and the physiological (genetic) capabilities of the fungus. While some features of the environment can be described, particularly the major carbon and nitrogen sources and the temperature, often we are ignorant of the local conditions sensed by a fungus. Modern technology should allow us to probe more successfully than hitherto those local (microscopic) conditions (5), but even so there are considerable practical difficulties to achieving a realistic assessment of the chemical and physical characteristics of the environment in which a fungus is living.

Therefore, it is not surprising that those studying how fungi adapt to their environment should focus on those situations in which a change in a major feature (a "stress") brings about clear and distinct changes in the fungal population. It is this factor that has led to the choice of some of the issues addressed in this volume. Essentially, the environment is assumed: that nutrients are not limiting and the growth of fungi is limited or inhibited by the presence of one particular factor, which might be either salinity (Chapter 9) or low water activity (Chapter 10) or toxic metals or sulfur dioxide and related chemical species (Chapter 7), or by very low (Chapter 6) and high temperatures. Nevertheless, the assumption just made must at some stage be addressed; Wainwright argues forcibly in Chapter 5 that, as seems to be the case in soil, fungi can be living in environments with an insufficiency of nutrients, when considered in terms of the rapid production of biomass.

There is a corollary to this point I am making about the environment of a fungus and it concerns the choice of species for experimental investigation. There have been numerous studies in which the fungi of a particular habitat have been isolated and their taxonomy established. It is not appropriate to dwell here on the great problems of quantifying the biomass of a fungus within that habitat. At the least, lists of species can be produced and the temporal change of species composition can be determined. It is arguable whether any of the species so isolated should be selected for experimental investigation. Certainly, important physiological investigations have been made on fungal species isolated from natural environments. Much of our knowledge of the effect of salinity on fungi has come from the study of three species so isolated, *Debaryomyces hansenii, Dendryphiella salina,* and *Zygosaccharomyces rouxii* (Chapter 9). But these fungi have been studied now for several decades. There is now sufficient known about each of these three species to produce a coherent, if simplistic, picture of the mechanisms that underpin the ability of these three species to grow in a medium of high salinity (6,7). However, it needs to be remembered that our understanding of the physiology of these three fungi with respect to salinity owes much to studies on *Neurospora crassa* and *Saccharomyces cerevisiae*, for which fungi we have the most detailed knowledge about how potassium and sodium move between the medium and the cytoplasm across the plasma membrane. That knowledge has been gained by a range of biophysical, biochemical, molecular, and, to a much lesser extent, genetic techniques. Some of the techniques used can be transferred readily for use with other fungi. Others are much more difficult to transfer, an outstanding example being the electrophysiological procedures used with *N. crassa*, which have depended on the availability of particularly wide-diameter hyphae and the production of large cells, both capable of allowing the insertion of one or more microelectrodes. Of course, had the studies on *N. crassa* and *S. cerevisiae* involved detailed genetic analysis, the information obtained could not be readily applied to another fungus without the presence of considerable knowledge about the genetics of that other fungus.

Thus selection of a fungus from a range of species isolated from a habitat in which a particular chemical or physical feature or set of features is thought to govern which fungi are able to live there should not be made lightly. There has to be much investment of effort before sufficient physiological information of relevance to ecological understanding is available. As just indicated, much effort will be needed because techniques used on one fungus are not necessarily readily applicable to the fungus of choice. There is, however, another and even more important reason. If we are thinking about a stressful environment, the use of the term "stressful" implies comparison with the ability of other environments to support the growth of fungi. Any proper comparison requires quantitative information because ability of a fungus to function in one environment rather than another will depend as much on the rates at which particular biochemical or physiological processes function as on the presence in the fungus of processes or properties not found in other species.

There are, of course, particular environments that support fungal growth with unique features, and progress can only be made by intensive study of the fungi that live there. Those fungi which are symbiotically (*sensu* de Barry) associated with green plants are one such group. Plant pathologists have devoted much effort to understanding how such fungi function. Here, of course, one can be dealing with not only physiological, but also morphological, specialization, the haustorium, the structure used by a fungus to gain nutrients from its green plant host, being a notable example. Striking examples of morphological specialization in a symbiotic association are given in Chapter 12 dealing with the ability of lichens to withstand drying and rewetting. There again, if we are considering fungi living in specialized environments, outstanding is the rumen, since those fungi that can live there do so in the absence of oxygen. These obligately anaerobic species demand study in their own right and their biology is discussed further in Chapter 11.

If we return to the matter of comparing fungi that grow saprophytically and in environments that do not differ very much in kind but do so in degree, then there are considerable virtues to studying the capability of a laboratory organism to adapt to a range of environmental conditions. I have indicated already the power of such an approach via studies on *N. crassa* and *S. cerevisiae* for enhancing our understanding of how fungi might cope with saline conditions. Several of the chapters (7, 8, 9, 10 and 11) in this volume are devoted to this approach. In particular, I would like to draw attention to the genetics and molecular biology of metal tolerance, which concern almost exclusively those fungi that have been used as laboratory organisms, with little if any attention being devoted to their ecology. The virtues of using laboratory organisms for the study of metal tolerance are well highlighted in Chapter 3. Essentially, such study has allowed us to obtain a good idea of the genetic underpinning of the process. The techniques and method of approach are capable of being applied to those fungi that are able to tolerate high concentrations of metals in the natural environment. But more important, DNA technology will allow us to screen the same fungi for those genes or members of the same family of genes that have been identified in the laboratory organisms.

It is a truism to say that the ability of a fungus to grow in a seemingly hostile environment is dependent on its genetic makeup. In reality, it is the phenotype that is the interface between the genome and the environment in which it is present. The environment is not constant, and if the genome is to evolve, there must be the capability of moving in both time and space between environments, the degree of movement being governed not only by the degree of difference between the environments, but also by the ability of the genome to allow the phenotype to vary in response to environmental conditions. Variation of the phenotype in this manner is termed "phenotypic plasticity."

Phenotypic plasticity is well known to botanists (8–11) but rarely considered explicitly by mycologists. That neglect may be because of the focus on genetic variation that is so readily observed. The neglect may have been encouraged also by the knowledge that, when a fungus faces unfavorable environmental conditions, more resistant structures such as spores and sclerotia are formed, removing the need for the fungus to adapt vegetatively. But a major reason for the lack of emphasis on phenotypic plasticity lies in the infrequent use of steady-state conditions for growing fungi. Most of our knowledge of phenotypic plasticity in fungi comes from those studies concerned with the presence or absence of nutrients in the medium. The use of such conditions has allowed us to discern how genes are switched on or off. Indeed such studies have produced a powerful body of information of considerable importance for understanding of phenotypic plasticity in fungi, as is demonstrated in Chapter 1. There, we are reminded that it is not only the phenotypic response, but also accurate perception of environmental change, which is important for the survival of the individual fungus. This is, of course, the important theme uppermost in the mind of Pickering (2) quoted at the start of this chapter.

Chapter 1 indicates how environmental change can be perceived within the protoplast by the genetic material. However, the plasma membrane is another important locus of perception (12). The location there of selective transport systems governs the entry of molecules into the protoplasm. Order of utilization of a range of similar substrates is governed by the affinity of the carrier (13). There are strong arguments for the plasma membrane being the locus for perceiving the vectorial stimulus for a change in the direction of growth. While oxygen gradients primarily control direction, it seems more likely that a hypha, via proton cotransport systems at the apex, senses gradients of protons consequent on metabolism governed in form and rate by the external oxygen concentration (14). We need to know much more about the role of the plasma membrane in the perception and monitoring of the environment by fungi.

While the plasma membrane can be important in perceiving the environment, we need to remember that the actual environment perceived by that membrane is governed by the properties of the wall. It is well established that the wall can bind metals (15). The wall of the marine fungus *Dendryphiella salina* has been shown to be 40% of the hyphal volume. It has been suggested that the wall maintains a pH outside the plasma membrane of *D. salina* lower than that in the bulk seawater to favor the proton economy on which transport in this fungus is based (7). So we can think of the wall as buffering against the vagaries of the external environment. Thus the wall may help to ensure the signal reaching the site of perception, i.e., the plasma membrane, is not obscured by too much noise. But, of course, the plasma membrane is not only a major site for perceiving the external environment. The membrane, via the various transport processes within it, is of key importance in helping to maintain a relatively constant internal milieu for the fungus, in spite of unfavorable or variable concentrations of solutes in the external medium (see Chapters 4 and 9).

Now I want to return to a matter alluded to earlier, namely the lack of studies of fungal growth under steady-state conditions or the response of a fungus as it is exposed to a change from one steady state to another. There are, of course, problems of growing filamentous fungi in continuous culture, but even for yeasts the number of studies involving this method of growth is limited. There ought to be more, because it is only by the use of continuous culture that one can obtain unambiguous information about the effect of an environmental change. Indeed, it is more possible now to obtain information for filamentous fungi, because there are species available, such as *Fusarium graminearum*, which can readily be grown in a chemostat (16).

Of the little we know about the growth of fungi in continuous culture and exposed to stress, one finding stands out. In studies on the marine yeast, it has been shown that *Debaryomyces hansenii* grown under glucose-limiting conditions and salinity moves from carbon limitation to energy limitation as the growth rate increases (17). The relationship between yield of biomass and growth rate is such that it is clear that carbon is being channeled away from anabolic to catabolic reactions. Under such conditions, energy generation can still occur at a high rate even though cell synthesis is curtailed. This means that there will be an increase in what have been termed "slip-mechanisms" (18) or "growth-unassociated (energy-spilling)" reactions (19). Probably the best term is "energy-spilling reactions" (20). The essential point, well known from studies of the growth of prokaryotes, is that there is uncoupling of growth and energy production. From the point of view of this volume, energy spillage becomes more important with an increasing degree of stress.

There are a number of ways in which energy spillage can take place (Table 1). Here I would like to focus attention on energy spillage by futile cycles, which are part of the metabolic machinery of the fungus. Futile cycles occur when two compounds can be interconverted by two different irreversible reactions. When the enzymes responsible for both reactions are functional, there is a futile cycling in which there is dissipation of energy. To date, four metabolic cycles have been suggested to have the role of energy dissipation in fungi (Fig. 1), namely: (a) a trehalose cycle in heat-shocked cells of *S. cerevisiae* (21); (b) the mannitol cycle in many fungi (22); (c) a glycerol cycle in *D. hansenii* growing under saline conditions (22,23); (d) the glutamine cycle in *N. crassa* under carbon limitation (24). It is arguable whether these cycles are operating in a strictly futile mode, since there can be net synthesis of the molecules named above. Indeed the accumulation of these compounds, certainly this is so for glycerol in *D. hansenii* under saline conditions, may be important in maintaining the integrity of the fungus. In that respect it is intriguing that the four compounds are all so-called compatible solutes. Be

Table 1 Possible Processes by Which Fungi Might Expend Energy on Nongrowth Processes Under Conditions of Stress (See Ref. 20), with Some Indication of the Area of Relevance, Particularly for This Volume

Process	Area of relevance
Maintenance of ion gradients	See Chapter 9
Futile cycles	
Metabolic	This chapter
Membrane-associated	See Chapters 4 and 9
Production of metabolites that are uncouplers	Production of organic acid Production of ethanol? (see Chapter 8)
Modification of respiratory chain and reduction of number of molecules of ATP synthesized	
Metabolic uncoupling	Catabolism of glucose by fermentation

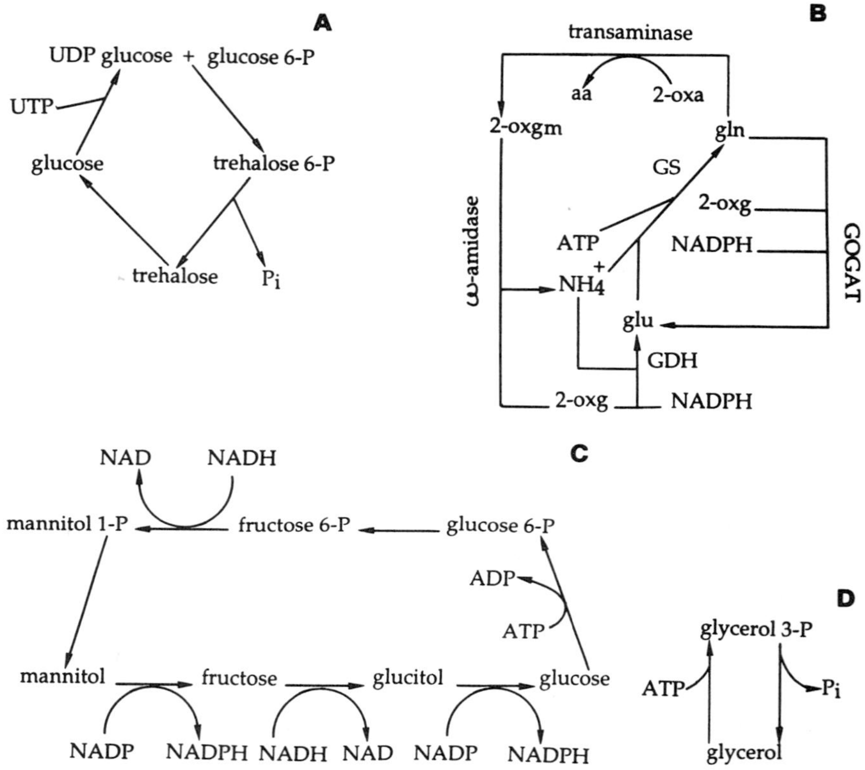

Figure 1 Futile cycles that have been proposed to be important for energy dissipation in fungi (see text for details). (A) Trehalose cycle. (B) Glutamine cycle (glu, glutamate; gln, glutamine; 2-oxg, 2-oxoglutarate; 2-oxa, 2-oxoacid; aa, amino acid; 2-oxgm, 2-oxoglutaramate). (C) Mannitol cycle. (D) Glycerol cycle.

that as it may and while the cycling may not be strictly futile, there is considerable cyclical turnover with concomitant dissipation of energy. Other metabolic processes associated with responses to environmental stress may also be involved in dissipation of energy. Thus the production of phytochelatins by such fungi as *Schizosaccharomyces pombe* in response to the presence of toxic metals involves metabolic cycles that could well act in a futile manner (22) (Chapter 3).

Considerable reduction in the growth rate of a fungus under nutrient unbalanced conditions is frequently accompanied by the production of secondary metabolites, those compounds with no obvious role in the internal economy of the organism (25–27). Many of these compounds have a physiological role and thus may protect the fungus from other organisms. This has yet to be properly demonstrated for fungi. Secondary metabolites may have a more subtle protective function. Certainly, some secondary metabolites such as melanins protect fungi possessing them against ultraviolet radiation (28). A number of secondary metabolites, particularly peptides, are known to bind metal ions. It is possible that some peptides have a role like siderophores with respect to iron (29); namely, the peptide is able to capture the metal and transport it more readily into the fungus. It can be argued also that, in being bound in this way, metal ions are removed from the local environment and as such cannot be used as cofactors in the metabolism of competing microorganisms (30).

Differentiation is at its simplest a change in cell chemistry. Genes functioning in the growing organism may be switched off and others switched on. The change in cell chemistry that takes place as a result of a change in the active genome underlies change in morphology. Secondary metabolism can be considered to be akin to differentiation and, for the biosynthesis of antibiotics such as penicillin and cephalosporin, a picture is beginning to emerge as to how the various processes are regulated both biochemically and genetically (31).

Low growth rates or cessation of growth often leads to spore formation. Reduction or cessation of growth can be the result of insufficient nutrients, exposure to light, or a change in temperature (32–34). Spores are not only reproductive propagules and agents of dispersion, but they are much more tolerant to environmental extremes. Thus two consequences of reduced growth rate, secondary metabolite production and spore formation, can be seen as mechanisms by which a fungus protects itself against an unfavorable environment. This interplay of themes that I have alluded to in considering the relationship between growth and stress appears to be becoming more pointed from the studies of heat shock proteins. Chapter 2, which deals with such proteins, casts new light on some of the topics just considered. The theme of heat shock proteins is embroidered in Chapter 8, in which the production of these proteins in yeast in the presence of ethanol is given detailed consideration.

It is implicit in the use of the term stress that an organism exposed to various kinds of stress will suffer damage. Damage to membranes can lead to movement of solutes to unfavorable locations both within and out of the protoplasm, as evidenced by the loss of polyols from lichens on dehydration (Chapter 12). Such damage is repaired on rewetting. Damage to DNA has both short-term and important long-term consequences. But fungi, like other organisms, have developed mechanisms to repair damage to DNA, which are considered in Chapter 13. The notion of repair mechanisms has wider currency among those studying prokaryotics (35). We need to know whether some of the repair processes operative in bacteria have their counterparts among the fungi. Chapter 6 describes the extent to which fungi can be subjected to low temperatures. The need now is to discover, at the biochemical and molecular level, what lesions might be generated at low temperatures, or indeed in any other seemingly unfavorable environment, and whether repair is possible.

REFERENCES

1. Harper, J. L. (1982). After description, in *The Plant Community as a Working Mechanism* (E. I. Newman, ed.), Blackwell Scientific Publications, Oxford, p. 11.
2. Pickering, G. (1961). Language: the lost tool of learning in medicine and science, *Lancet 2*: 115.
3. Edwards, C. (1990). Introduction, in *Microbiology of Extreme Environments* (C. Edwards, ed.), Open University Press, Milton Keynes, p. ix.
4. Jennings, D. H. (1987). Translocation in fungi, *Biol. Rev. 62*: 215.
5. Jennings, D. H. (1990). Nutrient acquisition by fungi—the relation between physiological understanding and ecological reality, *Proc. Indian Acad. Sci. 100*: 153.
6. Brown, A. D. (1990). *Microbial Water Stress Physiology*, Wiley, Chichester.
7. Clipson, N. J. W., and Jennings, D. H. (1992). *Dendryphiella salina* and *Debaryomyces hansenii*: models for ecophysiological adaptation to salinity by fungi which grow in the sea, *Can. J. Bot. 70*: 2097.
8. Bradshaw, A. D. (1965). Evolutionary significance of phenotypic plasticity in plants, *Adv. Genet. 13*: 115.
9. Jennings, D. H. (1977). Introduction, *Symp. Soc. Exp. Biol. 31*: 1.
10. Jennings, D. H., and Trewavas, A. J. (1986). *Plasticity in Plants*, 40th Symposium of the Society for Experimental Biology, Company of Biologists, Cambridge.
11. Smith, H. (1990). Signal perception, differential expressions within multigene families and the molecular basis of phenotypic plasticity, *Plant Cell Environ. 13*: 585.
12. Jennings, D. H. (1974). Sugar transport in fungi: an essay, *Trans. Br. Mycol. soc. 62*: 1.
13. Hofer, M., and Dahle, P. (1972). Glucose repression of inducible enzyme synthesis in the yeast *Rhodotorula gracilis*. Effect of cell-membrane transport, *Eur. J. Biochem. 29*: 326.
14. Jennings, D. H. (1991). The spatial aspects of fungal growth, *Sci. Prog. Oxford 75*: 141.
15. Gadd, G. M., and White, C. (1989). Heavy metal and radionuclide accumulation and toxicity in fungi and yeasts, in *Metal-Microbe Interactions* (R. K. Poole and G. M. Gadd, eds.), IRL Press, Oxford, p. 19.
16. Trinci, A. P. J. (1992). Myco-protein: a twenty-year overnight success story, *Mycol. Res. 96*: 1.

17. Burke, R. M., and Jennings, D. H. (1990). The effect of sodium chloride on the growth characteristics of the marine yeast *Debaryomyces hansenii* in batch culture and in continuous culture under carbon and under potassium limitation, *Mycol. Res. 94*: 378.
18. Neijssel, O. M., and Tempest, D. W. (1976). Bioenergetic aspects of aerobic growth of *Klebsiella aerogenes* NCTC 418 in carbon-limited and carbon-sufficient chemostat culture, *Arch. Microbiol. 107*: 215.
19. Tempest, D. W., and Neijssel, O. M. (1980). Growth yield values in relation to respiration, in *Diversity of Bacterial Respiratory Systems*, Vol. 1 (C. J. Knowles, ed.), CRC Press, Boca Raton, FL, p. 1.
20. Tempest, D. W., and Neijssel, O. M. (1984). The status of Y_{ATP} and maintenance energy as biologically interpretable phenomena, *Annu. Rev. Microbiol. 38*: 459.
21. Hottiger, T., Schmutz, P., and Wiemken, A. (1987). Heat-induced accumulation and futile cycling of trehalose in *Saccharomyces cerevisiae, J. Bacteriol. 169*: 5518.
22. Jennings, D. H., and Burke, R. M. (1990). Compatible solutes—the mycological dimension and their role as physiological buffering agents, *New Phytol. 116*: 277.
23. Adler, A., Blomberg, A., and Nilsson, A. (1985). Glycerol metabolism and osmoregulation in the salt-tolerant yeast *Debaryomyces hansenii, J. Bacteriol. 162*: 300.
24. Mora, J. (1990). Glutamine metabolism and cycling in *Neurospora crassa, Microbiol. Rev. 54*: 293.
25. Turner, W. B. (1971). *Fungal Metabolites*, Academic Press, London.
26. Turner, W. B., and Aldridge, D. C. (1983). *Fungal Metabolites II*, Academic Press, London.
27. Bu'lock, J. (1974). Secondary metabolism in fungi and its relationship to growth and development, in *The Filamentous Fungi*, Vol. I, *Industrial Mycology* (J. E. Smith and D. R. Berry, eds.), Edward Arnold, London, p. 33.
28. Arpin, N., and Bouillant, M-L. (1988). Métabolites secondaires fungiques: diversité de structures mais unité de fonction, la protection, *Cryptogam. Mycol. 9*: 267.
29. Winkelmann, G. (1991). Importance of siderophores in fungual growth, sporulation and spore germination, in *Frontiers in Mycology* (D. L. Hawksworth, ed.), C.A.B. International, Kew, Surrey, p. 49.
30. Jennings, D. H. (1987). The medium is the message, *Trans. Br. Mycol. Soc. 89*: 1.
31. Sanchez, S., Flores, M. E., and Demain, A. L. (1988). Nitrogen regulation of penicillin and cephalosporin fermentations, in *Nitrogen Source Control of Microbial Processes* (S. Sanchez-Esquival, ed.), CRC Press, Boca Raton, FL, p. 122.
33. Smith, J. E. (1977). Asexual sporulation in filamentous fungi, in *The Filamentous Fungi*, Vol. III, *Developmental Mycology* (J. E. Smith and D. R. Berry, eds.), Edward Arnold, London, p. 214.

34. Lysek, G. (1984). Physiology and ecology of rhythmic growth and sporulation in fungi, in *The Ecology and Physiology of the Fungal Mycelium* (D. H. Jennings and A. D. M. Rayner, eds.), Cambridge University Press, Cambridge, p. 323.
35. Hurst, A., and Nasim, A. (1984). *Repairable Lesions in Microorganisms*, Academic Press, London.

1

Perception and Response: Phenotypic Plasticity in Fungi

M. X. Caddick

The University of Liverpool
Liverpool, England

I. INTRODUCTION

Fungi represent a very diverse kingdom; consequently, some will have evolved to very specific environments, while others will actively colonize environments that are physically, chemically, and biologically diverse. Irrespective of this, all fungi will encounter variation in their environment and will therefore have evolved to some extent to cope with such change. The response will vary depending on the organism, and the environment and the rate of change, and may include the initiation of specific developmental processes, the induction or repression of anabolic or catabolic pathways, and so forth. Ultimately the survival of an organism will depend on its ability to accurately perceive the environment and respond appropriately.

Stress is a relative term, and to some extent the organism that can best obviate stress by perceiving environmental change and respond, both rapidly and appropriately, will be the one most likely to survive. Consequently the mechanisms underlying such processes will have had severe selection pressures applied to them. Not surprisingly, the systems that have evolved are found to be both precise and sensitive and, as studies continue, are being found to be both intricate and multifaceted.

The phenotype of any organism is defined by the interaction of the genome with the environment. With few exceptions, the genome is not altered in response to environmental change (see Section III.A), but an organism may appear phenotypically distinct in different environments. Much of the variation seen, the phenotypic plasticity of the organism, is the consequence of altered gene expression. Studies into the regulation of gene expression in fungi represent a major area of genetic research. Such studies are not merely of interest to mycologists, but represent an invaluable model for investigating similar processes in the full range of eukaryotes.

I have structured this chapter with the intention of giving a general view of regulation. To give an insight into the intricacies of specific regulatory responses, I have concentrated on just a few of the best-documented examples. Of the systems chosen, I have not restricted myself to those that are primarily concerned with the avoidance or limitation of stress. Rather I have concerned myself with mechanisms that can account for phenotypic change, as the underlying mechanisms are likely to have been adapted for different roles. Inevitably this area of study is dominated by a few fungi, principally *Saccharomyces cerevisiae, Schizosaccharomyces pombe, Aspergillus nidulans*, and *Neurospora crassa*, and in general they provide the basis of our current understanding. As my aim is to give a general perspective, describing common themes and not attempting the impossible task of giving a definitive view of current knowledge, this review is in general limited to these organisms. An unavoidable fact is that regulatory systems, though diverse and intricate, have been well conserved in many respects, not merely within species and genera, but within the eukaryotes as a whole. The consequence of this, for those concerned with a particular fungus, is that much of the information gleaned from another well-studied organism may readily be translated to their own particular one.

II. THE COMPONENTS OF REGULATION

Three elements are essential for an organism in order for it to correctly modulate gene expression in response to environmental change: perception, transduction, and response. Monitoring of an organism's environment, whether it be internal or external, is by necessity the first step in any regulated response. This in some cases requires the existence of specific receptors, some of which have been characterized. The initial perception must then be communicated by the receptor, through a process of transduction, to a biological signal. This transfer of information may require separate molecules, sometimes arranged in a cascade, or it may occur within the regulatory molecule itself. Similarly, the monitoring of the environment may in fact be the role of a domain or domains within the regulator. Ultimately the signal is translated into a regulatory response that reflects the perceived environmental change. This response may ultimately act to alter the environment, which will in turn induce an altered response in the organism. The whole process must therefore be regarded as dynamic, requiring continuous monitoring, communication, and response. Delineation between the three elements is not always possible, and the variety of mechanisms is enormous.

III. GENE EXPRESSION

Expression of a gene entails a complex series of events, including transcription, RNA maturation and transport, translation, posttranslational modification, and transport. Each of the steps involves a number of separate activities and inevitably results in possible points of regulation. What has become increasingly apparent is that regulation can and does act throughout. This allows for a precision and coordination that could not be achieved if a single point was used to regulate gene expression.

A. DNA Rearrangement

Until recently, it was generally felt that the basic structure of the genome remained essentially unaltered despite phenotypic change. What has become increasingly clear through both genetic and molecular analysis is that this dogma is not universally true. Increasingly, new examples of a genomic rearrangement are being observed. In addition to viruses and transposable elements that can integrate, delete, and rearrange the DNA and in some cases alter gene expression (e.g., 1), there are specific examples of gene rearrangements occurring in a reversible manner as the basis of specific regulatory systems. A number of well-documented examples have been demonstrated in fungi. The most relevant examples with respect to phenotypic plasticity, and possibly an adaptive response to the environment and stress (2,3), is the morphological switching found in a number of organisms, most notably *Candida albicans* (4). Such switching events are presumed to involve DNA rearrangement allowing the reversible transition from one morphological form to another. Different strains of *C. albicans* vary in their frequency of transition, and the number of distinct morphological forms that the organism can exhibit. The frequency of switching seems to be dependent on the environment and the strain. In one laboratory strain 17 distinct forms have been observed and the frequency of switching between forms varies but is generally between 10^{-2} and 10^{-4} (4). Among the most likely mechanisms for this "switching" would appear to be a reversible rearrangement of the DNA, but there are other possible explanations including aneuploidy. The electrophoretic karyotype of these organisms seems to alter with switching, which would be consistent with gross chromosomal changes (5). It is perhaps worth noting that the gross variation in karyotype between different fungal isolates is common to a number of species (e.g., 6, 7), but whether this is in any way a reversible or controlled process is difficult to ascertain.

By far the best-characterized example of reversible DNA rearrangement is the mating-type switch in *S. cerevisiae* (Fig. 1) (8). In the haploid growth phase, *S. cerevisiae* occurs in two morphologically distinct forms; these are described as the α and *a* mating types. These can readily be distinguished under the light microscope using mating-type-specific pheromones and represent two specific phenotypic forms exhibiting distinct expression of a number of genes. The particularly interesting aspect of this system is that the haploid organism can switch from one mating type to the other, the process being fully and equally reversible. The budding yeast at cell division forms a smaller daughter cell from the larger mother cell. Approximately 80% of the time the mother cell undergoes the mating-type switch, while the daughter cell remains unaltered. Both the progeny of the "switched" haploid cell exhibit the switched phenotype. The ability of a single haploid cell to produce a population of both *a* and α mating-type cells, which are then capable of undergoing cell fusion to produce a diploid, explains the apparent homothalic nature of the organism. An exception to this is the occurrence of mutations in the *HO* gene, which lead to an inability to switch (9). These

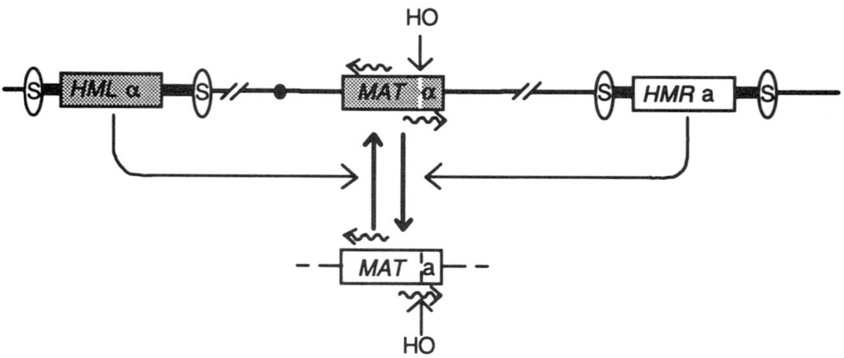

Figure 1 The mating-type switch in *S. cerevisiae*. The *MAT* loci can bear two alleles, *MATa* or MATα, which in the haploid determine the mathing type. Both alleles are expressed (⇝) and are substrates for cleavage by the sequence-specific endonuclease HO. *HML*α and *HMRa* are silent copies of the α and *a* alleles, respectively, being neither expressed nor digested by HO. This is accomplished owing to the presence of flanking sequences known as silencers (S) which are complexed with a number of proteins and alter the local chromatin structure (heavy line). To switch the mating type *HML*α acts as a donor to *MATa* and *HMRa* acts as a donor to *MAT*α. *HML*α is on the left arm and both *MAT* and *HMRa* are on the right arm of chromosome 3.

strains are heterothalic as they remain in either the α or *a* mating type. The *HO* gene has been shown to encode a restriction endonuclease that has a 20-bp recognition sequence (10). Expression of this gene occurs only in the haploid (11) and only in the mother cell (12), late in G1. Its expression leads to a double-stranded break in the *MAT* loci (13), which is the critical step in initiating the morphological switch (8). *HO* expression is therefore under complex regulation, relating to both the cell cycle and cell type, and as such has been studied extensively as a model of developmental regulation (14). The *HO* product has three potential sites of action in the genome, only one of which is "accessible," and this is at the *MAT* locus. The *MAT* locus contains one of two alleles, *a* or α, and the phenotype of the organism is defined by the allele being expressed at this locus. The other two sites, which can be described as "inaccessible," occur in the *HML* and *HMR* loci, which bear silent copies of α and *a*, respectively (15). These loci contain all the information necessary for the expression of the genes, but are not expressed owing to flanking DNA sequences described as silencers (16). The silencers represent binding sites for at least two proteins, but additionally require four specific gene products to suppress gene expression (17). This is apparently achieved, in part at least, by altering the chromatin structure, thus making the DNA inaccessible to the proteins necessary for gene expression or restriction (see Section III.C).

The transiently expressed *HO* gene product produces a double-strand cut in the *MAT* locus (13). This appears to catalyze a gene conversion event that involves degradation of the resident *MAT* allele and subsequent replacement by the alternative allele, which is achieved by utilizing the silent copy as the template (18). *MATa* expression leads to repression of the α-specific genes and expression of the *a*-specific genes, while *MAT*α acts conversely. Derepression of diploid-specific loci is dependent on simultaneous expression of both *MATa* and *MAT*α, which, in the wild type, occurs only in the diploid. Switching mechanisms with varying numbers of mating-type alleles have also been demonstrated in other fungi, e.g., *S. pombe* (8).

It is therefore apparent that rearrangement of DNA is a feasible mechanism for altering gene expression and morphology, but these rearrangements are only one part of the regulatory mechanism involved. Although there are a number of proven examples and even more putative examples, I suggest that these types of mechanism are rare in respect to other more conventional systems, which primarily act at, or subsequent to, transcription.

B. DNA Modification

DNA is subject to chemical and physical modifications, in particular methylation of cytosine residues (19) and torsional stress caused by the unwinding and winding of DNA. The altered physical and chemical properties that result inevitably affect both the accessibility of the DNA for a wide range of proteins and the nature of any DNA-protein interactions. Such changes will therefore have consequences in gene expression and in some instances appear to be involved in the regulation of genes.

Methylation of the DNA does not occur in all fungi and affects a small proportion of the genome in comparison to higher eukaryotes (20,21); there seems to be a direct relationship to genome size, and this possibly relates to the amount of "nonfunctional" or redundant DNA in the genome. Generally methylation occurs in nonexpressed sequences (19) and is maintained through DNA replication (22). Whether methylation is ever primarily regulatory, or merely a consequence of sequences being redundant, is open to debate, but in the case of the ribosomal repeat of *Schizophyllum commune*, for example, there does appear to be a correlation with developmental regulation (23).

Torsion within the linear chromosomes of eukaryotes is made possible by the arrangement of the chromosomes into loops tethered to a protein scaffold. This will allow the production of negative DNA supercoiling as a consequence of transcription and altered chromatin structure. A third mechanism would be the presence of an equivalent to the prokaryotic gyrase, though no such activity has been identified. Such physical alteration would affect the accessibility of the DNA to proteins involved in transcription and may alter its suitability as a substrate for transcription, as has been demonstrated in prokaryotes (24). In higher eukaryotes there is evidence that negative supercoiling does occur in transcriptionally active genes (25) and may relate largely to chromatin structure (26).

It is probable that both chemical and physical modification of the DNA are employed as mechanisms for the repression or derepression of specific areas of the genome in fungi.

C. Chromatin Structure

The availability of a specific segment of DNA to the transcription machinery can be altered by the local chromatin structure. Consequently chromatin plays an important role in transcription and its regulation (26,27). The principal component of chromatin is an octomer composed of the four histones H2a, H2b, H3, and H4. The histones are well conserved throughout eukaryotes and have been characterised for a number of fungi (28,29). DNA is wrapped approximately 1.8 times around this structure forming the nucleosome, which represent the basis of the 10-nm chromatin fiber. The 10-nm fiber is the precursor for higher orders of packaging, initially being arranged into the 30-nm solenoid fiber, which is transcriptionally inactive.

It has been demonstrated that elongation of the transcript is only marginally affected by the presence of nucleosomes in vitro (30), and therefore in vivo, where other factors are present, it may be of little or no consequence. The mechanism by which the RNA polymerase proceeds through chromatin may involve dissociation and reassociation of the nucleosomes (31). The location of nucleosomes at critical positions within a promoter can be correlated with the repression of transcription initiation in vitro (32), and in vivo depletion of the histones leads to the activation of some genes (33,34). This is consistent with the transcription complex competing with the histones for sites in the promoter. In *S. cerevisiae* derepression of the acid phosphatase structural gene *PHO5*, which occurs under conditions of limiting inorganic phosphate, requires the products of two regulatory genes, *PHO2* and *PHO4* (35). In the absence of either or both of these no expression is achieved. Their presence leads to repositioning of nucleosomes, displacing them from both a *PHO2* and a *PHO4* binding site, which lie immediately upstream of the structural gene (36). This nucleosome displacement appears to be a critical element in activating transcription, and the specific location of the nucleosomes in the regulatory region probably prevents constitutive expression of the *PHO5* gene.

The interaction of the histones with a specific region of DNA can be altered by their chemical modification. In particular, acetylation of the histones appears to loosen the DNA/histone interaction (37); this leads to negative supercoiling of the DNA linking the nucleosomes and may therefore mimic the role of the prokaryotic gyrase (26). In addition to acetylation, the histones are subject to a variety of other chemical modifications, which may well be regulated and associated, at least in part, with gene expression. A second important factor with respect to altered chromatin structure is the association of the fifth histone H1, or related histone proteins, which interact with both the nucleosomes and the linker DNA. H1 is generally depleted in transcriptionally active regions, is known to protect DNA from protein interactions, and is likely to be directly involved in gene regulation (26,27).

Heterochromatin is associated with specific regions of the chromosomes, particularly the centromeres and telomeres. For eukaryotes with sufficiently large chromosomes, the heterochromatin can be visualized using differential staining. Genes located within heterochromatic regions are generally repressed for much of the cell cycle (26). Similarly, loci in other regions seem to be corepressed, for example the silent mating type genes in *S. cerevisiae* (*HMLα* and *HMRa*). In *S. cerevisiae*, this repression is dependent on a number of genes designated *SIR* (17) and also the N-terminus of histone H4 (38), which has been shown to be acetylated at four residues including Lys-16. From genetic analysis it appears that *SIR3* interacts directly with Lys-16 of H4, preventing both acetylation and consequently the loosening of the chromatin structure (39).

Not only is the N-terminus of H4 necessary for repression, but it is also important for the activation of some genes. In particular, the induction of galactose metabolism in *S. cerevisiae* (which is mediated by *GAL4*, see below) is inhibited by deletion of the N-terminus, or substitution of the N-terminal acetylated residues, of H4 (40). This again implicates the histones and their modification in modulating gene expression. In addition to the apparent interaction between *GAL4* and the histones, the GAL system also includes a protein (GRF2 or protein Y), which appears to specifically position the nucleosomes in the regulatory domain adjacent to the structural genes (41,42). Whether this merely facilitates the regulatory mechanism or represents another level of regulation has not been determined.

It is therefore becoming increasingly apparent that chromatin is not merely a potential barrier to transcription but is an integral part of regulatory mechanisms. These range from the precise location of nucleosomes and the nature of specific histone-protein interactions to the gross chromatin structure in a specific region of the chromosome.

D. Transcription

The initiation of transcription (43–45) is regarded as the major point of gene regulation. This has a logic as well as the support of a wide array of genetic and biochemical data. By regulation at initiation you achieve a good economy of resources; i.e., the transcription, RNA maturation, and translation machinery as well as the pools of nucleotides and amino acids are not utilized unnecessarily. The speed of the response is not insignificant, but can be within a few minutes, and the level of regulation achieved is potentially in the thousandfold. The regulatory signals often, though not necessarily, occur in the DNA adjacent to the coding region. This minimizes the possible complications of dual function within the coding region. Finally, the flanking DNA provides an excellent region for coordinating a variety of signals, which can then be comprehensively communicated. Regulation at this level also allows for initiation to occur at different points resulting in distinct transcripts, and in a number of cases different protein products result.

1. Gene Structure and Arrangement

The majority of genes are highly regulated, while the remainder are considered to be "housekeeping" genes. The latter are regarded as being both constitutive and uniformly expressed, though it is debatable whether this is strictly true of any gene. All the genes encoding proteins, and some of the structural RNA genes, are transcribed by RNA polymerase II, the ribosomal RNA genes being expressed by RNA polymerase I and the small structural RNA genes (e.g., tRNA and 5sRNA) being transcribed by RNA polymerase III. Thus the vast majority of regulatory systems act to regulate the initiation of transcription by RNA polymerase II.

In general, the region immediately upstream of the structural gene is involved in regulating the initiation of transcription (Fig. 2). This is composed

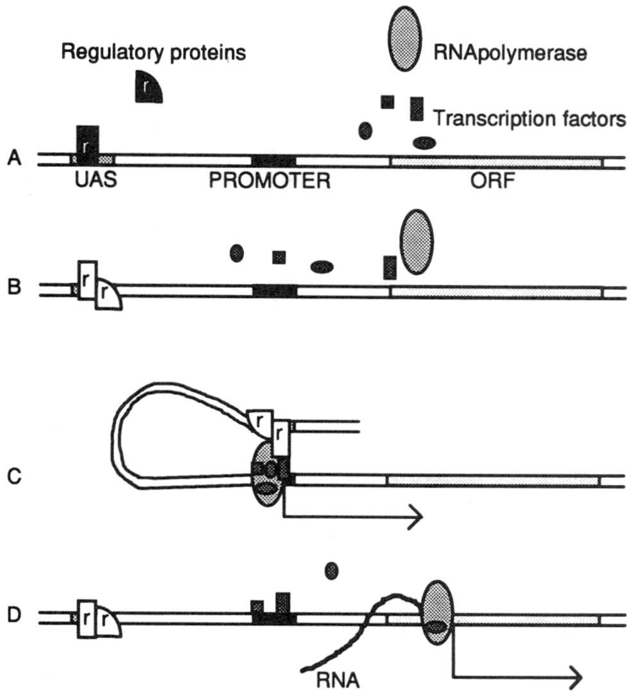

Figure 2 Transcriptional activation. (A) Generally upstream of the open reading frame (ORF) lies both a promoter and regulatory sequences such as an upstream activator sequence (UAS) or enhancer. (A and B) Regulatory proteins bind at the UAS; this may be the case under all growth conditions or on activation of the regulatory protein. The regulatory protein may be altered from an inactive (black) to an active (white) form. The regulatory proteins may act individually or in concert. (C) The binding of positive-acting regulatory proteins in some way assists in forming the transcription initiation complex, which involves both transcription factors and RNA polymerase. The regulatory proteins site of action may be the histones, the transcription factors, or RNA polymerase, or they may act indirectly via an adapter. Communication between proteins at the UAS and the promoter can be achieved by looping of the DNA. (D) Initiation of transcription releases the promoter for the recruitment of a second RNA polymerase molecule.

of the promoter, which is in close proximity to the point of initiation, and a number of short sequences, most notably upstream activator sequences (UAS) (46–49). The latter represent binding sites for specific regulatory proteins and can lie within the promoter or at a distance of a few kilobases. The UASs are generally regarded as the units that combine to form what are described as enhancers in animals and plants. The ability of regulatory proteins to act at a distance from the site of initiation is related to the strength of the activator domain(s), and in many cases the proteins act synergistically (43,44). Communication is achieved by the DNA looping round and consequently bringing the regulatory protein into close proximity with the promoter and the basic transcription machinery (43) (Fig. 2). In *S. cerevisiae*, and perhaps fungi in general, the regulatory sequences are not usually at a great distance from the promoter. This may reflect some basic differences in transcription machinery or it may reflect the constraints imposed by a small genome. *S. cerevisiae* also seems unable to utilize regulatory regions that are placed downstream from the promoter (50), which may be consistent with a limitation in the transcription machinery of fungi (51), though the principal component, TFIID, seems to function effectively in a mammalian transcription system (52) and is generally well conserved (45).

The promoter itself is composed of a number of regions that interact with transcription factors. The best characterized among these is the TATA box (53,54), which is found throughout the eukaryotes and shows strong similarity to the −10 region found in the majority of prokaryotic promoters. The TATA box is known to bind a protein complex TFIID (55), which in turn is stabilized by the binding of TFIIA and/or -B (56–58). This structure provides a point at which RNA polymerase II can be recruited, orientated, and transcription initiated. Although the TATA box is well conserved, it is not necessarily essential to gene expression, but in such cases its removal commonly leads to a diversity of initiation points (59). A variety of other sequence motifs are commonly found in the promoters of fungi, one particular example being an extended CT-rich region (48,49). These sequences are generally considered to have a role similar to that of the TATA box, and may in fact be interchangeable. The TATA box can also be replaced by specific UAS motifs or even a range of randomly synthesized sequences, without a loss of function (60,61). The AT-rich sequence of the TATA box may also facilitate melting of the DNA, which is essential for the initiation of transcription.

These sequences are the basis of the minimal promoters, which in the absence of regulatory proteins will often allow for a low basal level of expression in vitro (56,62). In vivo this is often not the case; it is not certain whether this variation is defined by specific elements within the surrounding sequence or the location of nucleosomes (63), but it should be noted that the location of nucleosomes is to some degree defined by the DNA sequence (26). The apparent similarity between the consensus core sequences within the promoter may in fact disguise a complexity not immediately apparent; for example, TATA boxes do not always function in a similar way (64,65). The identification of a specific motif may be misleading, as distinct regulatory proteins may distinguish between similar sequences (66).

Many coregulated genes are clustered in fungi (67), though in the majority of cases this is probably not the case. Where they are found in close proximity, it is not necessarily essential for their expression that they remain so (68), but often they share a common regulatory domain. In a number of cases some elements of the system are clustered while others are unlinked. The structural genes may be expressed divergently and/or contiguously (69,70) though they do not, generally at least, produce polycistronic messengers.

2. Structure and Function of Regulatory Proteins

In the last 5 years regulatory systems from fungi have been utilized successfully to answer a number of fundamental questions regarding the molecular mechanisms underlying transcription and its regulation. Much of this work has focused on two regulatory genes of *S. cerevisiae*: *GAL4*, which acts positively in the induction of at least five structural genes involved in galactose metabolism (71,72), and *GCN4*, which regulates amino acid biosynthesis (73). Molecular analysis of these and a variety of other regulatory genes has revealed that they generally have functionally distinct domains (43,74). The principal domains are concerned with the binding of DNA, quite often dimerization, activation of transcription, regulation of the protein, and nuclear localization.

The DNA-binding domains are particularly sensitive to mutation, a large proportion of loss of function mutations occurring within them (75). This is consistent with physical analysis, which has demonstrated that specific amino acid–nucleotide interactions underlie the binding specificity (76). Second, they represent complex structures within which minor alterations may well lead to altered function (77). The DNA-binding domains also appear to have had relatively few distinct progenitors. This has led to large families of related regulatory genes throughout eukaryotes whose functions have generally diverged, though there are often striking similarities (78–80). A number of distinct families of DNA-binding domain have been identified, many of which have been characterized structurally (76). The most common classes include the zinc finger proteins, which certainly contain a number of distinct subgroups (77,81), the leucine or B-zipper (82) and the helix-turn-helix (83).

In the case of *GAL4* the DNA-binding domain occurs in the N-terminus (84) and entails a six-cystine binuclear Zn^{2+} cluster (85). This domain spans approximately 50 residues and binds to a 17-bp consensus DNA sequence that shows dyad symmetry (86). This represents quite a long recognition sequence. The consequence of this, assuming random distribution of nucleotides, is that it will not occur at a high frequency within the genome, which is ideal for pathway-specific regulatory protein with only a few targets for regulation. This type of DNA-binding structure has been found in a variety of other fungal regulatory genes in *N. crassa* (87,88), *A. nidulans* (89–92), *S. cerevisiae* (93–98), and *Kluyveromyces lactis* (99) (see Fig. 3), all of which have a narrow domain of activity but are functionally diverse. This underlines

```
NIT4     T A C I A C R R R R S K C D - - - - - - - - - - G A L P S C A A C A S V Y G T E C I Y D
         * * * * * * * * *   * * * *                     *   * * * * * * *   * * *   *   *   * *
NIRA     T A C I A C R R R K S K C D - - - - - - - - - - G N L P S C A A C S S V Y H T T C V Y D
           * *     * *     *   * *                             *   * *   *               *   *
GAL4     Q A C D I C R L K K L K C S - - - - - - - - - - K E K P K C A K C L K - N N W E C R Y S
         * * * *   * *   * *   * * *                     *     *   *     * * *     *     *   * *
LAC9     Q A C D A C R K K K W K C S - - - - - - - - - - K T V P T C T N C L K - Y N L D C V Y S
           * * *   * *   * *   * *                             *   *     *               *   *
QUTA     R A C D S C R S K K D K C D - - - - - - - - - - G A Q P I C S T C A S - L S R P C T Y R
         * * * *   * *         * * *                     *   * *   *     *   *       *   * * *
QA1F     R A C D Q C R A A R E K C D - - - - - - - - - - G I Q P A C F P C V S - Q G R S C T Y Q
           * *     * *         * * *                           *   *     *               *
PPR1     T A C K R C R L K K I K C D - - - - - - - - - - Q E F P S C K R C A K - L E V P C V S L
           * *     * *     *   * * *                               *     * * *       * * *
LEU3     F A C V E C R Q Q K S K C D A - - - - - - - H E R A P E P C T K C A K - K N V P C I L K
           * * *   *             * * *                           * *     *               *
AMDR     A A C V H C H R R K V R C D A - - - - - - - - R L V G L P C S N C R S - A K T D C Q I H
             *     *       *   * * * *                             * * * *             * *
ALCR     H S C D P C R K G K R R C D A P E N R N F A N E N G W V S C S N C K R - W N K D C T F N
         *   *     *       *   * *                                 *     *   *           *
UGA      H G C I T C K I R K K R C S - - - - - - - - - - E D K P V C R D C R R - L S F P C I Y I
           * *   * *     * *     *                             *   *     *             * *
ARGR2    T G C W T C R G R K V K C D - - - - - - - - - - L R H P H C Q R C E K - S N L P C G G Y
             *     * *   *   * * * *                           *   *   * *         * *   *
MAL63    Q S C D C C R V R R V K C D - - - - - - - - - - R N K P - C N R C I Q - R N L N C T Y L
           * *     * *   *   * * * *                           *   *     *
HAP1     L S C T I C R K R K V K C D - - - - - - - - - - K L R P H C Q Q C T K - T G V A H L C H
```

Figure 3 Alignment of the DNA-binding domains of 14 fungal regulatory proteins. The amino acid sequence of the DNA-binding domains from 14 fungal regulatory proteins, all of which have a six-cystine structure; NIT4 and QA1F from *N. crassa* (87,88); NIRA, QUTA, AMDR, and ALCR from *A. nidulans* (89–92); GAL4, PPR1, LEU3, UGA, MAL63, and HAP1 from *S. cerevisiae* (86,93–97); LAC9 from *Kluyveromyces lactis* (99). Perfect matches with the sequence following are indicated by *.

the extent of conservation. A consequence of this conservation is that similar functions in distinct organisms appear likely to be undertaken by structurally similar proteins (e.g., 77,100,101). This allows the information gleaned from these well-studied organisms to be directly applied to other fungi by the use of specific probes derived from a consensus sequence, or even the cloned gene itself may be used as a probe to clone related genes by cross-hybridization. A third possibility is the direct amplification of related sequences by the polymerase chain reaction (PCR) using the conserved sequences to design synthetic oligonucleotides (102). Related genes from one organism can also be tested for function and potentially characterized in a second organism that is more amenable to such techniques (e.g., 103).

In many cases the DNA-binding domains are also involved in dimerization, which is a common feature of many regulatory molecules. One reason is the potential for forming heterodimers, which gives a specific regulatory molecule more flexibility in its function (78). The more general observation is that many of the binding domains are symmetrical, and a dimer is therefore ideal to interact with such a sequence. The antiparallel nature of DNA allows the formation of structurally perfect palindromic sequences, and these can be distributed at a relatively low frequency in optimal positions.

Nuclear localization signals or nuclear tags have been characterized in a number of regulatory proteins (104). In some cases they represent distinct domains, while in others they are less readily defined as they extend over a wide area (105). In general, they include a high proportion of very basic residues. The nuclear tag is essential as it facilitates transport through the nuclear pores; the nucleus, like other membrane-bound structures, does not permit the passive diffusion of proteins. This also provides an opportunity for regulating the location and therefore the function of a transcriptional regulator. This has been demonstrated in a number of cases (104), including the yeast regulatory gene *SWI5* (see Section III.F).

Transcriptional activation essentially appears to be a complex process, considering the number and variety of components involved and the subtleties required. Surprisingly though, the domains responsible for transcriptional activation are generally very insensitive to disruption by amino acid substitution (e.g., 75). This implies that, in general, specific amino acids are not involved in precise interactions. These findings were enhanced by the observation that fungal regulatory proteins are capable of activating transcription in mammalian (106), insect (107), and plant cells (108). Similarly nonfungal regulatory genes seem to function in fungi, provided a suitable DNA consensus binding site is adjacent to a promoter (e.g., 109). The basic requirement for a transcriptional activator is a negatively charged amphipathic α-helix (110), and, generally, the stronger the negative charge, the stronger the activation (43). But this level of apparent simplicity may in fact conceal an underlying complexity and variety of activation mechanisms.

There is strong evidence that, in some cases at least, specific amino acid interactions are important; different activators will have very different functions in relation to a specific promoter; and the point of action may vary or may be various (44,45,52,111,112). It should also be noted that not all activator domains in eukaryotes fit this charged α-helix model (45).

The mechanism and point of regulation for these molecules is varied, which is inevitable considering their diverse function. There are four basic mechanisms:

1. Direct binding of the effector molecule
2. Regulated expression of the regulatory protein
3. Posttranscriptional modification of the protein
4. Protein protein interaction

A single element may be employed, but it has become increasingly apparent that a combination of mechanisms is often used. It is interesting to note that the conservation of function in respect to DNA-binding domains and activator domains may also extend to some of the regulatory domains. For example, the human estrogen receptor protein not only retains its function within *S. cerevisiae*, but it also retains its regulated response to the presence of the hormone (113).

In the case of *GAL4* the gene is constitutively expressed (114), and the regulatory protein appears to be bound to its UAS even under noninducing conditions (115). The regulatory domain has been localized to the C-terminus of the protein (116). This domain includes one of the two activation domains (117) and is involved in binding a protein designated GAL80 (118,119). GAL80 appears to mask the activation domains and prevent them from interacting with the transcription machinery. It is proposed that in the presence of the inducer, probably a derivative of galactose, GAL80 dissociates from GAL4, unmasking the activation domains. Though this appears to be the principal mechanism, other factors are involved, including GAL4 phosphorylation and carbon catabolite repression (120,121).

The regulation of transcriptional activators is obviously important to their function in regulating transcription, but a second consideration is that overexpression of activation domains can be very damaging to gene expression in general (43), the activation domains being able to sequester essential elements of the transcription machinery and generally inhibit gene expression (122).

In the case of *GAL4* the regulatory gene is positive acting; i.e., it enhances and is essential for transcription. A number of regulatory genes are known to act negatively; i.e., their role is to repress transcription. The situation becomes further complicated by the ability of some regulatory genes to both activate and repress gene expression, usually at distinct loci (e.g., 123). In some cases the negative action will be achieved merely by preventing the binding of an activating protein. This can be achieved if the respective binding sites are overlapping or closely associated (see below). It is likely that in other situations repression is a more active process (120,124).

3. Coordinated Regulation

Expression of a given gene will often be subject to a number of regulatory signals that must be coordinated. This may involve interaction between regulatory genes, allowing the specific signals to be incorporated in determining the response. An example of this would be a structural gene expressed under conditions of either carbon or nitrogen depletion provided the appropriate substrate is present to induce the gene. In this example three signals are implicit in the regulation of such a gene, at least two of which are required for expression of the gene.

This situation is found for a number of structural genes characterized in *A. nidulans* (67), such as the *prn* cluster, which includes at least three structural genes and a pathway-specific regulatory gene involved in proline metabolism (125,126), and *gabA*, the structural gene for the γ-amino-*n*-butyrate (GABA) permease (127). Both proline and GABA can be utilized by *A. nidulans* as the sole carbon and/or nitrogen source. Consequently, if they are present and required, i.e., no better carbon or nitrogen source is available, then the genes will be expressed. In *A. nidulans*, carbon catabolite repression is mediated by a negatively acting regulatory gene; i.e., its product acts as a repressor, *creA* (128), which exerts its regulation at the mRNA level (129) and has been cloned and sequenced (130). Nitrogen metabolite repression is mediated by a positive-acting regulatory gene *areA* (77,131). Both these genes are involved in regulating the transcription of many structural genes, probably hundreds (67,132). By necessity, these two signals must, to a large extent, act independently. A further requirement is that they do not regulate all the structural genes within their domain equally, but rather there is a hierarchy of derepression. This can be achieved simply by varying the concentration of the active regulatory molecule, provided the affinity of specific regulatory domains, the number of bound molecules required, and the effectiveness of the regulator vary in each case. With respect to the *areA* product, each promoter will require a slightly different concentration of the active protein to bind and derepress the gene. This will result in a hierarchy of derepression, allowing the more favored nitrogen sources to be used preferentially. This means that the system is dynamic and gradual rather than a crude on/off switch. A consequence of this is that mutations in the DNA-binding domain can alter the protein-binding specificity and radically change the ability of the fungus to grow in distinct environments (67,77,133). This also provides an insight into how related regulatory proteins may have evolved.

The second regulatory requirement is a pathway-specific signal to allow expression of the derepressed genes only if the substrate is present. In the case of the *prn* cluster in *A. nidulans*, a gene within the cluster, *prnA*, is responsible for its pathway-specific induction (134). Induction of *gabA* is mediated by an unlinked regulatory gene, *amdR* (*intA*) (91,128), which detects the presence of various compounds, including ω-amino acids, and activates expression of the gene. The *amdR* product is not technically pathway specific as it coregulates, or "integrates," the induction of at least five structural genes involved in utilizing distinct nitrogen sources. Whether this is a form of economy, utilizing a single regulatory gene to regulate a number of distinct pathways, or there is some undefined relationship between the various pathways that requires coordination is unknown. In both cases the narrow domain regulatory genes act positively; i.e., their product is required for expression.

Thus, there are two distinct groups of signals. One is required for determining whether the gene may be expressed (repression/derepression) and the second is required for induction. Their combined function must explain both the coordination and the independence of distinct signals. In both cases, derepression can be achieved by simple mutations within the regulatory domain (128,131). In the case of the prn^d alleles, the lesion apparently disrupts the binding site for the *creA* product (C. Scazzocchio, personal communication), which it is postulated will lead to its inability to bind and therefore repress transcription. In the case of other genes under *creA* regulation, the binding sites for pathway-specific regulatory genes can overlap with that of the *creA* product (B. Felenbok, personal communication). This implies that the bound repressor specifically excludes the activator. With respect to the *areA* product, the situation is more complex. Its positive action would suggest that, in the examples given, its binding is required for, but not sufficient to, activate transcription, though for some structural genes the *areA* product may be sufficient for activation. The requirement for two positive signals is not so readily explained and has not been determined; it could be related to the exclusion of a repressor, synergy between the activation domains of the *areA* product and the pathway-specific regulatory proteins, allowing the combined effect to progress beyond a minimum threshold required for transcription. It is also possible that specific protein/protein interactions between the two regulatory proteins are required, for example assisting in either DNA binding of the complex or formation of a functional activator domain. Certainly regions of the *areA* product deemed nonessential for general function (77) are essential for expression of some structural genes (A. Platt, M. N. Arst, and M. X. Caddick, unpublished data), and in the case of nitrate and nitrite utilization there is good evidence for a protein/protein interaction between the products of the pathway-specific regulatory gene *nirA* and *areA* (135).

For any given regulatory proteins with a wide domain of action, interactions may occur with a wide range of regulatory proteins and transcription factors. Consequently it may be that these cannot be localized to a distinct domain but could involve different surfaces of the protein; this would enable the protein to act appropriately with a range of distinct mechanisms. At this level the simplistic view of functionally distinct domains begins to break down, particularly if the full spectrum of activity is monitored rather than viewing a regulatory system merely as the on/off switch for a specific gene.

E. Transcript Formation

The procedure of RNA polymerase complex along the gene could potentially be regulated by DNA structure or associated proteins. Ultimately termination occurs, though in eukaryotes this is not a well-characterized process as the mature end to the transcript is produced by subsequent processing (136). In the majority of cases the mature transcript is capped at the 5′ end and polyadenylated at the 3′ end. Capping involves the addition of a 7-methylguanosine residue in the 5′-5′ condensation, and its presence is important in both translational efficiency (137) and transcript stability (138). The formation of the poly(A) tail requires an initial cleavage of the transcript and subsequently the addition of adenosine residues. This requires specific signals in the 3′ end of the transcript (48,49,139). The poly(A) tail is again important to RNA stability and involves the binding of a protein [the poly(A)-binding protein] (140).

The stability of transcripts is very variable both between genes and growth conditions (141; M. X. Caddick unpublished). It is in the interest of the organism to produce stable transcripts if their product is going to be required over a long period of time, but in some cases it is important that a transcript not only appears rapidly in response to a signal but can also be removed, and consequently the expression of the gene is curtailed. In particular, this is true of regulatory genes, where it is likely that the concentration of their product is often an important factor in the regulatory system. The stability of the transcript involves its sequence, secondary structure, and interactions with specific proteins (142,143).

Maturation of the transcript often involves the splicing out of introns to create the contiguous coding sequence. In fungi introns are generally both small and infrequent compared with those of animal and plant genes. This may reflect restraints on genome size specific to rapidly growing organisms, though the location and paucity of introns in *S. cerevisiae* may be a special case (144). The introduction of an additional step in the expression of a gene inevitably provides a possible point for further gene regulation. This can be at the level of splicing or not splicing a gene, or differential splicing. Consequently a gene may be turned on or off, or distinct products produced (145).

An important caveat is that the sequence of a gene is not merely reflected in its coding potential, but it will contain information relating to transcript maturation and stability, both of which are increasingly being viewed as important elements in gene regulation.

F. Translation

Translation of nuclear encoded genes occurs exclusively in the cytosol; consequently the transcripts are transported from the nucleus to the endoplasmic reticulum. The initial step is the recognition of the mRNA CAP (137); the efficiency of this will inevitably be influenced by the structure of the mRNA, which will in turn reflect the rate of initiation and therefore expression. Global regulation of protein synthesis can also be mediated at the level of mRNA recognition by the translational machinery. For example, heat shock leads to reduced phosphorylation of the CAP-binding protein and results in reduced protein synthesis (146).

A more specific example of regulation at the level of translation is well described for the expression of *GCN4* (147), the wide domain regulatory gene responsible for activating amino acid biosynthesis in *S. cerevisiae* (73). The gene is constitutively transcribed, but the level of protein is highly regulated and relates directly to the expression of the biosynthetic genes (147–149). Regulation of *GCN4* expression is achieved at the level of translation. Within the first 590 nucleotides of the *GCN4* transcript there are four short open reading frames (ORF) initiated by an AUG (Fig. 4). These lie upstream of the long ORF that encodes the regulatory protein. Generally translation in eukaryotes starts at the first AUG encountered by the ribosome, as it "scans" the transcript from the 5′ end (150), and termination occurs at the first subsequent stop codon. Therefore, expression of *GCN4* requires reinitiation of translation, which is generally inefficient, particularly when the termination and initiation signals occur in rapid succession. Removal of all the upstream ORFs leads to a very high, deregulated level of *GCN4* expression, while the fourth short ORF is sufficient to repress *GCN4* expression to the repressed levels of the wild type. The critical steps in the regulation of GCN4 lie both in the ability of the ribosomes to recognize and initiate translation at the upstream AUGs and in the ability of ribosomes to reinitiate translation at subsequent start codons.

Initiation of translation at the first two short ORFs prevents reinitiation at ORF 3 or 4 allowing higher rates of reinitiation at the long ORF, while initiation at ORF 3 or 4 reduces the rate of initiation at the long ORF. A large number of loci are involved in mediating this regulation (73). At least

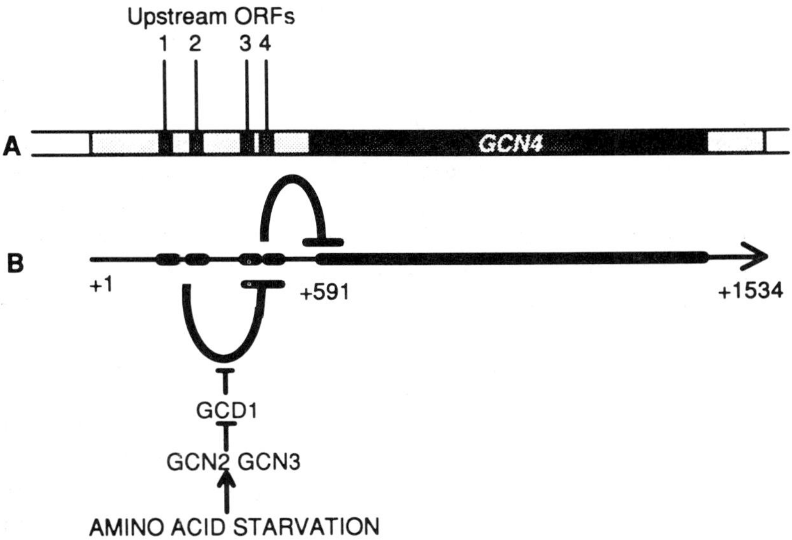

Figure 4 *GCN4* gene structure and regulation. GCN4 derepresses a number of structural genes involved in amino acid biosynthesis and its regulation is critical to amino acid biosynthesis. (A) The transcribed region of GCN4 contains four short open reading frames (ORF), and one long ORF that encodes GCN4 (the ORFs are indicated by the dark shading). All five ORFs are preceded by a start codon. The ribosomes proceed along the transcript (B) from the 5′ end (+1). For the majority of transcripts translation is initiated at the first AUG. Initiation at either ORF 1 or 2 is likely to prevent reinitiation at ORF 3 or 4 owing to their close proximity. GCD1, possibly in conjunction with GCD10–12, allows reinitiation to occur at ORF 3 and 4. GCN2 and 3 are able to inactivate GCD1 under conditions of amino acid starvation. Initiation at ORF 3 or 4 is likely to prevent reinitiation at the long, coding, ORF. The ability of the ribosomes to recognize the upstream ORFs and the level of reinitiation is subject to regulation by the intracellular levels of amino acids and involves the products of at least 14 genes; *GCN1, 2,* and *3* and *GCD1,2,3,4,5, 6,7,10,11,12,* and *13*.

11 loci, designated *GCD1,2,3,4,5,6,7,10,11,12,* and *13*, are required for repressing *GCN4* expression, some of which are under negative regulation by at least three loci, designated *GCN1, 2,* and *3*. The full intricacies of the regulation have not been characterized, but a model of some aspects has been forthcoming (151,152). *GCN2* has been characterized as a protein kinase, and it also bears a domain that has strong amino acid similarity to a histidyl-tRNA synthetase (HRS). Dominant mutations in *GCN2* leading to constitutive expression of GCN4 can occur in both the kinase and HRS domains. It is proposed that the HRS domain will bind uncharged tRNA mole-

cules (153); these are abundant under conditions of amino acid starvation and correlate with *GCN4* expression (73). This suggests that the binding of tRNA molecules is perceived by *GCN2* to indicate amino acid starvation and leads to activation of the kinase activity; this in turn phosphorylates and inactivates one or more of the GCD products. GCD mutants are generally highly pleiotropic and some may be directly associated with the ribosomes (73). Their inactivation increases the frequency of reinitiation at the long ORF, leaing to derepression of *GCN4* activity and amino acid biosynthesis. A similar use of an upstream ORF to repress gene expression has been characterized for the yeast gene *CPA1*, which is involved in arginine biosynthesis (153).

A distinct form of regulation at the level of translation has been identified in the expression of a number of retroviruses, retrotransposons, and prokaryotic genes (154); among the examples is the yeast transposable element Ty1 (155,156). In all these instances, the ORF is disrupted by a frameshift. Thus conventional translation would lead to a truncated polypeptide being formed. This can be obviated by the ribosome "slipping" into the second reading frame and producing the full-length polypeptide. Translational frameshifting in the case of Ty1 is dependent on the local nucleotide sequence around the frameshift (156) and is probably regulated via the ribosome complex.

Finally, the codon usage varies with the genes, such that highly expressed genes tend to have a strong codon bias while genes that are expressed at a low level do not. Codon bias is more likely to reflect rather than dictate the level of expression for a specific gene.

G. Posttranslational Modification

The extent and variety of posttranslational modification and transportation is beyond the scope of this review. Suffice it to say that the function of many proteins is dependent on, and regulated by, posttranslational mechanisms. In some respects, this is an inefficient point for regulation as all the resources entailed in protein synthesis will have been employed, but it provides additional flexibility, which may be superimposed onto other regulatory systems and may also allow a very rapid response.

With respect to regulatory protein (157), examples have already been given where phosphorylation is involved in modulating the regulatory proteins (e.g., GAL4 and GCN2/GCD). A further example is SWI5, which is required for expression of the *HO* gene and initiation of the mating-type switch in *S. cerevisiae* (see Section III.A). SWI5 is transcribed during S1, G2, and M phases of the cell cycle (158), but only on entry into G1 is the protein translocated into the nucleus (159). This cell-cycle-dependent transport involves a domain localized to the C-terminus and includes three Ser residues, which are essential for nuclear localization (160). The Ser residues are phosphorylated in a cell-cycle-dependent manner, probably by CDC28, a protein kinase involved in cell cycle regulation (161). Phosphorylation therefore controls the localization and function of SWI5 and is involved in cell cycle regulation in general (162).

In a number of cases a cascade of regulatory proteins has been identified (163); this may well represent the involvement of posttranslational modifications arranged to transduce and possibly amplify the signal. This will also allow for distinct points of regulation facilitating the coordination of specific signals. A third aspect is that this enables the transmission of a specific signal to a number of distinct regulatory proteins; this may be exemplified by the role of kinases (164) particularly in cell cycle regulation (162).

IV. CONCLUSIONS

Three principal themes recur when considering the mechanisms of regulation in fungi: diversity, flexibility, and conservation. The importance of regulation has meant that all aspects of a gene and its expression can and do play a role. The coordination of these various diverse mechanisms will vary for each gene, and the expression of each gene is therefore distinct. As regulation has to act through the various essential and complex components involved in processes ranging from the maintenance of a gene to the modification of its product, many of which are invariant or at least highly conserved within eukaryotes, limits are imposed on the ability of new regulatory systems and proteins to emerge. This has resulted in the conservation of established systems and has led to the isolation of similar genes exercising similar roles and an extraordinary conservation of function throughout eukaryotes. As a consequence, fungi not only provide a good indication of how gene regulation may occur in other eukaryotes, but the components characterized in one organism can be used to identify similar or equivalent genes and proteins from other organisms. The potential of molecular genetics should therefore open the regulatory systems of economically and medically important fungi to scrutiny and provides powerful techniques to characterize the control mechanisms for a variety of physiological phenomena.

REFERENCES

1. Errede, B., Cardillo, T. S., Sherman, F., Dubois, E., Deschamps, J., and Waime, J. M. (1980). Mating signals control expression of mutations resulting from insertion of transposable repetitive element adjacent to diverse yeast genes, *Cell 25*: 42.
2. Sarachek, A., Henderson, L. A., and Eddy, K. B. (1991). Genetic destabilization of *Candida albicans* by hydroxyurea, *Microbios 65*: 39.

3. Dutton, S., and Penn, C. W. (1989). Biological attributes of colony-type variants of *Candida albicans, J. Gen. Microbiol. 135*: 3363.
4. Soll, D. R. (1989). High-frequency switching in *Candida albicans*, in *Mobile DNA* (D. E. Berg and M. M. Howe, eds.), American Soc. of Microbiology, Washington, DC, p. 79.
5. Sukuki, T., Kobayashi, I., Mizuguchi, I., Banno, I., and Tanaka, K. (1988). Electrophoretic karyotypes in medically important *Candida* species, *J. Gen. Appl. Microbiol. 34*: 409.
6. Masel, A., Braithwaite, K., Irwin, J., and Manners, J. (1990). Highly variable molecular karyotypes in the plant pathogen *Colletotrichum glocosporioides, Curr. Genet. 18*: 81.
7. McDonald, B. A., and Martinez, J. P. (1991). Chromosome length polymorphisms in a *Septoria tritici* population, *Curr. Genet. 19*: 265.
8. Klar, A. J. S. (1989). The interconversion of yeast mating types: *Saccharomyces cerevisiae* and *Schizosaccharomyces pombe*, in *Mobile DNA* (D. E. Berge and M. M. Howe, eds.), American Soc. of Microbiology, Washington, DC, p. 671.
9. Russell, D. W., Jensen, R., Zoller, M., Burke, J., Errede, B., Smith, M., and Herskowitz, I. (1986). Structure of the *Saccharomyces cerevisiae HO* gene and analysis of its upstream regulatory region, *Mol. Cell. Biol. 6*: 4281.
10. Kostriken, R., and Heffron, F. (1984). The product of the *HO* gene is a nuclease: purification and characterisation of the enzyme, *Cold Spring Harbor Symp. Quant. Biol. 49*: 89.
11. Jensen, R., Sprague, F., and Herskowitz, I. (1983). Regulation of yeast mating type interconversion: feedback control of *HO* gene expression by the mating-type locus, *Proc. Natl. Acad. Sci. USA 80*: 3035.
12. Nasmyth, K. A. (1983). Molecular analysis of cell lineage, *Nature* (*Lond.*) *302*: 670.
13. Strathern, J. N., Klar, A. J. S., Hicks, J. B., Abraham, J. A., Ivy, J. M., Nasmyth, K. A., and McGill, C. (1982). Homothallic switching of yeast mating type cassettes is initiated by a double-stranded cut in the *MAT* locus, *Cell 31*: 182.
14. Nasmyth, K. A. (1987). The determination of mother cell cycle regulation of *SWI5* is required for motor-cell-specific mating type switching in yeast by a specific regulator of *HO* transcription, *EMBO J. 6*: 243.
15. Hicks, J. B., Strathern, J. N., and Klar, A. J. S. (1979). Transposable mating type genes in *Saccharomyces cerevisiae, Nature* (*Lond.*) *282*: 478.
16. Abraham, J., Nasmyth, K. A., Strathern, J. N., Klar, A. J. S., and Hicks, J. B. (1984). Regulation of mating type information in yeast. Negative control requiring sequences both 5′ and 3′ to the regulated region, *J. Mol. Biol. 176*: 307.
17. Rine, J., and Herskowitz, I. (1987). Four genes responsible for a position effect on expression from *HML* and *HMR* in *Saccharomyces cerevisiae, Genetics 116*: 9.
18. Klar, A. J. S., Hicks, J. B., and Strathern, J. N. (1982). Directionality of yeast mating type interconversion, *Cell 28*: 551.
19. Selker, E. U. (1990). DNA methylation and chromatin structure; a view from below, *Trends Biochem. Sci. 15*: 103.
20. Wagner, I., and Capesius, I. (1981). Determination of 5-methylcytosine from plant DNA by high-performance liquid chromatography, *Biochim. Biophys. Acta 654*: 52.

21. Russell, P. J., Rodland, K. D., Rachlin, E. M., and McCluskey, J. A. (1987). Differential DNA methylation during the vegetative life cycle of *Neurospora crassa, J. Bacteriol. 169*: 2902.
22. Holliday, R. (1987). The inheritance of epigenetic defects, *Science 238*: 163.
23. Buchner, B., Novotny, C. P., and Ullrich, R. C. (1988). Developmental regulation of the methylation of the ribosomal DNA in the basidiomycete fungus *Schizophllum commune, Curr. Genet. 14*: 105.
24. Pruss, G. J., and Drlica, K. (1989). DNA supercoiling and prokaryotic transcription, *Cell 56*: 521.
25. Leonard, M. W., and Patient, R. K. (1991). Evidence for torsional stress in transcriptionally active chromatin, *Mol. Cell. Biol. 11*: 6128.
26. Grunstein, M. (1990). Histone function in transcription, *Annu. Rev. Cell Biol. 6*: 643.
27. Felsenfeld, G. (1992). Chromatin as an essential part of the transcriptional mechanism, *Nature (Lond.) 355*: 219.
28. Maxson, R., Cohn, R., Kedes, L., and Muhun, T. (1983). Expression and organization of the histone genes, *Annu. Rev. Genet. 17*: 239.
29. Ehinger, A., Denison, S. H., and Gregory, S. M. (1990). Sequence organization and expression of the core histone genes in *Aspergillus nidulans, Mol. Gen. Genet. 222*: 416.
30. Izban, M. G., and Luse, D. S. (1991). Transcription on nucleosomal templates by RNA polymerase II in vitro: inhibition of elongation with enhancement of sequence-specific pausing, *Genes Dev. 5*: 683.
31. Thoma, F. (1991). Structural changes in nucleosomes during transcription: strip, split or flip? *Trends Genet. 7*: 175.
32. Lorch, Y., La Pointe, J. W., and Kornberg, R. D. (1987). Nucleosomes inhibit the initiation of transcription but allow chain elongation with the displacement of histones, *Cell 55*: 743.
33. Han, M., and Grunstein, M. (1988). Nucleosome loss activates yeast downstream promoters in vivo, *Cell 55*: 1137.
34. Han, M., Kim, U-J., Kayne, P. S., and Grunstein, M. (1988). Depletion of histone H4 and nucleosomes activates the *PH05* gene in *Saccharomyces cerevisiae, EMBO J. 7*: 2221.
35. Yoshida, K., Ogawa, N., and Oshima, Y. (1989). Function of the *PHO* regulatory genes for repressible acid phosphatase synthesis in *Saccharomyces cerevisiae, Mol. Gen. Genet. 217*: 40.
36. Straka, C., and Hoerz, W. (1991). A functional role for nucleosomes in the repression of a yeast promoter, *EMBO J. 10*: 361.
37. Norton, V. G., Imai, B. S., Yau, P., and Bradbury, E. M. (1989). Histone acetylation reduces nucleosome core particle linking number change, *Cell 57*: 499.
38. Kayne, P. S., Kim, U-J., Han, M., Mullen, J. R., Yoshizaki, F., and Grunstein, M. (1988). Extremely conserved histone H4 N-terminus is dispensable for growth but essential for respressing the silent mating loci in yeast, *Cell 55*: 27.
39. Johnson, L. M., Kayne, P. S., Kahn, E. S., and Grunstein, M. (1990). Genetic evidence for an interaction between SIR3 and histone H4 in the repression of the silent mating loci of *Saccharomyces cerevisiae, Proc. Natl. Acad. Sci. USA 87*: 6287.

40. Workman, J. L., Taylor, I. C. A., and Kingston, R. E. (1991). Activation domains of stably bound GAL4 derivatives activate repression of promoters by nucleosomes, *Cell 64*: 533.
41. Kornberg, R. D., and Stryer, L. (1988). Statistical distributions of nucleosomes: non-random localisations of a stochastic mechanism, *Nucleic Acids Res. 16*: 6677.
42. Fedcor, M. J., Neal, F. L., and Kornberg, R. D. (1988). Statistical positioning of nucleosomes by specific protein-binding to an upstream activating sequence in yeast, *J. Mol. Biol. 204*: 109.
43. Ptashne, M. (1988). How eukaryotic transcriptional activators work, *Nature (Lond.) 335*: 683.
44. Ptashne, M., and Gann, A. A. F. (1990). Activators and targets, *Nature (Lond.) 346*: 329.
45. Guarente, L., and Bermingham-McDonough, O. (1992). Conservation and evolution of transcriptional mechanisms in eukaryotes, *Trends Genet. 8*: 27.
46. Struhl, K. (1987). Promoters, activator proteins and the mechanism of transcriptional initiation in yeast, *Cell 49*: 295.
47. Guarente, L. (1987). Regulatory proteins in yeast, *Annu. Rev. Genet. 21*: 425.
48. Brown, A. J. P., and Lithgow, G. J. (1987). The structure and expression of nuclear genes in *Saccharomyces cerevisiae*, in *Gene Structure in Eukaryotic Microbes* (J. R. Kinghorn, ed.), IRL Press, Oxford, p. 9.
49. Gurr, S. J., Unkles, S. E., and Kinghorn, J. R. (1987). The structure and organization of nuclear genes in filamentous fungi, in *Gene Structure in Eukaryotic Microbes* (J. R. Kinghorn, ed.), IRL Press, Oxford, p. 93.
50. Struhl, K. (1989). Molecular mechanisms of transcriptional regulation in yeast, *Annu. Rev. Biochem. 58*: 1051.
51. Buratowski, S., Hahn, S., Sharp, P. A., and Guarente, L. (1988). Function of a yeast TATA element-binding protein in a mammalian transcription system, *Nature (Lond.) 334*: 37.
52. Gill, G., and Tjian, R. (1991). A highly conserved domain of TFIID displays species specificity in vivo, *Cell 65*: 333.
53. Breathnach, R., and Chambon, P. A. (1981). Organization and expression of eukaryotic split genes coding for proteins, *Annu. Rev. Biochem. 50*: 349.
54. Nakajima, N., Horikoshi, M., and Roeder, R. G. (1988). Factors involved in specific transcription by mammalian RNA polymerase II: purification, genetic specificity, and TATA box-promoter interactions of TFIID, *Mol. Cell. Biol. 8*: 4028.
55. Matsui, T., Segall, J., Weil, P. A., and Roeder, R. G. (1980). Multiple factors required for accurate initiation of transcription by purified RNA polymerase II, *J. Biol. Chem. 255*: 11992.
56. Lillie, J. W., and Green, M. R. (1989). Activators target in sight, *Nature (Lond.) 341*: 279.
57. Hahn, S., Buratowski, S., Sharpe, P. A., and Guarente, L. (1989). Identification of a yeast protein homologous in function to the mammalian general transcription factor TFIIA, *EMBO J. 8*: 3379.
58. Lin, Y. S., and Green, M. (1991). Mechanism of action of an acidic transcriptional activator in vitro, *Cell 64*: 971.

59. Ogden, J. E., Stanway, C., Kim, S., Mellor, J., Kingsman, A. J., and Kingsman, S. M. (1986). Efficient expression of the *Saccharomyces cerevisiae PGK* gene depends on an upstream activation sequence but does not require TATA sequences, *Mol. Cell. Biol. 6*: 4335.
60. Chen, W., and Struhl, K. (1989). Yeast upstream activator protein GCN4 can stimulate transcription when its binding site replaces the TATA element, *EMBO J. 8*: 261–268.
61. Singer, V. L., Wobbe, C. R., and Struhl, K. (1990). A wide variety of DNA sequences can functionally replace a yeast TATA element for transcriptional activation, *Genes Dev. 4*: 636.
62. Saltzman, A. G., and Weinmann, R. (1989). Promoter specificity and modulation of RNA polymerase II transcription, *FASEB J. 3*: 1723.
63. Brandl, C. J., and Struhl, K. (1990). A nucleosome-positioning sequence is required for GCN4 to activate transcription in the absence of a TATA element, *Mol. Cell. Biol. 10*: 4256.
64. Struhl, K. (1986). Constitutive and inducible *Saccharomyces cerevisiae* promoters: evidence for two distinct molecular mechanisms, *Mol. Cell. Biol. 6*: 3847.
65. Ponticelli, A. S., and Struhl, K. (1990). Analysis of yeast *HIS3* transcription in vitro: biochemical support for multiple mechanisms of transcription, *Mol. Cell. Biol. 10*: 2832.
66. Chodosh, L. A., Olesen, J., Hahn, S., Baldwin, A. S., Guarente, L., and Sharp, P. A. (1988). A yeast and a human CCAAT-binding proteins have heterologous subunits that are functionally interchangeable, *Cell 53*: 25.
67. Arst, H. N., Jr., and Scazzocchio, C. (1985). Formal genetics and molecular biology of the control of gene expression in *Aspergillus nidulans*, in *Gene Manipulations in Fungi* (J. W. Bennet and L. L. Lasure, eds.), Academic Press, Orlando, FL, p. 309.
68. Rand, K. N., and Arst, H. N., Jr. (1977). A mutation in *Aspergillus nidulans* which affects the regulation of nitrate reductase and is tightly linked to its structural gene, *Mol. Gen. Genet. 155*: 67.
69. Hopper, J. E., and Rowe, L. B. (1978). Molecular expression and regulation of the galactose pathway genes in *Saccharomyces cerevisiae, J. Biol. Chem. 253*: 7566.
70. Johnston, I. L., McCabe, P. C., Greaves, P., Gurr, S. J., Cole, G. E., Brau, M. A. D., Unkels, S. E., Clutterbuck, A. J., Kinghorn, A. J., and Innis, M. A. (1990). Isolation and characterization of the *crnA-niiA-niaD* gene cluster for nitrate assimilation in *Aspergillus nidulans, Gene 90*: 181.
71. Oshima, Y. (1982). Regulatory circuits for gene expression: the metabolism of galactose and phosphate, in *Molecular Biology of the Yeast* Saccharomyces cerevisiae: *Metabolism and Gene Expression* (J. N. Strathern, E. W. Jones, and J. R. Broach, eds.), Cold Spring Harbor Laboratory, Cold Spring Harbor, NY, p. 159.
72. Johnston, M. (1987). A model fungal gene regulatory mechanism: the *GAL* gene of *Saccharomyces cerevisiae, Microbiol. Rev. 51*: 458.
73. Hinnebusch, A. G. (1988). Mechanisms of gene regulation in the general control of amino acid biosynthesis in *Saccharomyces cerevisiae, Microbiol. Rev. 52*: 248.

74. Keegan, L., Gill, G., and Ptashne, M. (1986). Separation of DNA binding from the transcription-activating function in a eukaryotic regulatory protein, *Science 231*: 699.
75. Johnston, M., and Dover, J. (1988). Mutational analysis of the *GAL4*-encoded transcriptional activator protein of *Saccharomyces cerevisiae, Genetics 120*: 63.
76. Harrison, S. C. (1991). A structural taxonomy of DNA-binding domains, *Nature* (*Lond.*) *353*: 715.
77. Kudla, B., Caddick, M. X., Langdon, T., Martinez-Rossi, N. M., Bennett, C. F., Sibley, S., Davies, R. W., and Arst, H. N., Jr. (1990). The regulatory gene *are*A mediating nitrogen metabolite repression in *Aspergillus nidulans.* Mutations affecting specificity of gene activation alter a loop residue of a putative zinc finger, *EMBO J. 9*: 1355.
78. Chodosh, L. A., Olesen, J., Hahn, S., Baldwin, A. S., Guarente, L., and Sharp, P. A. (1988). A yeast and human CAAT-binding protein have heterologous subunits that are functionally interchangeable, *Cell 53*: 25.
79. Struhl, K. (1987). The DNA-binding domains of the jun oncoprotein and the yeast *GCN4* transcriptional activator protein are functionally homologous, *Cell 50*: 841.
80. Cunningham, T. S., and Cooper, T. G. (1991). Expression of the *DAL80* gene, whose product is homologous to the GATA factors and is a negative regulator of multiple nitrogen catabolic genes in *Saccharomyces cerevisiae*, is sensitive to nitrogen catabolite repression, *Mol. Cell. Biol. 11*: 6205.
81. Vallee, B. E., Coleman, J. E., and Auld, D. S. (1991). Zinc fingers, zinc clusters and zinc twists in DNA-binding protein domains, *Proc. Natl. Acad. Sci. USA 88*: 999.
82. Saver, R. T. (1990). Scissors and helical forks, *Nature* (*Lond.*) *347*: 514.
83. Aggarwal, A. K., and Harrison, S. C. (1990). DNA regulation by proteins with the helix-turn-helix motif, *Annu. Rev. Biochem. 59*: 933.
84. Keegan, L., Gill, G., and Ptashne, M. (1986). Separation of DNA binding from the transcription-activating function of a eukaryotic regulatory protein, *Science 231*: 699.
85. Pan, T., and Coleman, J. E. (1990). GAL4 transcription factor is not a "zinc finger" but forms a $Zn(II)_2Cys_6$ binuclear cluster, *Proc. Natl. Acad. Sci. USA 87*: 2077.
86. Giniger, E., Varnum, S. M., and Ptashne, M. (1985). Specific DNA binding of GAL4, a positive regulatory protein of yeast, *Cell 40*: 767.
87. Yuan, G-F., Fu, Y-H., and Marzluf, G. A. (1991). *nit-4* a pathway-specific regulatory gene of *Neurospora crassa*, encodes a protein with a putative binuclear zing DNA-binding domain, *Mol. Cell. Biol. 11*: 5735.
88. Baum, J. A., Geever, R., and Giles, N. (1987). Expression of qa-1F activator protein: identification of upstream binding sites in a *qa* gene cluster and localization of the DNA-binding domain, *Mol. Cell. Biol. 7*: 1256.
89. Burger, G., Strauss, J., Scazzocchio, C., and Franz, B. E. (1991). *nirA*, the pathway-specific regulatory gene of nitrate assimilation in *Aspergillus nidulans*, encodes a putative GAL4-type zinc finger protein and contains four introns in highly conserved regions, *Mol. Cell. Biol. 11*: 5746.

90. Beri, R. K., Whittington, H., Roberts, C. F., and Hawkins, A. R. (1987). Isolation and characterization of the positively acting regulatory gene *QUTA* from *Aspergillus nidulans, Nucleic Acids Res. 15*: 7991.
91. Ardrianopoulos, A., and Hynes, M. J. (1990). Sequence and functional analysis of the positively acting regulatory gene *amdR* from *Aspergillus nidulans, Mol. Cell. Biol. 10*: 3194.
92. Klumberg, P., Prange, T., Mathieu, M., Sequeval, D., Scazzocchio, C., and Felenbok, B. (1991). Correct intron splicing generates a new type of putative zinc-binding domain in a transcriptional activator of *Aspergillus nidulans, FEBS Lett. 280*: 11.
93. Kammerer, B., Guyonvarch, A., and Hubert, J. C. (1984). Yeast regulatory gene PPRI. I. Nucleotide sequence, restriction map and codon usage, *J. Mol. Biol. 180*: 239.
94. Friden, P., and Schimmel, P. (1987). *LEU3* of *Saccharomyces cerevisiae* encodes a factor for control of RNA levels of a group of leucine-specific genes, *Mol. Cell. Biol. 7*: 2708.
95. Andre, B. (1990). The *UGA3* gene regulating the GABA catabolic pathway in *Saccharomyces cerevisiae* codes for a putative zinc-finger protein acting on RNA amount, *Mol. Gen. Genet. 220*: 269–276.
96. Messenguy, F., Dubois, E., and Descamps, F. (1986). Nucleotide sequence of the *ARGRII* regulatory gene and amino acid sequence homologies between ARGRII, APRI and GAL4 regulatory proteins, *Eur. J. Biochem. 157*: 77.
97. Kim, J. K., and Mitchels, C. A. (1988). The *MAL63* gene of *Saccharomyces cerevisiae* encodes a cystine-rich finger protein, *Curr. Genet. 14*: 319.
98. Pfeifer, K., Kim, K.-S., Kogan, S., and Guarente, L. (1989). Functional dissection and sequence of yeast HAP 1 activator, *Cell 56*: 291.
99. Salmeron, J. M., Jr., and Johnston, S. A. (1986). Analysis of the *Kluyveromyces lactis* positive regulatory gene *LAC9* reveals functional homology to, but sequence divergence from, the *Saccharomyces cerevisiae GAL4* gene, *Nucleic Acids Res. 14*: 7767.
100. Fu, Y-H., and Marzluf, G. A. (1990). *nit-2*, the major nitrogen regulatory gene of *Neurospora crassa*, encodes a sequence specific DNA binding protein, *Proc. Natl. Acad. Sci. USA 87*: 5331.
101. Munchart, P. L., and Magasanik, B. (1991). Sequence and expression of *GLN3*, a positive nitrogen regulatory gene of *Saccharomyces cerevisiae* encoding a protein with a putative zinc finger DNA-binding domain, *Mol. Cell. Biol. 11*: 6216.
102. Compton, T. (1990). Degenerate primers for DNA amplification, in *PCR Protocols: A Guide to Methods and Applications* (M. A. Innis, D. H. Gelfand, J. J. Sninsky, and T. J. White, eds.), Academic Press, San Diego, p. 39.
103. Davis, M. A., and Hynes, M. J. (1987). Complementation of *areA*$^-$ regulatory mutations of *Aspergillus nidulans* by the heterologous regulatory gene *nit-2* of *Neurospora crassa, Proc. Natl. Acad. Sci. USA 84*: 3753.
104. Silver, P. A. (1991). How proteins enter the nucleus, *Cell 64*: 489.
105. Garcia-Bustos, J., Heitmann, J., and Hall, M. N. (1991). Nuclear protein localization, *Biochim. Biophys. Acta 1071*: 83.

106. Kakidani, H., and Ptashne, M. (1988). GAL4 activates expression in mammalian cells, *Cell 52*: 161.
107. Fischer, J., Giniger, E., Maniatis, T., and Ptashe, M. (1988). GAL4 activates transcription in *Drosophila, Nature* (*Lond.*) *332*: 853.
108. Ma, J., Przibilla, E., Hu, J., Bogorad, L., and Ptashe, M. (1988). Yeast activators stimulate plant gene expression, *Nature* (*Lond.*) *334*: 631.
109. Lech, K., Anderson, K., and Brent, R. (1988). DNA-bound FOS proteins activate transcription in yeast, *Cell 52*: 179.
110. Ginger, E., and Ptashne, M. (1987). Transcription in yeast activated by a putative amphipathic α-helix linked to a DNA binding unit, *Nature* (*Lond.*) *330*: 670.
111. Lewin, B. (1990). Commitment and activation at PolII promoters: a tail of protein-protein interactions, *Cell 61*: 1161.
112. Kelleher, R. J., III, Flanagan, P. M., and Kornberg, R. D. (1990). A novel mediator between activator proteins and the RNA polymerase II transcription apparatus, *Cell 61*: 1209.
113. Metzger, D., White, J. H., and Chambon, P. (1988). The human oestrogen receptor functions in yeast, *Nature* (*Lond.*) *334*: 31.
114. Perlman, D., and Hopper, J. E. (1979). Constitutive synthesis of the Gal4 protein, a galactose pathway regulator in *Saccharomyces cerevisiae, Cell 16*: 89.
115. Selleck, S. B., and Majors, J. E. (1987). In vivo DNA-binding properties of a yeast transcription activator protein, *Mol. Cell. Biol. 7*: 3260.
116. Ma, J., and Ptashne, M. (1987). The carboxy-terminal 30 amino acids of GAL4 are recognised by GAL80, *Cell 50*: 137.
117. Ma, J., and Ptashne, M. (1987). Deletion analysis of GAL4 defines two transcription activating segments, *Cell 48*: 847.
118. Lue, N. F., Chasman, D. I., Buchman, A. R., and Kornberg, R. D. (1987). Interaction of GAL4 and GAL80 regulatory proteins in vitro, *Mol. Cell. Biol. 7*: 3446.
119. Chasman, D. I., and Kornberg, R. D. (1990). GAL4 protein: purification, association with GAL80 protein, and conserved domain structure, *Mol. Cell. Biol. 10*: 2916.
120. Flick, J. S., and Johnston, M. (1990). Two systems of glucose repression of the *GAL1* promoter in *Saccharomyces cerevisiae, Mol. Cell. Biol. 10*: 4757.
121. Mylin, L. M., Johnston, M., and Hopper, J. E. (1990). Phosphorylated forms of GAL4 are correlated with ability to activate transcription, *Mol. Cell. Biol. 10*: 4623.
122. Gill, G., and Ptashne, M. (1988). Negative effect of the transcriptional activator GAL4, *Nature* (*Lond.*) *334*: 721.
123. Caddick, M. X., Brownlee, A. G., and Arst, H. N., Jr. (1986). Regulation of gene expression by pH of the growth medium in *Aspergillus nidulans, Mol. Gen. Genet. 203*: 346.
124. Mylin, L. M., Bhat, J. P., and Hopper, J. E. (1989). Regulated phosphorylation and dephosphorylation of GAL4, a transcriptional activator, *Genes Dev. 3*: 1157.

125. Jones, S. A., Arst, H. N., Jr., and MacDonald, D. W. (1981). Gene roles in the prn cluster of *Aspergillus nidulans, Curr. Genet. 3*: 49.
126. Hull, E. P., Green, P. M., Arst, H. N., Jr., and Scazzocchio, C. (1989). Cloning and physical characterisation of the L-proline catabolism gene cluster of *Aspergillus nidulans, Mol. Microbiol. 3*: 553.
127. Bailey, C. R., Penfold, H. A., and Arst, H. N., Jr. (1979). *Cis*-dominant regulatory mutations affecting the expression of GABA permease in *Aspergillus nidulans, Mol. Gen. Genet. 169*: 79.
128. Arst, H. N., Jr., and MacDonald, D. W. (1975). A gene cluster in *Aspergillus nidulans* with an internally located *cis*-acting regulatory region, *Nature* (*Lond.*) *254*: 26.
129. Lockington, R., Scazzocchio, C., Sequevel, D., Mathieu, M., and Felenbok, B. (1987). Regulation of *alcR*, the positive regulatory gene of the ethanol utilization regulon of *Aspergillus nidulans, Mol. Microbiol. 1*: 275.
130. Dowzer, C. E. A., and Kelly, J. M. (1991). Analysis of the *creA* gene, a regulator of carbon catabolite repression in *Aspergillus nidulans, Mol. Cell Biol. 11*: 5701–5709.
131. Wiame, J-M., Grenson, M., and Arst, H. N., Jr. (1985). Nitrogen catabolite repression in yeasts and filamentous fungi, *Adv. Microbial Physiol. 26*: 1.
132. Caddick, M. X., and Arst, H. N., Jr. (1990). Nitrogen regulation in *Aspergillus*: are two fingers better than one? *Gene 95*: 123.
133. Sharma, K. K., and Arst, H. N., Jr. (1985). The product of the regulatory gene of the proline catabolism gene cluster of *Aspergillus nidulans* is a positive-acting protein, *Curr. Genet. 9*: 299.
134. Arst, H. N., Jr. (1976). Integrator gene in *Aspergillus nidulans, Nature* (*Lond.*) *262*: 231.
135. Tollervey, D. W., and Arst, H. N., Jr. (1981). Mutation to constitutivity and derepression are separate and separable in a regulatory gene of *Aspergillus nidulans, Curr. Genet. 4*: 63.
136. Wickens, M. (199). How the messenger got his tail: addition of poly(A) in the nucleus, *Trends Biochem. Sci. 15*: 277.
137. Shatkin, A. J. (1985). mRNA CAP binding proteins: essential factors for initiating transcription, *Cell 40*: 223.
138. Furuichi, Y., La Fiandra, A., and Shatkin, A. J. (1977). 5′-Terminal structure and mRNA stability, *Nature* (*Lond.*) *266*: 235.
139. Proudfoot, N. (1991). Poly(A) signals, *Cell 64*: 671.
140. Jackson, R. J., and Standart, N. (1990). Do the poly(A) tail and 3′ untranslated region control mRNA translation? *Cell 62*: 15.
141. Herrick, D., Parker, R., and Jacobson, A. (1990). Identification and composition of stable and unstable mRNA's in *Saccharomyces cerevisiae, Mol. Cell. Biol. 10*: 2269.
142. Hunt, T. (1988). Controlling mRNA lifespan, *Nature* (*Lond.*) *334*: 567.
143. Brewerman, G. (1989). mRNA decay: finding the right targets, *Cell 57*: 9.
144. Fink, G. R. (1987). Pseudogenes in yeast? *Cell 49*: 5.
145. Smith, C. W. J., Patton, J. G., and Bernardo, N-G. (1989). Alternative splicing in the control of gene expression, *Annu. Rev. Genet. 23*: 527.

146. Duncan, R., Milburn, S. C., and Hershey, J. W. B. (1987). Regulated phosphorylation and low abundance of the HeLa cell initiation factor eIF-4F suggests a role in translational control: heat shock effect as eIF-4F, *J. Biol. Chem. 226*: 380.
147. Hinnebusch, A. G. (1984). Evidence for translational regulation of the activator of the general control of amino acid biosynthesis in yeast, *Proc. Natl. Acad. Sci. USA 81*: 6442.
148. Mueller, P. P., and Hinnebusch, A. G. (1986). Multiple upstream AUG codons mediate translational control of GCN4, *Cell 45*: 201.
149. Thireos, G., Driscoll-Penn, M., and Greer, H. (1984). 5′ uptranslated sequences are required for the translational control of a yeast regulatory gene, *Proc. Natl. Acad. Sci. USA 81*: 5096.
150. Kozak, M. (1984). Selection of initiation sites by eukaryotic ribosomes: effect of inserting AUG triplets upstream from the coding sequence of preproinsulin, *Nucleic Acids Res. 12*: 3873.
151. Roussou, I., Thireos, G., and Hauge, B. M. (1988). Transcriptional-translational regulatory circuit in *Saccharomyces cerevisiae* which involves the GCN4 transcriptional activator and the GCN2 protein kinase, *Mol. Cell. Biol. 8*: 2132.
152. Wek, R. C., Ramirez, M., Jackson, B. M., and Hinnebusch, A. G. (1990). Identification of positive-acting domains in GCN2 protein kinase required for translational activation of *GCN4* expression, *Mol. Cell. Biol. 10*: 2820.
153. Werner, M., Feller, A., Messenguy, F., and Pierard, A. (1987). The leader peptide of yeast gene *CPA2* is essential for the translational repression of its expression, *Cell 49*: 805.
154. Craigen, W. J., and Cashey, C. T. (1987). Translational frameshifting: where will it stop? *Cell 50*: 1.
155. Mellor, J., Wilson, J., Kingsman, S. M., and Kingsman, A. J. (1985). A retrovirus-like strategy for expression of a protein encoded by yeast transposon Ty1, *Nature (Lond.) 313*: 243.
156. Wilson, W., Malim, M. M., Mellor, J., Kingsman, A. J., and Kingsman, S. M. (1986). Expression strategies of the yeast retrotransposon Ty: a short sequence directs ribosomal frameshifting, *Nucleic Acids Res. 14*: 7001.
157. Berk, A. J. (19889). Regulation of eukaryotic transcription factors by posttranslational modification, *Biochim. Biophys. Acta 1009*: 103.
158. Nasmyth, K., Seddon, A., and Ammerer, G. (1987). Cell cycle regulation of SWI5 is required for mother-cell specific *HO* transcription in yeast, *Cell 49*: 549.
159. Nasmyth, K., Adolf, G., Lydall, D., and Seddon, A. (1990). The identification of a second cell cycle control on the HO promoter in yeast: cell cycle regulation of SWI5 nuclear entry, *Cell 62*: 631.
160. Moll, T., Tebb, G., Surana, V., Robitsch, M., and Nasmyth, K. (1991). The role of phosphorylation and the CDC28 protein kinase in cell cycle-regulated nuclear import of the *S. cerevisiae* transcription factor SWI5, *Cell 66*: 743.
161. Reed, S. I., Hardwiger, J. A., and Loricz, A. (1989). Protein kinase activity associated with the product of the yeast cell division cycle gene *CDC28, Proc. Natl. Acad. Sci. USA 82*: 4055.

162. Nurse, P. (1990). Universal control mechanism regulating onset of M-phase, *Nature* (*Lond.*) *344*: 503.
163. Metzenberg, R. L. (1979). Implications of some genetic control mechanisms in *Neurospora, Microbiol. Rev. 43*: 361.
164. Hoeksta, M. F., Demaggio, A. J., and Dhillon, N. (1991). Genetically identified protein kinases in yeast, *Trends Genet. 7*: 256.

2

Heat Shock Proteins in Fungi

Nora Plesofsky-Vig and Robert Brambl

University of Minnesota
St. Paul, Minnesota

I. INTRODUCTION

All organisms from bacteria to fungi, plants, and animals respond to elevated, but sublethal temperatures with a major redirection of gene expression that leads to the strong synthesis of a small group of proteins. These heat shock-induced proteins have been shown to help cells adapt to the inducing temperature and to survive exposure to higher, otherwise lethal temperatures. It was initially unclear what the functions of these newly discovered proteins were, but within the last 5 years, intensive investigations into the activities of individual heat shock proteins have produced important advances in our understanding of their cellular roles. It has been discovered that many of the heat shock proteins, or closely related proteins, are essential proteins that are also synthesized by cells at normal temperature. The functions of these proteins, especially in non-heat-shocked cells, have been dissected and normal cellular processes have been identified in which the heat shock proteins participate.

The heat shock proteins that are most strongly synthesized by all organisms fall into four general classes: hsp70, hsp83/90, hsp100, and the α-crystallin-related small heat shock proteins. hsp60 and ubiquitin are two other common heat shock proteins. Many of these heat shock proteins interact primarily with other cellular proteins to facilitate their proper conformational folding, assembly into protein complexes, and translocation into target organelles. Because of these effects on the conformation of other proteins, these heat shock proteins have been called molecular chaperones. In some organisms, the heat shock proteins are also induced to high levels during certain stages of development, often connected with sexual reproduction or embryogenesis. In certain fungi, heat shock itself is a signal for a developmental transition. Another area of active investigation is the mechanism by which heat shock gene transcription is induced in cells that experience a heat stress. Heat shock transcription factors from several types of organisms, including *Saccharomyces cerevisiae*, have been identified and characterized. These transcription factors generally preexist in cells, and their activation by heat shock may be due to their extensive phosphorylation at high temperature.

The heat shock response of *S. cerevisiae* has been more intensively studied than that of other fungi and forms a model to which others can be compared (1). Many of its heat shock genes have been sequenced and their regulated expression has been analyzed. The functions of the heat shock proteins of *S. cerevisiae* have been assayed in vitro and, most important, these functions have been assayed in vivo by the engineering of targeted mutations in specific heat shock genes. Although *S. cerevisiae* has proved to be extraordinarily useful in gaining an understanding of heat shock proteins and the expression of their genes, this yeast does not appear to be typical in all respects. In particular, *S. cerevisiae* differs from most other organisms in not regulating expression of heat shock genes at the translational level (2). In contrast, the filamentous fungus *Neurospora crassa* does exercise translational regulation during heat shock, discriminating against its preserved normal mRNAs (3). *S. cerevisiae* also differs from certain other fungi in the DNA-binding properties of its heat shock transcription factor and in the cellular localization of certain heat shock proteins, as will be discussed.

II. HEAT SHOCK PROTEINS: CHAPERONES

A. hsp70

hsp70 is an intensively studied heat shock protein that belongs to a family of proteins with different expression patterns; some hsp70s are expressed constitutively, while some are induced by high temperature or by other stresses. The hsp70s of eukaryotes have approximately 50% homology to a bacterial heat shock protein, DnaK of *Escherichia coli* (4). The hsp70 proteins have been found to bind other peptides, as well as ATP, and peptide release depends on ATP hydrolysis (5). One function of hsp70, originally proposed by Lewis and Pelham (6), is through this binding and release to disrupt aggregated or misfolded proteins that result from protein denaturation at elevated temperature or from other causes. All the hsp70s that have been investigated, cytoplasmic and organellar, have been found to play a role in protein folding and/or translocation during normal growth. During heat shock, the functions of hsp70 may be expanded to include the renaturation of proteins that have become improperly folded as a result of high temperature.

In *S. cerevisiae* nine genes for these proteins have been identified. They have overlapping expression patterns, with seven being constitutively expressed at normal temperature and six genes being heat-inducible (7). Craig and co-workers have tested the essentiality of some of these genes by creating targeted mutations in specific sequences, which are used to replace endogenous genes in transformed cells. They have found that certain hsp70 genes can substitute for others that are defective, while other hsp70 genes cannot complement the lost function. This indicates that there are separate functional groups of hsp70-related genes, due either to different functions or to different temperature optima of the encoded hsp70 proteins. Two constitutively expressed genes, *SSA1* and *SSA2*, which are 96% identical, can substitute for one another, although only *SSA1* is induced by heat shock. At least one of these genes is necessary for sustained growth and colony formation at 37°C (8). Cell viability at normal temperature (23°C) is abolished if double mutant *ssa1-ssa2* cells also lack functional genes for SSA3 and SSA4, which are not normally expressed constitutively. This indicates that the four hsp70-related proteins SSA1–4, which have different expression patterns, are functionally related (9). Two other hsp70-related genes *SSB1* and *SSB2*, whose expression is repressed by heat shock, can compensate for one another at low temperatures. Yeast cells with mutations in both genes grow slowly at temperatures below 30°C (8).

The smut fungus *Ustilago maydis* has four identified genes for hsp70-related proteins. All these genes are constitutively expressed, and two are induced by heat shock (10). Targeted disruption of one of the inducible genes, *ums2*, appeared to be recessively lethal, suggesting that this gene is essential for cell viability at normal temperature and that none of the other hsp70 genes can functionally complement it (10). Alternatively, *ums2* may be required specifically for *U. maydis* sporulation or germination.

Experiments to test the functions of the cytoplasmic hsp70 of *S. cerevisiae*, at normal temperature, were performed in a yeast strain disrupted in all four functionally related hsp70 genes, *SSA1–4*. The mutant strain contained a single introduced *SSA1* gene whose expression was inducible by galactose. hsp70 was found to be required for translocation of the α-factor precursor into the endoplasmic reticulum and for its secretion. Furthermore, it is implicated in mitochondrial transport, since the precursor of a mitochondrial protein, the β-subunit of ATPase, also accumulated in the absence of hsp70 (11). To show biochemically that hsp70 was needed for organelle translocation, in vitro protein import was assayed with isolated organelles. It was found that another protein in yeast extracts, in addition to hsp70 (SSA1), was needed for the translocation of α-factor precursor into microsomes (11, 12) and for the import of a mitochondrial protein precursor (13). It has not yet been determined whether hsp70 interacts directly with precursor proteins, keeping them in a translocation-competent state, or whether hsp70 may instead facilitate import indirectly by interacting with other, essential import factors (13).

Other hsp70-related proteins are located within the organelles themselves. Genes encoding hsp70 proteins that are imported into mitochondria have been isolated from *S. cerevisiae* (*SSC1*) (14) and from *Schizosaccharomyces pombe* (15); the proteins of these two fungi have 68% identity. Both these hsp70s are synthesized during normal growth, but their expression increases severalfold upon heat shock. The *SSC1* gene has been found, by gene disruption, to be essential for growth of *S. cerevisiae* at normal temperature (16). A temperature-sensitive defect in *SSC1* rendered this yeast unable to complete the transport of precursor proteins into mitochondria and to refold them (17). Although localized in the matrix of mitochondria, the mitochondrial hsp70 forms a complex with precursor proteins at membrane contact sites (18,19). Mitochondrial hsp70 may have an additional role in DNA recombination, since it has been identified in a protein complex with a DNA endonuclease, and its association is required for full endonucleolytic activity (20).

A gene (*KAR2*) for an hsp70 that is located in the endoplasmic reticulum has also been characterized in *S. cerevisiae* (21,22). This hsp70 has a carboxy-terminal signal for retention in the endoplasmic reticulum, as well as an amino-terminal signal sequence. It is proposed that the microsomal hsp70 facilitates the proper folding of secretory and transmembrane proteins during normal growth, by stabilizing partially unfolded proteins in the endoplasmic reticulum. Expression of the microsomal hsp70 is strongly induced by inhibitors of protein glycosylation and by mutations that cause secretory proteins to accumulate in the endoplasmic reticulum (22). *KAR2* expression in yeast is also induced by heat shock (21,22), a stimulus that does not induce expression of the corresponding mammalian gene, *BiP/GRP78*. Disruption of *KAR2* is lethal to *S. cerevisiae* (21,22). Mutations in the *KAR2* gene that lead to temperature-sensitive growth also block conjugative nuclear fusion, an effect on nuclei that may be an indirect result of a lesion in endoplasmic reticulum structure or function (21,22).

B. hsp83

hsp83 is an extremely abundant cellular protein at normal temperature that is also induced by heat shock. The eukaryotic hsp83 has 40% identity to the *E. coli* protein htpG (7). A molecular target for hsp83 in animal cells is uncomplexed steroid hormone receptors, to which hsp83 binds. The aquatic fungus *Achlya ambisexualis* secretes the steroid antheridiol as one of its mating hormones, and the complex interactions of this fungal steroid hormone and its receptor with hsp83 are remarkably similar to the interrelationships discovered in animals. The hsp83 of *A. ambisexualis* has been isolated in a stabilized complex with the fungal steroid receptor (23). Furthermore, as in animal cells, treatment with the steroid antheridiol leads to increased synthesis of hsp83 by this fungus, without inducing expression of most of the other heat shock proteins (24). *A. ambisexualis* has two genes encoding hsp83 (25), but their individual expression patterns are not known.

In *S. cerevisiae*, there are also two genes encoding hsp83, one of which is expressed more strongly during normal growth, while the other is induced to higher levels by heat shock (26). In accordance with their overlapping expression patterns, these genes can substitute for one another in gene disruption experiments. The presence of one functional gene for hsp83 is necessary and sufficient for cell viability, but growth at heat shock temperatures requires more than one gene (26). This appears to be due to a requirement for a greater accumulation of hsp83 at high temperature, rather than to any distinction between the two hsp83 proteins.

A steroid hormone receptor has not been identified in yeast, and the function of its hsp83 remains obscure. Nevertheless, *S. cerevisiae* has proved very useful for analyzing the interaction of hsp83 (of yeast) with a mammalian hormone receptor, after the gene for this receptor was introduced into yeast. Since the steroid hormone receptor does not bind hsp83 at the same time that it binds the hormone, it was generally assumed that hsp83 binding might prevent the steroid receptor from binding hormone and, thereby, block transcriptional activation by the hormone-receptor complex. However, Picard et al. (27) were able to determine, by varying the level of hsp83 expressed in *S. cerevisiae*, that hsp83 was in fact required for maximal hormonal activation of the steroid receptor. High levels of hsp83, in conjunction with constant amounts of the hormone and its receptor, activated transcription of a reporter gene (for β-galactosidase) driven by a steroid response element, which had been introduced into the yeast cells. These experiments suggest an important role for hsp83 in activating a protein, such as the steroid receptor, for subsequent interactions with other molecules. In animal cells, the binding of hsp83 to other proteins such as the dioxin receptor (28) and tyrosine kinases (29) may serve as an initial step in the activation of these proteins also.

C. hsp60

hsp60 was first found to be a mitochondrial-localized protein in *Tetrahymena thermophila* (30), but shortly thereafter it was identified genetically in

S. cerevisiae as a protein necessary for the assembly of peptides that were imported into mitochondria (31). Mitochondrial proteins that require a functional hsp60 for assembly into complexes include intermembrane proteins, such as cytochrome b_2 and the Rieske Fe/S protein of complex III (31), and proteins of the mitochondrial matrix, such as the β-subunit of F_1-ATPase (31) and hsp60 itself (32). hsp60 is synthesized constitutively, as well as being induced by heat shock, and it has been shown by gene disruption to be an essential protein in *S. cerevisiae* at normal temperature (33). hsp60 is closely related to the GroEL protein of *E. coli* (54% identity), which similarly participates in protein assembly and, in addition, is required for export of certain secretory proteins (34). GroEL is aided by GroES, a 10-kDa peptide, for which a homolog has been identified in mitochondria (35).

The hsp60 of *N. crassa*, like other hsp60s and GroEL (30), is assembled into a 14-subunit oligomeric structure consisting of two stacked rings with seven monomers each (36). This particle catalyzes, in an ATP-mediated reaction, the folding of proteins that have been unfolded during the normal process of translocation into mitochondria (37). Assembly of multicomponent protein complexes may also occur directly on the hsp60 particle or assembly may be indirectly facilitated by protein folding. The copurification of hsp60 with mitochondrial holoenzymes, such as the cytochrome reductase of *N. crassa* (36) and the yeast ATPase (38), suggests that hsp60 may directly participate in protein assembly as well as in protein folding.

D. DnaJ-Related Proteins

Three genes of *S. cerevisiae, SCJ1, MAS5(YDJ1),* and *SEC63*, encode proteins that are homologous to DnaJ of *E. coli*, a bacterial heat shock protein that interacts with the hsp70-homolog DnaK in several processes (39). The results of overexpressing SCJ1 in yeast suggest that this protein is involved in translocating proteins into mitochondria and in protein secretion via the endoplasmic reticulum. Nevertheless, SCJ1 itself is located mainly in mitochondria (40). *SCJ1* is not essential, according to gene disruption, for yeast growth at any temperature (40). The related *MAS5* gene, however, is essential for yeast growth at high temperature, and it is moderately induced by heat shock (41). The MAS5 protein is required for import of precursor proteins into mitochondria, especially at elevated temperature (41), and it is located outside of mitochondria (42). Another DnaJ-related protein, SEC63, is required for translocation of proteins into the endoplasmic reticulum (43). Yeast cells with a mutant *SEC63* accumulate secretory precursors in the cytoplasm, and at high temperature, they are unable to grow (44).

III. HEAT SHOCK PROTEINS: PROTEASES

A. hsp100

A protein of approximately 100 kDa is synthesized by most organisms in response to heat shock. The gene for hsp104 of *S. cerevisiae* has been sequenced (45) and a partial peptide sequence for hsp98 of *N. crassa* was determined (46). Both these proteins are closely related to the prokaryotic ClpA/ClpB protein, which is the regulatory subunit of an ATP-dependent protease. ClpB and the catalytic subunit of the protease, ClpP, are induced in *E. coli* by heat shock (47,48). Yeast hsp104 is not synthesized constitutively, nor is it essential for cell viability. However, it is required for induced thermotolerance. Deletion of *HSP104* abolished the thermotolerance to a lethal temperature that cells acquire from a moderate heat shock, although the deletion did not affect yeast growth at normal or heat shock temperature (49). This crucial role of hsp104 in induced thermotolerance requires functional ATP-binding sites that have been identified in hsp104 and homologous proteins (45). The function of this ClpB-related protein, therefore, might involve regulation of an ATP-dependent protease that has not yet been identified. Deleting *HSP104* of yeast led to reduced protein synthesis during prolonged heat shock (49) and delayed recovery of mRNA splicing, following a severe cellular heat shock (50). The homologous hsp98 of *N. crassa* is highly enriched in the microsomal fraction of cell extracts, and a portion of it associates with polyribosomes, particularly the large ribosomal subunit (46).

B. Ubiquitin

Ubiquitin is a small protein (76 amino acids) that becomes covalently attached to lysine residues of other proteins, thereby marking them for ATP-dependent proteolysis. An early indication of the connection between ubiquitin and the heat shock response was the discovery, first in mouse (51) and later in yeast cells (52), that a defect in conjugating ubiquitin to proteins leads to the production of heat shock proteins by the mutant cells. It was also found that the levels of ubiquitin increase in wild-type cells in response to cellular heat shock. Of the four genes for ubiquitin that have been characterized in *S. cerevisiae*, one (*UBI4*) encodes a polyubiquitin fusion protein consisting of five ubiquitin repeats; in the other three genes, the ubiquitin sequence is fused to an unrelated gene (53). Only the polyubiquitin *UBI4* gene is induced by heat shock (54), while all four ubiquitin genes are expressed by exponentially growing yeast cells (53). The *UBI4* gene was shown to be dispensable for normal vegetative growth, but it is essential for yeast resistance to various stresses, including chronic high temperature (38.5°C), starvation, and exposure to amino acid analogs (54). A functional *UBI4* gene is also required for yeast sporulation and for the maintenance of viable spores (54, 55). *N. crassa* also has a polyubiquitin gene, which is composed of four ubiquitin repeats, but its expression is not induced by heat shock or other stresses (56).

Also in the ubiquitin pathway, two enzymes that conjugate ubiquitin to proteins were found to be more strongly expressed in yeast in response to heat shock (52). Codisruption of genes for these two ubiquitin-conjugating enzymes led to impaired growth at normal temperature and cell inviability at the heat shock temperature (52). The mutant cells were unable to grow in the presence of an amino acid analog, and they were defective specifically in the proteolysis of short-lived and abnormal proteins. Since eight ubiquitin-conjugating enzymes have been identified in yeast, several of which are engaged in other cellular processes (52), these results suggest that the role of ubiquitin in mediating proteolysis of abnormal and short-lived proteins is crucial for cell survival during stress.

IV. HEAT SHOCK PROTEINS: UNKNOWN FUNCTION

A. α-Crystallin-Related Proteins

Relative to the other heat shock proteins, less progress has been made in understanding the function of a widespread group of divergent heat shock proteins of 15–30 kDa that have sequence homology to vertebrate lens α-crystallin (57,58). Despite their divergent sequences, these small heat shock proteins have similar hydropathy profiles, and they all appear to be organized into roughly spherical particles that consist of 20 or more monomers (59). Although multiple, related small heat shock proteins are synthesized by most plants and animals, only one gene for an α-crystallin-related protein has been identified in *S. cerevisiae* (60) and in *N. crassa* (61). The respective genes are not expressed constitutively in either fungus.

The intracellular location of these small heat shock proteins varies. The *N. crassa* hsp30 is predominantly associated with mitochondria, according to cell fractionation, but this association requires continued cell exposure to high temperature. Shift-down of the cells to normal temperature leads hsp30 to dissociate from mitochondria (61). This in vivo pattern of localization is reflected by in vitro studies, which show that hsp30 binds with strong preference to mitochondria isolated from heat-shocked cells, compared to those from unstressed cells (61). Near its amino-terminus, *N. crassa* hsp30 has a sequence that can form an amphiphilic α-helix. This structure may regulate the in vivo mitochondrial binding of hsp30, since engineered deletion of this sequence destroyed the in vitro binding selectivity shown by hsp30 (61). In contrast to this dependence on heat shock, the location of yeast hsp26 varies with the carbon source, regardless of temperature. When *S. cerevisiae* cells are growing fermentatively on glucose, hsp26 concentrates in the nucleus. However, hsp26 becomes distributed throughout the cells when they are growing on respiratory substrates, such as galactose or acetate (62).

In contrast to other heat shock proteins of *S. cerevisiae*, deletion or disruption of the gene for hsp26 does not affect yeast growth at normal or heat shock temperatures, nor does this mutation interfere with the acquisition of heat shock–induced thermotolerance (60). Engineered constitutive expression of hsp26, however, does produce a small increase in cell survival at 50°C, a lethally high temperature (59,63).

B. Other Heat Shock Proteins

hsp12 and STI1 are heat shock proteins of yeast that do not have clear homologs in other organisms. hsp12 shows a small amount of homology to an α-crystallin-related protein of *Drosophila melanogaster*, hsp26, but this similarity does not include the characteristic α-crystallin-related region (64). Disruption of the gene for hsp12 does not affect growth or thermotolerance, nor does a double mutation in hsp12 and hsp26 appear to make yeast cells more heat-sensitive than a single mutation (64). The yeast STI1 is a heat shock–inducible protein of 66 kDa that is unrelated in sequence to the hsp70 family of proteins, but may have some overlapping functions (65). STI1 appears to be important for yeast growth at temperature extremes, since disruption of its gene reduced cell growth at temperatures below 23°C and above 30°C (65).

V. HEAT SHOCK PROTEINS: GLYCOLYSIS AND RESPIRATION

The synthesis of certain glycolytic enzymes and respiratory proteins has been found to increase during heat shock. An isoform of enolase was one of the first heat shock proteins to be identified as a glycolytic enzyme; this isoform, enolase 1, is also strongly expressed during stationary phase in *S. cerevisiae* (66). Heat shock also induces one of the three isoforms of glyceraldehyde-3-phosphate dehydrogenase, which is an abundant protein during yeast growth at normal temperature (7). The glycolytic enzyme phosphoglycerate kinase, similarly, is induced severalfold at the transcriptional level by heat shock, but only if the yeast cells are growing fermentatively on glucose, not in cells growing on the respiratory substrates glycerol or ethanol (67). An isoform of cytochrome *c* (iso-2 cytochrome *c*) that is expressed during stationary phase was also found to be strongly induced transcriptionally by heat shock, under both aerobic and anaerobic conditions (68). The heat shock–induced synthesis of these glycolytic and respiratory proteins may help heat-stressed cells continue to generate sufficient ATP for essential metabolic processes.

VI. HEAT SHOCK PROTEINS: OXIDATIVE DAMAGE

High temperature stress has been found to increase oxidative damage to cells (69), possibly by accelerating the formation of reactive oxygen intermediates and/or by increasing their reactivity. Enzymes that convert these oxygen species are induced by heat shock. In fungi, enzymes that metabolize and, thereby, detoxify hydrogen peroxide are induced by high temperature. Heat shock produces high levels of peroxidase activity in *N. crassa*, although there is no increase in superoxide dismutase activity (70). The activities of two of the three catalases that have been identified in *N. crassa* increased dramatically in response to heat shock (71). Heat shock also strongly induced the cytosolic catalase T of *S. cerevisiae*, while catalase A, a peroxisomal enzyme, was not induced (72). Yeast strains that were disrupted in the catalase T gene showed reduced survival at lethally high temperature, indicating that this peroxide-detoxifying activity does enhance thermotolerance (72).

The metallothioneins are small cysteine-rich peptides that, in fungi, appear to be induced mainly by copper, but which bind various heavy metals. It has also been reported that metallothionein is extremely efficient at scavenging free hydroxyl radicals, which may account for its ability to protect cells against radiation damage (73). In *S. cerevisiae* the metallothionein gene is moderately, but consistently, induced by heat shock. The normal induction of metallothionein by copper is mediated by a copper-dependent activator of metallotionein gene transcription. In yeast strains that are deleted in the activator gene, it was found that the resultant sensitivity to copper could be suppressed by a point mutation in the heat shock transcription factor (HSF, discussed below) that affected its binding to DNA. The mutant HSF bound to a region within the promoter of the metallothionein gene that contains a consensus heat shock element (74,75). Interestingly, this mutation of HSF decreased the induction of heat shock gene expression, while it increased the basal level of metallothionein gene transcription and also made it inducible by cadmium (75).

DNA damage can be produced by mutagenic oxidants such a 4-nitroquinoline-1-oxide and by ultraviolet irradiation. Treatment of *S. cerevisiae* with these agents leads to the expression of a set of DNA-damage-responsive (DDR) genes. Two of these genes, *DDRA2* and *DDR48*, are rapidly induced by heat shock as well as by treatment with mutagens (76). Other DNA damage–induced genes, however, including those involved directly in DNA synthesis or repair, are not induced by heat shock (76,77). Surprisingly, induction of *DDRA2* by heat shock depends on promoter sequences that differ from the standard consensus elements and do not bind the heat shock transcription factor (see below). The heat shock–induced ubiquitin gene *UBI4* has a similar promoter element (78), as well as the standard heat shock element (53).

VII. THE HEAT SHOCK TRANSCRIPTION FACTOR

The transcription of heat shock genes depends on activation of a transcription factor (HSF), which binds to the heat shock consensus element (HSE) found in the upstream promoter region of heat shock genes. The basic unit of the HSE is the sequence -GAA- repeated every five nucleotides on alternating strands of DNA (79). The gene for the 93.3-kDa HSF of *S. cerevisiae* has been sequenced, and the domain of the protein that binds to DNA has been identified experimentally (80,81). Although this domain has no motifs that are characteristic of other eukaryotic DNA-binding proteins, it has a small region of similarity to the sigma factors of bacterial RNA polymerases (82). Oligomerization of the transcription factor, which forms a trimer in solution, requires leucine zipper-like sequences within the protein that can form a three-stranded α-helical coiled-coil (83). When bound to DNA, especially that containing an extended HSE, the transcription factor may form hexamers as well as trimers.

The *S. cerevisiae* HSF binds to its target DNA in the absence of a heat shock (84), but a heat shock is required to fully activate the factor for promoting transcription (84). The mechanism by which heat shock activates HSF, thereby enhancing transcription of heat shock genes, probably involves the increased phosphorylation of HSF that results from a heat shock (81, 85). As in *S. cerevisiae*, the HSF of another budding yeast, *Kluyveromyces lactis*, also binds to DNA constitutively (86). However, such binding does not occur in the fission yeast *S. pombe*, whose HSF requires a heat shock to bind the consensus HSE (87); the heat shock factor of animals also does not bind DNA constitutively (88). The binding capabilities of the HSF of budding yeasts may, therefore, be an exception to the general pattern of heat shock–dependent binding of the transcription factor.

The gene for HSF has been shown, by disruption, to be essential for the viability of *S. cerevisiae* at normal temperatures, as well as at heat shock temperatures (80,81), and the transcription factor shows a low level of transcriptional activation at normal temperatures (84). It has been determined that the constitutive expression of hsp70 (SSA1) and hsp83 of *S. cerevisiae* is, in fact, regulated by HSF. One of the consensus elements present in the *SSA1* promoter was shown by deletion analysis to be much more important than the other HSEs for heat shock–inducible transcription. It was found that the same HSE is also required for constitutive expression of *SSA1* (89). Similarly, mutation of a TATA-proximal HSE in the hsp83 gene strongly reduced both heat shock–induced transcription of hsp83 and the basal level of its transcription, at the same time that HSF binding decreased (90). Despite the similar requirement for HSEs that bind the HSF, there are apparently different mechanisms that govern basal transcription versus heat shock–induced transcription. Based on deletion analyses of the HSF, the protein domain that is required for heat shock–induced transcriptional activation is distinct from the regions of HSF that make it constitutively active (85, 91). Furthermore, mechanisms that repress transcription are operative at normal temperatures. The hsp70 gene *SSA1* has an adjacent, upstream site in the DNA that, in conjunction with a yeast cell factor, represses constitutive transcription (89). Also identified is a domain within the HSF that represses strong transcriptional activity at normal temperature (85,92); repression may be mediated by a conserved heptapeptide within this domain (86).

These results indicate that the constitutive binding of HSF to HSEs, which has only been described for budding yeasts, serves an important function for these cells that is unrelated to heat shock. This binding apparently mediates the constitutive expression of heat shock proteins that are essential for cell survival and growth at normal temperatures. The importance of HSF and, hence, the heat shock proteins specifically during heat shock is demonstrated by a mutant strain of *S. cerevisiae* with a temperature-sensitive defect in HSF. The two cellular functions that were most affected by this defect are the posttranslational import of mitochondrial proteins and cell progression through the G2 phase of the cell cycle (92). During heat shock, these functions may require a higher level of the constitutively synthesized heat shock proteins, or these activities may require other proteins whose synthesis is limited to heat shock-induction. By itself, increased expression of hsp70 (SSA1) did not correct the defect in mitochondrial protein translocation (92); nor was protein secretion, another hsp70-requiring function (11,12), interrupted by the temperature-sensitive defect in HSF (92).

VIII. CELLULAR EFFECTS OF HEAT SHOCK

A. Respiration and Energy Generation

The heat shock induction of several proteins that, under normal conditions, participate in the import and assembly of mitochondrial proteins suggests that mitochondrial functions and/or structures are vulnerable to cellular heat shock. hsp60, the mitochondrial hsp70, and possibly the DnaJ-related

SCJ1 are three identified heat shock proteins that are themselves located in mitochondria and that function in protein folding and unfolding reactions.

One of the immediate physiological responses of *N. crassa* to heat shock is a 30% decline in respiratory oxygen consumption, occurring within the first 20 min of exposure to high temperature (93). Energization of the mitochondrial inner membrane is also dramatically reduced by cellular heat shock, as determined by fluorescence microscopy with a potentiometric reporter dye (93). These initial effects on mitochondria would be expected to lead to a reduction in respiratory ATP production and a lag in protein import, which requires an energized mitochondrial membrane. These deleterious initial effects appear to be reversed in *N. crassa* during a prolonged heat shock (93). Increased temperature also affects the respiration of fungi that are human pathogens and that convert from mycelial growth to a yeast form under the stimulus of high temperature. Heat shock initially stimulated respiration in these fungi, but after 20 min oxygen consumption declined (94). High temperature also caused partial uncoupling of oxidative phosphorylation in *S. cerevisiae* (95) and *Blastomyces dermatitidis*; but uncoupling did not occur in *Paracoccidioides brasiliensis* (94), and it only occurred in an attenuated, but not in a virulent, strain of *Histoplasma capsulatum* (96).

B. Splicing of mRNA Precursors

Although expression of most normal genes is reduced or discontinued during heat shock, the synthesis of heat shock proteins requires functioning cellular machinery for the transcription and translation of heat shock genes. A vulnerable step in gene expression has been found to be RNA splicing. Many heat shock genes do not contain introns, possibly because an extreme heat shock inhibits splicing of precursor RNAs and the consequent production of mature mRNAs (4). In *S. cerevisiae* the actin gene, which contains an intron, continues to be transcribed during cellular heat shock. Mature actin mRNA is produced by yeast at 37°C, whereas a larger precursor RNA accumulates at 41°C, indicating that RNA processing is blocked (50). In contrast, the pathogenic, dimorphic fungus *H. capsulatum* does not display an inhibition of RNA splicing during extreme heat shock. The *H. capsulatum* gene for hsp83 contains two introns, but the introns were spliced out of its RNA at the high heat shock temperature of 42°C, as well as at 37°C (97). Processing of the mRNA precursor for β-tubulin, which contains several introns, appeared similarly resistant to extreme heat shock. It seems reasonable that the splicing apparatus of this fungus might have unique resistance to extreme high temperature, since *H. capsulatum* is subjected to heat shock continuously as a condition for infection of its mammalian host (97).

IX. HEAT SHOCK AND DEVELOPMENT

A. Differentiation

Heat shock proteins are induced in many organisms as they undergo sexual or embryonic development. High levels of RNAs for hsp26 and hsp83 accumulate in *S. cerevisiae* cultures that are induced to sporulate, and these RNAs are retained in the mature spores (98); hsp26 itself also accumulates to high levels during sporulation (99). It has been shown that yeast sporulation requires the heat-inducible ubiquitin gene, *UBI4* (55). The induction of heat shock genes, however, is not tightly coupled to sporulation. Diploid cells that are nonsporulating, due to homozygosity at the mating-type locus, also accumulate high levels of hsp26 and hsp83 mRNA on the nitrogen-deficient sporulation medium (98). Furthermore, yeast strains carrying a disrupted or deleted hsp26 gene sporulate normally (60). Unlike in *S. cerevisiae*, RNA for the *Neurospora* hsp30 has not been detected in mature ascospores of *Neurospora tetrasperma* (N. Plesofsky-Vig, A. Paulson, E. P. Hill, and R. Brambl, unpublished data).

Upon infection of a mammalian host, dimorphic fungi undergo a transition from a filamentous to a yeast morphology. A high temperature that corresponds to the host temperature, by itself, can induce this transition in culture. There is a strong correlation between the heat-induced transition of these dimorphic, pathogenic fungi and the synthesis of heat shock proteins (100). There is no evidence, however, that the heat shock proteins themselves are involved in the developmental change to a yeast form. They may instead be associated with fungal adaptation to a higher temperature. This is suggested by the finding that an avirulent, heat-sensitive strain of *H. capsulatum* maximally synthesized heat shock proteins at a lower temperature (34°C) than more virulent strains (37°C), and the avirulent strain maintained heat shock protein synthesis for a longer period of time (100).

The rust fungi, which are pathogens of plants, constitute another group of fungi whose differentiation can be induced in culture by heat shock. Unlike the human pathogens, high temperature is not a normal part of the differentiation of the rust fungi on their plant hosts. Nevertheless, a heat shock of 28.5°C caused cultured uredospore germlings of the bean rust fungus, *Uromyces appendiculatus*, to develop infection structures or appressoria, and vesicle development continued after the temperature was lowered to 19°C (101). This heat shock–induced differentiation occurred with the same kinetics as the differentiation induced thigmotropically, which is considered the natural stimulus. The heat-induced germlings clearly synthesized heat shock proteins, but despite the similarity of the ensuing differentiation, heat shock proteins were not synthesized during thigmotropically induced development (101). In the case of these rust fungi, it seems likely that a heat shock induces differentiation independently of the production of heat shock proteins.

Similarly, heat shock proteins do not appear to be involved in the heat-induced activation of *Neurospora* ascospores. Brief exposure to an extremely high heat shock (10 min at 60°C) is the signal, both in nature and in the laboratory, for mature *N. tetrasperma* ascospores to resume active metabolism and elaborate germination structures. This treatment, however, does not elicit the synthesis of heat shock proteins or the accumulation of RNA for hsp30. The germinating ascospores, in fact, do not become capable of a strong heat shock response until after 5 hr of germination (102). An inability to express hsp70 genes in response to heat shock is also shown by *S. cerevisiae* cultures in the later stages of ascospore formation (99).

B. Trehalose Accumulation

A metabolic activity that is shared by developing fungal spores and heat-stressed cells is the synthesis of the disaccharide trehalose, a nonreducing sugar that is a source of stored carbohydrate (103). Trehalose accumulates in *S. cerevisiae* cells during stationary phase and when cells are exposed to high temperature, ethanol, or heavy metals (104). The germinating conidiospores of *N. crassa* also accumulate trehalose when subjected to a heat shock (105), and furthermore, they degrade the alternate storage carbohydrate glycogen (93,105). The accumulation of trehalose correlates well with the induction of thermotolerance in *S. cerevisiae*, and a protective function has been suggested for trehalose both in spores and in vegetative cells subjected to stress (106,107). There is conflicting evidence, however, whether trehalose is protective in the absence of heat shock protein synthesis (107,108), and an independent requirement for trehalose has not been demonstrated.

C. The Cell Cycle

When *S. cerevisiae* cells are subjected to a heat shock, they transiently arrest at the G1 (postmitotic) phase of the cell cycle (109). Correspondingly, when nutrient-depleted yeast cells enter stationary phase, or G1 arrest, they show increased synthesis of heat shock proteins, and thermotolerance is strongly enhanced. The heat shock proteins that are synthesized in stationary phase and, hence, cell cycle–regulated include hsp26 (60), hsp102 (49), hsp83 (26), *UBI4*-encoded ubiquitin (55), hsp12 (64), *STI1* RNA (65), catalase T (72), enolase 1 (66), and iso-2 cytochrome *c* (68).

In yeast, this growth arrest is regulated, at least partly, by cAMP. Strains of *S. cerevisiae* that have constitutively low levels of cAMP, due to a defect in adenylate cyclase, are arrested in G1. Conversely, nutrient depletion does not induce G1 arrest in strains with cAMP-independent protein kinase activity, due to a defect in the cAMP-binding regulatory subunit of the kinase (110). cAMP also appears to regulate the expression of certain heat shock proteins during growth arrest. An adenylate cyclase mutant synthesized three heat shock proteins (hsp72A, hsp72B, and hsp41) constitutively, and these mutant cells were extremely resistant to elevated temperature (110,111). cAMP affects the heat shock inducibility of the same subset of heat shock proteins. Mutant yeast cells, with a cAMP-independent protein kinase, could not synthesize these three heat shock proteins upon exposure to high temperature. These cAMP-independent cells were also more sensitive than wild-type cells to elevated temperature, and they failed to arrest in G1 following heat shock (110). These characteristics of the mutant cells were mimicked in wild-type yeast cells by the addition of cAMP; exogenous cAMP prevented the cells from arresting in G1 following heat shock and it abolished the thermotolerance that is acquired from heat shock (110). An adenylate cyclase-defective strain of *N. crassa* also showed increased heat resistance relative to the wild-type strain, and its thermotolerance was destroyed by the addition of cAMP (112).

These experiments suggest that low cAMP levels, whether resulting from nutrient depletion, heat shock, or a defective adenylate cyclase, induce thermotolerance by two possible mechanisms. One mechanism is the increased production of a subset of heat shock proteins and the other is the induction of a transient growth arrest. cAMP is probably not the only signal, however, that mediates growth arrest and the associated heat resistance. Mutant yeast cells that lack the protein kinase regulatory subunit (which makes the kinase cAMP-dependent), but have an attenuated catalytic subunit, are nevertheless able to enter a thermotolerant, arrested state in response to nutrient depletion (113).

Complementary to the question of cell cycle regulation of heat shock protein synthesis is the possible role of the heat shock proteins in cell cycle arrest. A striking cell cycle phenotype is found in a yeast strain that cannot synthesize heat shock proteins at elevated temperature, owing to a temperature-sensitive defect in the heat shock transcription factor (HSF). The mutant cells are blocked or delayed in their progression through the G2 (premitotic) phase of the cell cycle (92). One or more of the heat shock proteins may be esssential for cells to complete the cell division cycle at elevated temperature and progress to the G1 phase, where they transiently arrest (92). *S. cerevisiae*, therefore, has intersecting pathways in which heat shock proteins and growth arrest in G1 reciprocally influence one another and make cells resistant to high temperature stress.

X. CONCLUSION

The discovery that organisms exposed to the stress of high temperature synthesize a unique group of proteins, together with the characterization of these proteins, has opened up new areas of biological understanding. The molecular basis for adaption of organisms to high temperature and other stresses can now be analyzed. Of fundamental importance is the realization that many heat shock proteins are essential for cell growth and metabolism at normal temperature, and they provide functions that were hitherto unrecognized. The complete functions of the heat shock proteins are not yet understood, and their elucidation will give us expanded insight into the complex behavior of cells during normal growth and under stress. In these investigations of heat shock proteins, fungi will continue to serve as appropriate and useful models, as well as being inherently interesting.

REFERENCES

1. Plesofsky-Vig, N., and Brambl, R. (1985). The heat shock response of fungi, *Exp. Mycol. 9*: 187–194.
2. Lindquist, S. (1981). Regulation of protein synthesis during heat shock, *Nature 293*: 311–314.
3. Plesofsky-Vig, N., and Brambl, R. (1987). Two developmental stages of *Neurospora crassa* utilize similar mechanisms for responding to heat shock but contrasting mechanisms for recovery, *Mol. Cell. Biol. 7*: 3041–3048.
4. Lindquist, S. (1986). The heat-shock response, *Annu. Rev. Biochem. 55*: 1151–1191.
5. Gething, M-J., and Sambrook, J. (1992). Protein folding in the cell, *Nature 355*: 33–45.
6. Lewis, M. J., and Pelham, H. R. B. (1985). Involvement of ATP in the nuclear and nucleolar functions of the 70 kd heat shock protein, *EMBO J. 4*: 3137–3143.
7. Lindquist, S., and Craig, E. A. (1988). The heat-shock proteins, *Annu. Rev. Genet. 22*: 631–677.
8. Craig, E. A., and Jacobsen, K. (1985). Mutations in cognate genes of *Saccharomyces cerevisiae* hsp70 result in reduced growth rates at low temperatures, *Mol. Cell. Biol. 5*: 3517–3524.
9. Werner-Washburne, M., Stone, D. E., and Craig, E. A. (1987). Complex interactions among members of an essential subfamily of hsp70 genes in *Saccharomyces cerevisiae, Mol. Cell. Biol. 7*: 2568–2577.
10. Holden, D. W., Kronstad, J. W., and Leong, S. A. (1989). Mutation in a heat-regulated hsp70 gene of *Ustilago maydis, EMBO J. 8*: 1927–1934.
11. Deshaies, R. J., Koch, B. D., Werner-Washburne, M., Craig, E. A., and Schekman, R. (1988). A subfamily of stress proteins facilitates translocation of secretory and mitochondrial precursor polypeptides, *Nature 332*: 800–805.
12. Chirico, W. J., Waters, M. G., and Blobel, G. (1988). 70K heat shock related proteins stimulate protein translocation into microsomes, *Nature 332*: 805–810.
13. Murakami, H., Pain, D., and Blobel, G. (1988). 70-kD heat shock-related protein is one of at least two distinct cytosolic factors stimulating protein import into mitochondria, *J. Cell Biol. 107*: 2051–2057.

14. Craig, E. A., Kramer, J., Shilling, J., Werner-Washburne, M., Holmes, S., Kosic-Smithers, J., and Nicolet, C. M. (1989). SSC1, an essential member of the yeast HSP70 multigene family, encodes a mitochondrial protein, *Mol. Cell. Biol. 9*: 3000–3008.
15. Powell, M. J., and Watts, F. Z. (1990). Isolation of a gene encoding a mitochondrial HSP70 protein from *Schizosaccharomyces pombe, Gene 95*: 105–110.
16. Craig, E. A., Kramer, J., and Kosic-Smithers, J. (1987). *SSC1*, a member of the 70-kDa heat shock protein multigene family of *Saccharomyces cerevisiae*, is essential for growth, *Proc. Natl. Acad. Sci. USA 84*: 4156–4160.
17. Kang, P. J., Ostermann, J., Shilling, J., Neupert, W., Craig, E. A., and Pfanner, N. (1990). Requirement for hsp70 in the mitochondrial matrix for translocation and folding of precursor proteins, *Nature 348*: 137–143.
18. Scherer, P. E., Krieg, U. C., Hwang, S. T., Vestweber, D., and Schatz, G. (1990). A precursor protein partly translocated into yeast mitochondria is bound to a 70 kd mitochondrial stress protein, *EMBO J. 9*: 4315–4322.
19. Ostermann, J., Voos, W., Kang, P. J., Craig, E. A., Neupert, W., and Pfanner, N. (1990). Precursor proteins in transit through mitochondrial contact sites interact with hsp70 in the matrix, *FEBS Lett. 277*: 281–284.
20. Morishima, N., Nakagawa, K., Yamamoto, E., and Shibata, T. (1990). A subunit of yeast site-specific endonuclease *SceI* is a mitochondrial version of the 70-kDa heat shock protein, *J. Biol. Chem. 265*: 15189–15197.
21. Normington, K., Kohno, K., Kozutsumi, Y., Gething, M.-J., and Sambrook, J. (1989). *S. cerevisiae* encodes an essential protein homologous in sequence and function to mammalian BiP, *Cell 57*: 1223–1236.
22. Rose, M. D., Misra, L. M., and Vogel, J. P. (1989). *KAR2*, a karyogamy gene, is the yeast homolog of the mammalian BiP/GRP78 gene, *Cell 57*: 1211–1221.
23. Riehl, R. M., Sullivan, W. P., Vroman, B. T., Bauer, V. J., Pearson, G. R., and Toft, D. O. (1985). Immunological evidence that the nonhormone binding component of avian steroid receptors exists in a wide range of tissues and species, *Biochemistry 24*: 6586–6591.
24. Brunt, S. A., Riehl, R., and Silver, J. C. (1990). Steroid hormone regulation of the *Achlya ambisexualis* 85-kilodalton heat shock protein, a component of the *Achlya* steroid receptor complex, *Mol. Cell. Biol. 10*: 273–281.
25. Brunt, S. A., and Silver, J. C. (1991). Molecular cloning and characterization of two distinct *hsp85* sequences from the steroid responsive fungus *Achlya ambisexualis, Curr. Genet. 19*: 383–388.
26. Borkovich, K. A., Farrelly, F. W., Finkelstein, D. B., Taulien, J., and Lindquist, S. (1989). hsp82 is an essential protein that is required in higher concentrations for growth of cells at higher temperatures, *Mol. Cell. Biol. 9*: 3919–3930.
27. Picard, D., Khursheed, B., Garabedian, M. J., Fortin, M. G., Lindquist, S., and Yamamoto, K. R. (1990). Reduced levels of hsp90 compromise steroid receptor action in vivo, *Nature 348*: 166–168.
28. Wilhelmsson, A., Cuthill, S., Denis, M., Wikstrom, A.-C., Gustafsson, J-A., and Poellinger, L. (1990). The specific DNA binding activity of the dioxin receptor is modulated by the 90 kd heat shock protein, *EMBO J. 9*: 69–76.

29. Oppermann, H., Levinson, W., and Bishop, J. M. (1981). A cellular protein that associates with the transforming protein of Rous sarcoma virus is also a heat-shock protein, *Proc. Natl. Acad. Sci. USA 78*: 1067–1071.
30. McMullin, T. W., and Hallberg, R. L. (1988). A highly evolutionarily conserved mitochondrial protein is structurally related to the protein encoded by the *Escherichia coli groEL* gene, *Mol. Cell. Biol. 8*: 371–380.
31. Cheng, J. Y., Hartl, F.-U., Martin, J., Pollock, R. A., Kalousek, F., Neupert, W., Hallberg, E. M., Hallberg, R. L., and Horwich, A. L. (1989). Mitochondrial heat-shock protein hsp60 is essential for assembly of proteins imported into yeast mitochondria, *Nature 337*: 620–625.
32. Cheng, M. Y., Hartl, F.-U., and Horwich, A. L. (1990). The mitochondrial chaperonin hsp60 is required for its own assembly, *Nature 348*: 455–458.
33. Reading, D. S., Hallberg, R. L., and Myers, A. M. (1989). Characterization of the yeast *HSP60* gene coding for a mitochondrial assembly factor, *Nature 337*: 655–659.
34. Kusukawa, N., Yura, T., Ueguchi, C., Akiyama, Y., and Ito, K. (1989). Effects of mutations in heat-shock genes *groES* and *groEL* on protein export in *Escherichia coli, EMBO J. 8*: 3517–3521.
35. Lubben, T. H., Gatenby, A. A., Donaldson, G. K., Lorimer, G. H., and Viitanen, P. V. (1990). Identification of a groES-like chaperonin in mitochondria that facilitates protein folding, *Proc. Natl. Acad. Sci. USA 87*: 7683–7687.
36. Hutchinson, E. G., Thichelaar, W., Hofhaus, G., Weiss, H., and Leonard, K. R. (1989). Identification and electron microscopic analysis of a chaperonin oligomer from *Neurospora crassa* mitochondria, *EMBO J. 8*: 1485–1490.
37. Ostermann, J., Horwich, A. L., Neupert, W., and Hartl, F.-U. (1989). Protein folding in mitochondria requires complex formation with hsp60 and ATP hydrolysis, *Nature 341*: 125–130.
38. Gray, R. E., Grasso, D. G., Maxwell, R. J., Finnegan, P. M., Nagley, P., and Devenish, R. J. (1990). Identification of a 66 KDa protein associated with yeast mitochondrial ATP synthase as heat shock protein hsp60, *FEBS Lett. 268*: 265–268.
39. Ang, D., Liberek, K., Skowyra, D., Zylicz, M., and Georgopoulos, C. (1991). Biological role and regulation of the universally conserved heat shock proteins, *J. Biol. Chem. 266*: 24233–24236.
40. Blumberg, H., and Silver, P. A. (1991). A homologue of the bacterial heat-shock gene *DnaJ* that alters protein sorting in yeast, *Nature 349*: 627–630.
41. Atencio, D. P., and Yaffe, M. P. (1992). MAS5, a yeast homolog of DnaJ involved in mitochondrial protein import, *Mol. Cell. Biol. 12*: 283–291.
42. Caplan, A. J., and Douglas, M. G. (1991). Characterization of YDJ1: a yeast homologue of the bacterial dnaJ protein, *J. Cell Biol. 114*: 609–621.
43. Sadler, I., Chiang, A., Kurihara, T., Rothblatt, J., Way, J., and Silver, P. (1989). A yeast gene important for protein assembly into the endoplasmic reticulum and the nucleus has homology to DnaJ, an *Escherichia coli* heat shock protein, *J. Cell Biol. 109*: 2665–2675.
44. Rothblatt, J. A., Deshaies, R. J., Sanders, S. L., Daum, G., and Schekman, R. (1989). Multiple genes are required for proper insertion of secretory proteins into the endoplasmic reticulum in yeast, *J. Cell Biol. 109*: 2641–2652.

45. Parsell, D. A., Sanchez, Y., Stitzel, J. D., and Lindquist, S. (1991). Hsp104 is a highly conserved protein with two essential nucleotide-binding sites, *Nature 353*: 270–273.
46. Vassilev, A., Plesofsky-Vig, N., and Brambl, R. (1992). Isolation, partial amino acid sequence, and cellular distribution of heat shock protein hsp98 from *Neurospora crassa, Biochim. Biophys. Acta 1156*: 1–6.
47. Squires, C. L., Pedersen, S., Ross, B. M., and Squires, C. (1991). ClpB is the *Escherichia coli* heat shock protein F84.1, *J. Bacteriol. 173*: 4254–4262.
48. Kroh, H. E., and Simon, L. D. (1990). The ClpP component of Clp protease is the σ^{32}-dependent heat shock protein F21.5, *J. Bacteriol. 172*: 6026–6034.
49. Sanchez, Y., and Lindquist, S. L. (1990). HSP104 required for induced thermotolerance, *Science 248*: 1112–1115.
50. Yost, H. J., and Lindquist, S. (1991). Heat shock proteins affect RNA processing during the heat shock response of *Saccharomyces cerevisiae, Mol. Cell. Biol. 11*: 1062–1068.
51. Finley, D., Ciechanover, A., and Varshavsky, A. (1984). Thermolability of ubiquitin-activating enzyme from the mammalian cell cycle mutant ts85, *Cell 37*: 43–55.
52. Seufert, W., and Jentsch, S. (1990). Ubiquitin-conjugating enzymes UBC4 and UBC5 mediate selective degradation of short-lived and abnormal proteins, *EMBO J. 9*: 543–550.
53. Ozkaynak, E., Finley, D., Solomon, M. J., and Varshavsky, A. (1987). The yeast ubiquitin genes: a family of natural gene fusions, *EMBO J. 6*: 1429–1439.
54. Finley, D., Ozkaynak, E., and Varshavsky, A. (1987). The yeast polyubiquitin gene is essential for resistance to high temperatures, starvation, and other stresses, *Cell 48*: 1035–1046.
55. Tanaka, K., Matsumoto, K., and Toh-e, A. (1988). Dual regulation of the expression of the polyubiquitin gene by cyclic AMP and heat shock in yeast, *EMBO J. 7*: 495–502.
56. Taccioli, G. E., Grotewold, E., Aisemberg, G. O., and Judewicz, N. D. (1989). Ubiquitin expression in *Neurospora crassa*: cloning and sequencing of a polyubiquitin gene, *Nucleic Acids Res. 17*: 6153–6165.
57. Ingolia, T. D., and Craig, E. A. (1982). Four small *Drosophila* heat shock proteins are related to each other and to mammalian α-crystallin, *Proc. Natl. Acad. Sci. USA 79*: 2360–2364.
58. Plesofsky-Vig, N., Vig, J., and Brambl, R. (1992). Phylogeny of the α-crystallin-related heat shock proteins, *J. Mol. Evol. 35*: 537–545.
59. Tuite, M. F., Bentley, N. J., Bossier, P., and Fitch, I. T. (1990). The structure and function of small heat shock proteins: analysis of the *Saccharomyces cerevisiae* Hsp26 protein, *Antonie van Leeuwenhoek 58*: 147–154.
60. Petko, L., and Lindquist, S. (1986). Hsp26 is not required for growth at high temperatures, nor for thermotolerance, spore development, or germination, *Cell 45*: 885–894.
61. Plesofsky-Vig, N., and Brambl, R. (1990). Gene sequence and analysis of hsp30, a small heat shock protein of *Neurospora crassa* which associates with mitochondria, *J. Biol. Chem. 265*: 15432–15440.

62. Rossi, J. M., and Lindquist, S. (1989). The intracellular location of yeast heat-shock protein 26 varies with metabolism, *J. Cell Biol. 108*: 425–439.
63. Susek, R. E., and Lindquist, S. L. (1989). hsp26 of *Saccharomyces cerevisiae* is related to the superfamily of small heat shock proteins but is without a demonstrable function, *Mol. Cell. Biol. 9*: 5265–5271.
64. Praekelt, U. M., and Meacock, P. A. (1990). *HSP12*, a new small heat shock gene of *Saccharomyces cerevisiae*: analysis of structure, regulation and function, *Mol. Gen. Genet. 223*: 97–106.
65. Nicolet, C. M., and Craig, E. A. (1989). Isolation and characterization of *STI1*, a stress-inducible gene from *Saccharomyces cerevisiae, Mol. Cell. Biol. 9*: 3638–3646.
66. Iida, H., and Yahara, I. (1985). Yeast heat-shock protein of M_r 48,000 is an isoprotein of enolase, *Nature 315*: 688–690.
67. Piper, P. W., Curran, B., Davies, M. W., Lockheart, A., and Reid, G. (1986). Transcription of the phosphoglycerate kinase gene of *Saccharomyces cerevisiae* increases when fermentative cultures are stressed by heat-shock, *Eur. J. Biochem. 161*: 525–531.
68. Pillar, T. M., and Bradshaw, R. E. (1991). Heat shock and stationary phase induce transcription of the *Saccharomyces cerevisiae* iso-2 cytochrome *c* gene, *Curr. Genet. 20*: 185–188.
69. Freeman, M. L., Spitz, D. R., and Meredith, M. J. (1990). Does heat shock enhance oxidative stress? Studies with ferrous and ferric iron, *Radiat. Res. 124*: 288–293.
70. Kapoor, M., and Sveenivasan, G. M. (1988). The heat shock response of *Neurospora crassa*: stress-induced thermotolerance in relation to peroxidase and superoxide dismutase levels, *Biochem. Biophys. Res. Commun. 156*: 1097–1102.
71. Chary, P., and Natvig, D. O. (1989). Evidence for three differentially regulated catalase genes in *Neurospora crassa*: effects of oxidative stress, heat shock, and development, *J. Bacteriol. 171*: 2646–2652.
72. Wieser, R., Adam, G., Wagner, A., Schuller, C., Marchler, G., Ruis, H., Krawiec, Z., and Bilinski, T. (1991). Heat shock factor-independent heat control of transcription of the *CTT1* gene encoding the cytosolic catalase T of *Saccharomyces cerevisiae, J. Biol. Chem. 266*: 12406–12411.
73. Thornalley, P. J., and Vasak, M. (1985). Possible role for metallothionein in protection against radiation-induced oxidative stress. Kinetics and mechanism of its reaction with superoxide and hydroxyl radicals, *Biochim. Biophys. Acta 827*: 36–44.
74. Silar, P., Butler, G., and Thiele, D. J. (1991). Heat shock transcription factor activates transcription of the yeast metallothionein gene, *Mol. Cell. Biol. 11*: 1232–1238.
75. Yang, W., Gahl, W., and Hamer, D. (1991). Role of heat shock transcription factor in yeast metallothionein gene expression, *Mol. Cell. Biol. 11*: 3676–3681.
76. McClanahan, T., and McEntee, K. (1986). DNA damage and heat shock dually regulate genes in *Saccharomyces cerevisiae, Mol. Cell. Biol. 6*: 90–96.
77. Sebastian, J., Kraus, B., and Sancar, G. B. (1990). Expression of the yeast *PHR1* gene is induced by DNA-damaging agents, *Mol. Cell. Biol. 10*: 4630–4637.

78. Kobayashi, N., and McEntee, K. (1990). Evidence for a heat shock transcription factor-independent mechanism for heat shock induction of transcription in *Saccharomyces cerevisiae, Proc. Natl. Acad. Sci. USA 87*: 6550–6554.
79. Amin, J., Anathan, J., and Voellmy, R. (1988). Key features of heat shock regulatory elements, *Mol. Cell. Biol. 8*: 3761–3769.
80. Wiederrecht, G., Seto, D., and Parker, C. S. (1988). Isolation of the gene encoding the *S. cerevisiae* heat shock transcription factor, *Cell 54*: 841–853.
81. Sorger, P. K., and Pelham, H. R. B. (1988). Yeast heat shock factor is an essential DNA-binding protein that exhibits temperature-dependent phosphorylation, *Cell 54*: 855–864.
82. Clos, J., Westwood, J. T., Becker, P. B., Wilson, S., Lambert, K., and Wu, C. (1990). Molecular cloning and expression of a hexameric *Drosophila* heat shock factor subject to negative regulation, *Cell 63*: 1085–1097.
83. Sorger, P. K., and Nelson, H. C. M. (1989). Trimerization of a yeast transcriptional activator via a coiled-coil motif, *Cell 59*: 807–813.
84. Jakobsen, B. K., and Pelham, H. R. B. (1988). Constitutive binding of yeast heat shock factor to DNA in vivo, *Mol. Cell. Biol. 8*: 5040–5042.
85. Sorger, P. K. (1990). Yeast heat shock factor contains separable transient and sustained response transcriptional activators, *Cell 62*: 793–805.
86. Jakobsen, B. K., and Pelham, H. R. B. (1991). A conserved heptapeptide restrains the activity of the yeast heat shock transcription factor, *EMBO J. 10*: 369–375.
87. Gallo, G. J., Schuetz, T. J., and Kingston, R. E. (1991). Regulation of heat shock factor in *Schizosaccharomyces pombe* more closely resembles regulation in mammals than in *Saccharomyces cerevisiae, Mol. Cell. Biol. 11*: 281–288.
88. Sorger, P. K., Lewis, M. J., and Pelham, H. R. B. (1987). Heat shock factor is regulated differently in yeast and HeLa cells, *Nature 329*: 81–84.
89. Park, H.-O., and Craig, E. A. (1989). Positive and negative regulation of basal expression of a yeast HSP70 gene, *Mol. Cell. Biol. 9*: 2025–2033.
90. McDaniel, D., Caplan, A. J., Lee, M.-S., Adams, C. C., Fishel, B. R., Gross, D. S., and Garrard, W. T. (1989). Basal-level expression of the yeast *HSP82* gene requires a heat shock regulatory element, *Mol. Cell. Biol. 9*: 4789–4798.
91. Nieto-Sotelo, J., Wiederrecht, G., Okuda, A., and Parker, C. S. (1990). The yeast heat shock transcription factor contains a transcriptional activation domain whose activity is repressed under nonshock conditions, *Cell 62*: 807–817.
92. Smith, B. J., and Yaffe, M. P. (1991). A mutation in the yeast heat-shock factor gene causes temperature-sensitive defects in both mitochondrial protein import and the cell cycle, *Mol. Cell. Biol. 11*: 2647–2655.
93. Habel, D., Plesofsky-Vig, N., and Brambl, R. (1991). The respiratory response to heat shock in *Neurospora crassa, FEMS Microbiol. Lett. 81*: 317–322.
94. Medoff, G., Painter, A., and Kobayashi, G. S. (1987). Mycelial- to yeast-phase transitions of the dimorphic fungi *Blastomyces dermatitidis* and *Paracoccidioides brasiliensis, J. Bacteriol. 169*: 4055–4060.
95. Patriarca, E. J., and Maresca, B. (1990). Acquired thermotolerance following heat shock protein synthesis prevents impairment of mitochondrial ATPase activity at elevated temperatures in *Saccharomyces cerevisiae, Exp. Cell Res. 190*: 57–64.

96. Lambowitz, A. M., Kobayashi, G. S., Painter, A., and Medoff, G. (1983). Possible relationship of morphogenesis in pathogenic fungus, *Histoplasma capsulatum*, to heat shock response, *Nature 303*: 806–808.
97. Minchiotti, G., Gargano, S., and Maresca, B. (1991). The intron-containing *hsp82* gene of the dimorphic pathogenic fungus *Histoplasma capsulatum* is properly spliced in severe heat shock conditions, *Mol. Cell. Biol. 11*: 5624–5630.
98. Kurtz, S., and Lindquist, S. (1984). Changing patterns of gene expression during sporulation in yeast, *Proc. Natl. Acad. Sci. USA 81*: 7323–7327.
99. Kurtz, S., Rossi, J., Petko, L., and Lindquist, S. (1986). An ancient developmental induction: heat-shock proteins induced in sporulation and oogenesis, *Science 231*: 1154–1157.
100. Shearer, G., Birge, C. H., Yuckenberg, P. D., Kobayashi, G. S., and Medoff, G. (1987). Heat-shock proteins induced during the mycelial-to-yeast transitions of strains of *Histoplasma capsulatum, J. Gen. Microbiol. 133*: 3375–3382.
101. Staples, R. C., Hoch, H. C., Freve, P., and Bourett, T. M. (1989). Heat shock-induced development of infection structures by bean rust uredospore germlings, *Exp. Mycol. 13*: 149–157.
102. Plesofsky-Vig, N., Paulson, A., Hill, E. P., Glaser, L., and Brambl, R. (1992). Heat shock gene expression in germinating ascospores of *Neurospora tetrasperma, FEMS Microbiol. Lett. 90*: 117–122.
103. Thevelein, J. M. (1984). Regulation of trehalose mobilization in fungi, *Microbiol. Rev. 48*: 42–59.
104. Attfield, P. V. (1987). Trehalose accumulates in *Saccharomyces cerevisiae* during exposure to agents that induce heat shock response, *FEBS Lett. 225*: 259–263.
105. Neves, M. J., Jorge, J. A., Francois, J. M., and Terenzi, H. F. (1991). Effects of heat shock on the level of trehalose and glycogen, and on the induction of thermotolerance in *Neurospora crassa, FEBS Lett. 283*: 19–22.
106. Van Laere, A. (1989). Trehalose, reserve and/or stress metabolite? *FEBS Microbiol. Rev. 63*: 201–210.
107. De Virgilio, C., Piper, P., Boller, T., and Wiemken, A. (1991). Acquisition of thermotolerance in *Saccharomyces cerevisiae* without heat shock protein hsp104 and in the absence of protein synthesis, *FEBS Lett. 288*: 86–90.
108. Winkler, K., Kienle, I., Burgert, M., Wagner, J.-C., and Holzer, H. (1991). Metabolic regulation of the trehalose content of vegetative yeast, *FEBS Lett. 291*: 269–272.
109. Johnston, G. C., and Singer, R. A. (1980). Ribosomal precursor RNA metabolism and cell division in the yeast *Saccharomyces cerevisiae, Mol. Gen. Genet. 178*: 357–360.
110. Shin, D.-Y., Matsumoto, K., Iida, H., Uno, I., and Ishikawa, T. (1987). Heat shock response of *Saccharomyces cerevisiae* mutants altered in cyclic AMP-dependent protein phosphorylation, *Mol. Cell. Biol. 7*: 244–250.
111. Iida, H. (1988). Multistress resistance of *Saccharomyces cerevisiae* is generated by insertion of retrotransposon Ty into the 5′ coding region of the adenylate cyclase gene, *Mol. Cell. Biol. 8*: 5555–5560.

112. Cruz, A. K., Terenzi, H. F., Jorge, J. A., and Terenzi, H. F. (1988). Cyclic AMP-dependent, constitutive thermotolerance in the adenylate cyclase-deficient *cr-1* (crisp) mutant of *Neurospora crassa, Curr. Genet. 13*: 451–454.
113. Cameron, S., Levin, L., Zoller, M., and Wigler, M. (1988). cAMP-independent control of sporulation, glycogen metabolism, and heat shock resistance in *S. cerevisiae, Cell 53*: 555–566.

3

Genetics and Molecular Biology of Metal Tolerance in Fungi

A. Brian Tomsett

The University of Liverpool
Liverpool, England

I. INTRODUCTION

The primary source of metals in all ecosystems is the underlying bedrock of the planet. At the surface, there can be considerable variation in metal content, which results from three major sources: differences in the underlying bedrock, the atmosphere, and the biosphere (for review, see 1). In terrestrial systems, ore deposits will greatly increase metal content in comparison with most soils, which have low concentrations of metal. Atmospheric inputs can result from natural sources, e.g., volcanoes and forest fires, as well as pollution as a result of man's activities. In some ecosystems, the input of metal ions from biotic sources is not insignificant; this can result either from bioaccumulation by some species or from leaching (i.e., solubilization) of metals from soils by others. Such movements of metals through ecosystems are obviously increased by water.

Metal ions are thus a primary component of all ecosystems and, hence, all organisms must have evolved mechanisms to cope with varying levels of these elements. In many cases, insufficient levels of metal can result in stress responses just as severe as those resulting from excess metal, and hence cells must have the means to avoid both extremes. This may be justification for the view that the mechanisms of metal tolerance are likely to be adaptations of the processes of normal metal homeostasis; i.e., if the cell already has processes to monitor and maintain intracellular concentrations of metals, adaptation of these processes could raise the threshold at which the organism can survive. Genetic and molecular analysis of the response of organisms to metals will be required to prove any such interrelationship.

A. The Response of Organisms to Metals

1. Heavy Metals and Metal Metabolism

Some metals are essential, i.e., they are required for normal cellular growth, while others are nonessential, i.e., they fulfill no known cellular role. Essential metals are required as micronutrients (trace elements) in the cultivation of plants, fungi, and bacteria, which in turn act as major sources of metals for animals.

A major group of toxic elements in biological systems are the "heavy metals," which have been recently defined as the class B/borderline heavy metals (2). This group has a specific gravity greater than 4, forming complexes primarily with ligands containing nitrogen and sulphur centers, rather than oxygen: it includes copper, zinc, nickel, lead, silver, cadmium, and mercury. Each of these is toxic when present in excess, but copper, zinc, and nickel are essential micronutrients.

Metal homeostasis must involve uptake of sufficient essential metals for the optimal functioning of metabolic processes, while providing protection against toxicity from excess metal. These processes rely on the ability of the cellular apparatus to distinguish both individual elements and their relative concentrations. This molecular recognition involves proteins that can bind certain metal ions but not interact with others. Such molecular recognition allows an organism to differentiate between ions and, if necessary, partition them in different ways.

2. Protection Against Metal Toxicity and the Evolution of Tolerance

There are two major strategies organisms can adopt to protect themselves against heavy metal toxicity (see 3). The first, *avoidance*, restricts metal entry into the cell, either by reduced uptake/active efflux or by the formation of complexes outside the cell. The second, *sequestration*, reduces levels of free ions in the cytosol, either by the synthesis of ligands to achieve intracellular chelation or by compartmentation into vacuoles. Both avoidance and sequestration protect the wide array of macromolecules known to be sensitive to metal binding.

The evolution of metal tolerance is thus most likely to occur by adaptation of such cellular processes to provide increased resistance under extremes of metal stress.

3. The Genetic Control of Metal Tolerance

Higher plants remain the best-studied example of the evolution of metal tolerance in natural populations (4). From these studies, it is still not possible to determine whether adaptation to tolerance is simple or polygenic (5,6). This is perhaps not too surprising given that there must be a considerable number of genes involved in metal homeostasis, which *potentially* could be mutated to confer tolerance to a greater or lesser extent. Adaptation in a natural environment is likely to result from several mutagenic events accumulated over time. Some will confer tolerance, others will enable metabolism to compensate for those changes (see Section II.B.3). Whether in plants or fungi, unraveling the contribution of each of these is complex. For this reason, the analysis of artificially isolated mutants can shed light on the areas of cellular activity that can be modified to confer tolerance.

4. The Importance of Metal Tolerance in Fungi

Fungi play a central role in the biosphere, in the degradation of organic matter and the recycling of essential elements. One index of the "health" of a terrestrial ecosystem could be, therefore, the diversity of the microflora, especially the fungi. Increased stress from heavy metals, and the concomitant decreased growth and activity, could limit the survival of the whole food chain in that location. Mycorrhizal fungi have also been implicated in increased metal tolerance of higher plants (see 7). For example, zinc accumulation in fungal walls and extra hyphal slime have been shown to enhance tolerance in *Betula* sp. (8).

Metal tolerance in fungi has recently been reviewed (7). It is clear that the diversity of fungi in soil is reduced by heavy metal pollution (9), and that tolerant fungi can be isolated from such environments (see 7). However, just as in higher plants, metal-tolerant fungi occur at sites without increased levels of metal (10). Brown and Hall (7) point out that some fungi accumulate heavy metals to concentrations that would cause severe toxicity in higher plants; is this an indication of tolerance, or merely a demonstration of a greater toxicity threshold? As will be discussed below, some components of such species may provide this increased basal tolerance. Since, at the cytosolic level, many enzymes are likely to be (almost) equally sensitive across wide taxonomic groups, organisms exhibiting increased basal tolerance are of interest in understanding tolerance mechanisms in adapted populations of both fungi and more complex eukaryotes.

The biotechnological aspects of the interaction of fungi with metals have also been reviewed recently (11). In general terms, these comprise three major areas: the reclamation of land by the use of mycorrhizas; the development of new antifungal compounds, both for the preservation against fungal decay and in the control of pathogenic fungi; the use of fungal biomass for detoxification or recovery of valuable metals (see 11).

B. Mechanisms of Metal Tolerance

A variety of mechanisms can result in metal tolerance (see 12–14). As discussed in Section I.A.2, these are of two types, avoidance and sequestration. Potentially, any cellular component involved in the transport or binding of a metal ion could be modified, by mutation, to increase the tolerance of the organism; Table 1 outlines some of these possibilities. Since each of these mechanisms has been observed in a range of microorganisms (see 12–14), do naturally occurring metal-tolerant organisms use more than one?

Most naturally occurring fungi present enormous difficulties for genetic research, and hence, work in my laboratory (and in others) has attempted to explore the range of metal tolerance mechanisms available to a single species. While such studies may not be an entirely accurate reflection of isolates from natural ecosystems, it does enable specific questions to be posed when studying natural populations.

In this review, I shall discuss the data obtained by a variety of research groups, adopting genetic and molecular approaches, to unravel the molecular mechanisms underlying tolerance to a variety of metals.

Table 1 Mechanisms of Metal Tolerance

Cellular constituent/process	Result of mutation	Tolerance mechanism
Secreted protein or organic molecule or inorganic ion	Increased production resulting in increased extracellular precipitation	Avoidance
Fungal cell wall	Increased metal-binding capacity	Avoidance
Permeases	Reduced transport into the cell or increased efflux	Avoidance
Vacuoles	Increased transport into subcellular compartments	Sequestration
Inorganic chelators, e.g., sulfides	Altered metabolism	Sequestration
Organic chelators, e.g., citrates	Altered metabolism	Sequestration
Metal-binding proteins	Increased synthesis	Sequestration

II. GENETIC ANALYSIS OF METAL TOLERANCE

A. The Isolation of Mutants

1. Metal-Resistant Mutants

The isolation of mutants resistant to a toxic compound can be achieved by direct selection, which is by far the simplest approach. In some cases, culture collections have been tested to examine the differences between natural ("wild-type") isolates, as well as genetically marked strains isolated by mutagenesis (for example, 15). In such cases, a strain is sought that can grow at a significantly higher concentration of metal; Tohoyama et al. (15) tested 29 strains of *Saccharomyces cerevisiae*, one of which was able to grow on 0.2 mM cadmium sulfate, a concentration at which all others were totally inhibited.

In most cases, however, metal-resistant mutants have been isolated by direct selection from a metal-sensitive strain, with or without treatment with a mutagen; cells are plated onto a medium containing a single concentration of metal that is highly inhibitory, and often lethal, to the metal-sensitive "parent" strain. This has the advantage that the spontaneous or induced mutation is likely to be a single event, and hence the subsequent analysis to identify individual components of cellular processes will be more straightforward. A danger of screening stock collections is that resistance may have resulted from mutation of more than one gene.

A second approach is to select for metal resistance by stepwise culture: cells are repeatedly subcultured onto increasing concentrations of the metal, such that at each stage, only the most resistant survive. While this also selects for multiple mutagenic events, it has proven very successful for the isolation of a particular class of resistant mutants (see Section II.B.1).

In the isolation of resistant mutants, care must be taken to ensure that the strains either exhibit heritable resistance or have altered DNA sequences. Resistant variants can be identified among cell populations that do not result from mutation, but rather from an inducible cellular response. For example, *Rhizopus stolonifer* and *Cunninghamella blakesleeana* were "trained" to tolerate high levels of copper (16) but this metal tolerance was neither stable nor heritable. The metals may have simply activated genes that coded for biochemical processes that confer tolerance to copper, an approach that can be used for molecular analysis but not for the isolation of mutants.

Similar "induction" variants can be selected after mutagenesis, but can be eliminated by allowing a period of growth on metal-free media before subsequent analysis. For example, Phelan et al. (17) excluded 25 putative copper-resistant mutants of *Aspergillus nidulans* (out of 40 isolates) after growth on complete medium before undertaking further genetic analysis: in each of these strains, tolerance was lost once the organism was grown in the absence of metal. Ultimately, however, mutants are only proven by genetic tests of their heritability, or by molecular analysis of changes to DNA.

2. Metal-Sensitive Mutants

Just as the comparison of metal-resistant mutants to a normal "sensitive" strain enables examination of tolerance and normal metal homeostasis, mutants that are hypersensitive to metal can provide additional insight since they may include functions (genes) incapable of mutation to resistance. In most cases, the isolation of hypersensitive mutants is labor-intensive because individual colonies growing on unsupplemented media must be screened on supplemented media to identify the hypersensitivity (total isolation).

A metal-sensitive mutant of *A. nidulans* has been reported that grew at a slower rate on cadmium and copper than the wild-type strain from which it was derived and failed to grow in the presence of nickel, cobalt, zinc, and lead (18). This phenotype resulted from mutation of a single gene located on chromosome VIII. Unfortunately, this strain was difficult to maintain in culture and has now been lost.

Two copper-sensitive mutants of *S. cerevisiae* (*cup2* and *ace1-1*) have been isolated, which have subsequently been shown to be alleles of the same gene (19–21); these mutants have led to the description of a novel regulatory protein involved in copper homeostasis (see Section III.B.4).

Two research groups have used cadmium hypersensitivity to identify mutants of *Schizosaccharomyces pombe* defective in glutathione (GSH) metabolism, a key element of cadmium tolerance in a number of organisms (see Section III.C). Mutoh and Hayashi (22) isolated 20 mutants using total isolation, whereas Glaeser et al. (23) first selected mutants that were resistant to *N*-methyl-*N'*-nitro-*N*-nitrosoguanidine (MNNG) and subsequently tested these for cadmium hypersensitivity. Sixty GSH-deficient mutants were identified. The latter group exploited the observation that the mutagen MNNG is activated by intracellular thiol groups (mostly GSH), and hence cells deficient in GSH would be more resistant to MNNG (allowing direct selection) but more sensitive to cadmium.

B. Genetic Analysis of Mutants

The genetic analysis of mutants aims to identify which mutations are allelic, how many different genes are involved, and their location on the genetic map. Allelism is usually determined by complementation tests, but as in many cases metal-resistant mutants are partially dominant, this approach is not possible. Instead, recombination frequencies are often examined in sexual crosses between the mutant strains to examine how closely linked they may be; this process can also be used to find their location on the genetic map by determining their proximity to other known genetic markers.

1. The Genetics of Copper Resistance

The first description of the genetic control of copper resistance was in *S. cerevisiae* (24), in which resistant and sensitive alleles were shown to segregate in a Mendelian fashion. The gene identified, *CUP1*, encodes a copper-binding metallothionein (see 25). The properties of this protein will be discussed below (Sections III.B.2 and III.B.4), together with its transcriptional activator, the product of the *CUP2* (ACE1) gene identified by copper hypersensitivity. In another yeast, *Candida glabrata*, copper-resistant mutants were isolated after stepwise selection on copper sulfate–containing media (26); most of these mutants were also shown to result from increased metallothionein levels, by amplification of a single gene, although some showed no such changes.

In *A. nidulans*, 15 copper-resistant mutants have been isolated (17). When backcrossed to wild-type strains, copper resistance segregated among the progeny in a Mendelian fashion. By intercrossing these strains, two genes, *cupA* and *cupB*, have been identified. Thirteen mutants map to the *cupA* locus: crosses between these yield progeny that are all copper-resistant; similar results were obtained for the two mutants located within the *cup*B locus. Crosses between *cup*A^r and *cup*B^r mutants, however, segregated copper-resistant:wild-type in a 3:1 ratio, indicating no genetic linkage. Furthermore, *cup*A^r *cup*B^r double mutants could be recognized as one quarter of these progeny (as expected) owing to their increased resistance in comparison with the two single mutants. Unlike the yeasts described above, these mutants do not appear to result in increased synthesis of metal-binding proteins, despite preliminary evidence that they are present in this organism (17). It seems more likely that these mutants have restricted uptake of copper ions, especially in *cupB^r* mutants, because germination is unaffected by the presence of the metal and resistance is metal-specific.

In *Neurospora crassa*, a copper-binding protein has been described, but when copper-resistant mutants have been isolated, these also do not have increased synthesis of this protein (27–29).

2. The Genetics of Cadmium and Zinc Resistance

Cadmium has been used extensively to analyze metal tolerance in a wide variety of organisms because it is highly toxic and nonessential. It has a similar ionic radius and chemical properties to zinc and cobalt with which it can compete for uptake, and it replaces zinc in enzymes involved in DNA replication, repair, and recombination, leading to a variety of mutagenic and carcinogenic effects (30–32).

In *S. cerevisiae*, three cadmium resistance loci have been reported. There has been little characterization of the CAD1 gene, except that it is located on chromosome 4 (see 15). The *CAD2* gene, a dominant nuclear marker on chromosome 2, was identified among a series of laboratory stocks (15). The resistance of this mutation does not result from increased synthesis of metal-binding proteins, rather most of the cadmium appears to be bound to structural components including the cell wall. The third gene, *ZRC1*, is responsible for both zinc and cadmium tolerance in *S. cerevisiae* (33). It has been identified by recombinant DNA techniques, and since its chromosomal location has not yet been determined, it is formally possible that it is an allele of *CAD1* or *CAD2*. The molecular mechanism of this resistance is discussed below (Section III.A).

In *A. nidulans*, 13 cadmium-resistant mutants have been selected, which can grow on higher levels of cadmium than wild-type strains (34). Three mutants were located in the *cad*A locus on chromosome IV; the other 10 have been mapped to the *cad*B locus on chromosome VI. As with copper, *cad*A^r *cad*B^r double mutants were recognized among the progeny of crosses.

Phenotypic tests of *cad*A^r and *cad*B^r mutants showed that they possessed elevated growth rates in the presence of cadmium, and to a lesser extent copper, but did not exhibit resistance to nickel, cobalt, zinc, or lead (34). In addition to this increased growth rate, *cad*B^r strains showed an increased ability to germinate in the presence of cadmium. *cad*A^r strains, on the other hand, showed a similar response to wild-type (sensitive) strains in this respect; this suggests that the resistance of these mutants is apparent only after a period of growth; i.e., resistance results from a cadmium-inducible gene, and hence conidia are not inherently resistant. There is no doubt that these are true mutants because of their heritable nature, only that resistance is not developed until the strains are under cadmium stress.

Little is known of the basis of cadmium resistance in *A. nidulans*. The *cadA* and *cad*B genes determine different mechanisms of resistance because their effect is additive (rather than epistatic). There is no evidence that either produces cadmium-binding proteins (35), although in *cad*B^r mutants, cadmium may induce a cysteine-rich protein that is normally only copper-inducible in the wild-type.

The location of the *cad*B locus, however, is of interest. This gene is sufficiently tightly linked to the *meth*B locus to suggest that they are either continuous, or *cad*B^r could be alleles of the *meth*B locus. Since there are a number of examples of clustering of related genes in *A. nidulans*, it is possible that *cad*B, like *meth*B, is involved in sulfur metabolism. *meth*B encodes cystathionine γ-synthase (E.C.4.2.99.9), an enzyme in the pathway for the conversion of cysteine, eventually to methionine (36). This suggests that *cad*B^r may alter the thiol pools to confer resistance.

To determine whether any more loci could be identified, 14 additional cadmium-resistant mutants were isolated; all mapped to *cad*B (37).

Two cadmium resistance loci have also been identified in *N. crassa* (38). Of four mutants isolated, one mapped to a gene on linkage group II, while three were located on chromosome VII. These mutants did not demonstrate cotolerance to copper, zinc, or cobalt, nor did they exhibit any difference in cadmium uptake in comparison with the wild-type sensitive strain. No increase in cytosolic cadmium-binding proteins was observed. The only clue to a possible basis of their tolerance was that these strains exhibited morphological variations consistent with alterations to the cell wall and/or perhaps the plasma membrane.

3. The Genetics of Cobalt Tolerance

Cobalt is an essential element for many organisms and is unusual in that only microorganisms can use it in an inorganic form; animals rely on cobalamin synthesized by microbes. In *N. crassa*, cobalt toxicity produced changes similar to those of iron deficiency (39), and the effect of cobalt on iron metabolism has been well documented. Relatively little attention has been paid to cobalt tolerance. In *S. cerevisiae*, zinc and cobalt ions were transported to the vacuole for storage and detoxification, and a cobalt-resistant strain has been shown to have a higher vacuolar cobalt concentration than the sensitive parental strain (40). This is a good example of compartmentation, in which metal ions are stored separately from metabolically active cellular components. In *N. crassa*, cobalt resistance has been described that does not involve changes to iron metabolism, and magnesium ions reversed the toxicities of nickel and cobalt with parallel suppression of toxic ion uptake (41, 42).

In *A. nidulans*, cobalt tolerance was first described by Elorza (43), but all the mutants were lost. Eight additional cobalt-tolerant strains have been isolated and characterized in my laboratory. Of these, six mapped to locus *cbt*A, and two to *cbt*B; both genes appear to be located on chromosome VI (18,44,45). As with the copper and cadmium studies described above, *cbt*A^r *cbt*B^r double mutants were identified from crosses.

In each of these strains, germinating spores exhibited cobalt tolerance; i.e., a period of growth was not necessary for tolerance. *cbt*A^r strains were approximately three times more tolerant than wild-type strains; *cbt*B^r mutants and *cbt*A^r *cbt*B^r double mutants were almost fivefold more tolerant. Tolerance during germination was not additive, because *cbt*B^r single and *cbt*A^r *cbt*B^r double mutants were equally tolerant. In growing hyphae, however, *cbt*A^r *cbt*B^r double mutants were more resistant than either of the single mutants, indicating some degree of additivity. In tests for tolerance to other metals, *cbt* mutants showed some resistance to nickel, cadmium, and manganese but not to copper or zinc. Strains of *N. crassa* have also been described that are cotolerant to nickel and cobalt (41).

Because of the proximity of *cbt*A and *cbt*B to *cad*B on chromosome VI, chromosome mapping was attempted: no linkage was found to *cad*B or *meth*B. During these studies, Hodges (44) noted that *meth*B$^-$ strains appeared slightly more resistant to cobalt during germination and slightly more sensitive to manganese. Further studies revealed that *all cbt*A^r strains carried a second mutation, located within a novel gene involved in methionine catabolism, which we have designated *mec*D. *mec*D$^-$ mutants do not confer cobalt tolerance and cannot be detected phenotypically in the absence of *cbt*A^r alleles. By sexual crossing to wild-type, *cbt*A^r and *mec*D$^-$ alleles can be separated to produce *cbt*A^r *mec*D$^+$ strains. These accumulate *mec*D$^-$ mutations spontaneously even when maintained on complete media containing no cobalt. The *mec*D$^-$ allele does not appear to enhance cobalt tolerance per se; rather it is thought to alleviate a toxic side effect resulting from changes in metabolism caused by mutations in the *cbt*A gene. There is evidence that the toxic side effect results from increased homocysteine pools, which indirectly inhibit a range of methylation reactions: excess homocysteine leads to excess *S*-adenosyl homocysteine, which is a potent inhibitor of methyl transferases (46). *mec*D$^-$ mutations are thus thought to prevent the conversion of excess homocysteine to *S*-adenosyl homocysteine. Homocysteine accumulation is a possible basis for the cobalt tolerance of *cbt*A^r strains.

4. Resistance to Other Metals

In this section, I have outlined some genetic studies on metal-resistant mutants of fungi that are amenable to genetic and/or molecular analyses. It was not appropriate here to extensively review the field, covering all fungi and all metals. However, there are very good examples from these other studies that indicate the scope for research in this field. For instance, Elorza (43) isolated zinc-, cobalt-, nickel-, iron-, manganese-, and tin-resistant mutants of *A. nidulans*, indicating that this organism has an enormous amount still to teach us. An alternative approach is to use a specific metal: for example, vanadate-resistant mutants of yeast have been used to identify five genes involved in protein glycosylation (47). There is a wealth of literature in this field and many unanswered questions.

III. MOLECULAR ANALYSES OF METAL TOLERANCE

The molecular analysis of metal tolerance has centered on two major approaches. First, the identification of metal-binding polypeptides has enabled the identification of important proteins, which function in both metal tolerance and metal homeostasis. If a protein can be purified to homogeneity, and sequenced, then parts of the DNA sequence can be deduced to start recombinant DNA analysis. Second, DNA sequences can be isolated directly by complementation of mutants, differential screening to identify metal-regulated sequences, or random "shotgun" cloning to identify metal resistance among transformants. The identification of a clone, and subsequent DNA sequencing, can lead to predictions of the primary structure of the protein product: experiments can then be designed to identify the function of the protein.

In this section, I shall outline some studies that have led to the isolation from *S. cerevisiae* of the *ZRC1* gene product and to the description of metal-binding proteins/peptides and their regulation by metals. A more detailed and excellent review has been published recently (48).

A. The ZRC1 Gene Product

The *ZRC1* gene of *S. cerevisiae* has been isolated by shotgun-cloning random fragments of DNA into a recipient cell and selecting transformants exhibiting enhanced resistance to zinc (49). This approach will work only if the DNA encodes a dominant, selectable marker. Several DNA fragments conferring zinc resistance were isolated, one of which defines the *ZRC1* locus (33).

The *ZRC1* gene has been sequenced to reveal a single open reading frame (ORF) of 1326 bp, which has the potential to encode a protein of 48.3 kDa (442 amino acids) (33). This protein is not cysteine-rich like metallothioneins (see below) nor does it have zinc finger motifs, which are characteristic of certain transcription factors (50). This may indicate resistance is not conferred by increased binding of zinc ions to the protein. Analysis of its primary sequence reveals six regions characteristic of membrane-spanning proteins, and there is significant homology to the *cysB* gene product of both *Escherichia coli* and *Salmonella typhimurium*. Since *cysB* acts as a regulator of cysteine biosynthesis in *E. coli* (51), it is possible that ZRC1 acts to increase intracellular cysteine pools (or some other thiol compound), hence conferring resistance. This explanation is, however, unlikely since these strains show no increased resistance to copper, cobalt, or nickel (33).

The ability of *ZRC1* to confer resistance was elegantly demonstrated by introducing a frameshift mutation into the ORF, which abolished zinc and cadmium resistance (33). In addition, while multiple copies of this gene confer tolerance to 40 mM Zn^{2+}, disruption of the chromosomal *ZRC1* gene in the recipient makes cells more sensitive than wild-type to zinc and cadmium ions. These *zrc1*$^{-}$ mutants were not lethal and did not result in cysteine auxotrophy.

Although the mechanism of zinc tolerance is not known in this case, the available evidence suggests that ZRC1 encodes a membrane protein that increases the efflux of zinc and cadmium ions from these yeast cells (33).

B. Metallothioneins

Three classes of metallothioneins (MTs) have been described (see 52). Class I and II MTs are the products of genes, synthesized by transcription and translation via an mRNA template. Class III MTs (phytochelatins) are enzyme-synthesized polypeptides and are not gene-encoded. Here, I describe gene-encoded MTs; phytochelatins (class III MTs) will be discussed in Section III.C.

1. Fungal Class I MTs

MTs are low-molecular-mass, cysteine-rich, metal-binding proteins (52). Their classification is based on the arrangement of cysteine residues within the protein. The first isolate, equine renal MT, has characteristic cys-cys and cys-x-cys (where x is an amino acid other than cysteine) clusters. The cysteine clusters of class I MTs align with those of the horse. Whereas mammalian MTs are 61 or 62 amino acids long, fungal class I MTs have only 25 residues and approximate to half the mammalian molecule. Fungal class I MTs thus form one metal-binding domain, containing 6 moles of Cu(I) bound to 6 cysteines (27,53), whereas mammalian MTs have two domains, binding 6 Cu(I) to 11 cysteines in the α domain and 6 Cu(I) to 9 cysteines in the β domain (54). In addition to this M_{12} structure with the monovalent cations [i.e., 12 Cu(I) or Ag(I) bound in two clusters], mammalian MT forms an M_7 structure with divalent metal ions: 4 divalent ions bind to the α cluster, while 3 bind to the β cluster (54). These two structures may provide differential specificity (or partitioning) for Zn and Cu, the two major essential metals in cells.

In contrast, the two well-characterized fungal class I MTs, namely *N. crassa* and *Agaricus bisporus*, are both copper-specific. Despite evidence that these proteins can bind divalent cations in vitro, this has not been detected in vivo (27,53,55).

The *N. crassa* MT was purified by traditional chromatographic techniques and sequenced (28). The MT gene was isolated by synthesizing an oligonucleotide to part of the amino acid sequence, which was subsequently used to screen a *N. crassa* genomic DNA library (55). Because of a lack of mutants, the MT gene was mapped to chromosome 6, by using restriction fragment length polymorphisms as genetic markers (27). Transcription of the MT gene is only induced by copper, which presumably accounts for the metal specificity, even though in vitro the protein can bind other metals. Unlike the yeast class II MTs described below (Sections III.B.2 and III.B.3), copper-resistant *N. crassa* strains do not have increased MT gene copy number (see 27), suggesting that this protein has a primary role in copper homeostasis but not in copper resistance.

2. Fungal Class II MTs: *S. cerevisiae* CUP1 Gene Product

The most extensively studied class II MT is that of *S. cerevisiae* (see 25,56). It was first described in 1974 (57,58). The protein sequence was initially elucidated by cloning and sequencing the gene, *CUP1*. Fogel and Welch (59) first cloned the *CUP1* gene by transforming a copper-sensitive strain, *cup1*^{s}, using a genomic DNA library, and isolating the DNA fragment that transformed this strain to copper resistance. Butt and his co-workers (60,61) also cloned this gene, but by using differential hybridization. They initially isolated copper-resistant and copper-sensitive strains by replica plating onto medium containing $CuSO_4$. mRNA from these strains was used to prepare hybridization probes that were tested on gels containing restriction-digested DNA from the copper-resistant strain. A 1350-bp DNA fragment was identified that strongly hybridized to the probe from the copper-grown, copper-resistant strain but that had little cross-reaction to the probe made from the copper-sensitive strain grown in the absence of $CuSO_4$. This fragment was cloned and, on transformation of the copper-sensitive strain, generated copper-resistant cells.

The sequence of the gene was determined by both research groups (61, 62). The deduced protein sequence indicated that *CUP1* encoded a 6573-Da protein, containing an arrangement of 12 cysteines consistent with other metal-binding proteins, but sufficiently divergent to be described as a class II MT. When the *CUP1* protein was subsequently purified, it was found to lack the first eight amino acids relative to the deduced gene product (63). These residues are presumably removed by proteolysis and may indicate a transit function, such as sequestration of the protein in the vacuole, although as yet this has not been proven. The mature protein is thus 5655 Da, containing 53 amino acid residues, which will bind 8 moles of Cu(I) [or Ag(I)] to 12 cysteine residues. Cd(II), Zn(II), and Co(II) also bind in vitro with a maximal stoichiometry of 4 metal ions per mole of protein (63–65).

There can be no doubt about the involvement of the *CUP1* locus and its gene product in copper resistance of *S. cerevisiae*. There have been a number of reports of the transformation of copper-sensitive strains to copper resistance using the *CUP1* gene (59–61,64,66). Ecker and his co-workers (64) constructed a yeast strain which the MT locus was deleted (*cup1*Δ): this strain was transformed with high-copy-number plasmids containing the MT gene, under either normal or constitutive regulatory control. The untransformed *cup1*Δ strain could grow uninhibited (on solid media) on 1 μM $CuCl_2$, 1 μM $CdCl_2$, and 5 mM $ZnCl_2$. When transformed with normally regulated *CUP1* gene constructs at high copy number, resistance to copper was increased 1000-fold (uninhibited growth on 1 mM) specifically. With the constitutive gene constructs, however, uninhibited growth occurred on 500

μM $CuCl_2$ (500-fold increase in resistance) and on 1 mM $CdCl_2$ (1000-fold). There was no increased resistance to silver, mercury, zinc, cobalt, nickel, tin, gold, platinum, uranium, or lanthanum. In addition to these genetically manipulated strains, naturally occurring copper-resistant yeast strains also have multiple copies of the *CUP1* locus. Strains carrying 2–14 copies of the *CUP1* locus have been described (67).

The regulation of *CUP1* gene expression appears to be mediated only by copper (see 25). Zinc, cadmium, nickel, iron, tin, lead, and gold, for example, do not activate its transcription (60,61,68). This specificity of copper induction presumably dictates the specificity of copper resistance: even though the MT can bind other metals in vitro, resistance will not occur in vivo if the expression of the protein is not stimulated. Because of similar ionic properties, silver [Ag(I)] also activates transcription of *CUP1*, but silver resistance has not been documented in this species, even in strains constitutively synthesizing the MT protein (see above).

Recently, some evidence has emerged for cadmium regulation of the *CUP1* gene (69). By screening a genomic library for sequences capable of conferring cadmium resistance on *S. cerevisiae*, a clone was isolated containing two complete copies of the *CUP1* gene. Subsequent analysis showed that *CUP1* gene expression could be induced in these transformants with 0.02 μM cadmium, but that at concentrations of 2 μM or higher, the transcript level was depressed to below that of the untreated control. Despite being on a high-copy-number plasmid, this clone was unable to confer cadmium resistance on a *cup1*Δ strain, even though it restored copper resistance: the clone only confers cadmium resistance when transformed into a strain already possessing 12 chromosomal copies of this gene. As described above, it appears that this MT can only confer cadmium tolerance when present at high (constitutive) levels.

The activation of transcription only by low levels of cadmium is curious. In *A. nidulans*, low levels of cadmium stimulate germination of some conidia that otherwise appear inviable (34). The most likely explanation of this is that chelation of Cd^{2+} to constituents of the medium releases a greater free pool of an essential trace metal (such as Cu^{2+}). Perhaps the stimulation of transcription of *CUP1* by Cd^{2+} in yeast is also indirect; that is, micronutrient levels of copper are released from chelated sites resulting in the *CUP1* transcripts.

Recently, the *CUP1* gene of *S. cerevisiae* has been shown to hybridize to genes of *Mucor rouxii* and *Mucor racemosus*; the equivalent gene from *M. racemosus* has also been cloned (70).

3. *Candida glabrata* MTs

Studies of *Candida glabrata* (*Torulopsis glabrata*) have yielded novel information on metal-binding polypeptides, because it is the first documented example of an organism containing both gene-encoded MTs and enzyme-synthesized class III MTs (see Section III.C) (71). This is in complete contrast to the data currently available for *N. crassa, S. cerevisiae,* and *Sch. pombe*, which appear to possess one or the other of these molecules, but not both.

The gene-encoded MTs, MT-I and MT-II, of *C. glabrata* were identified initially by fractionation of cell-free extracts of copper-grown cells (71). These molecules have been purified and amino acid sequences have been obtained (71,72). Their respective genes were cloned by screening genomic libraries with degenerate oligonucleotides prepared against the amino acid sequences (72). The amino acid sequences of MT-I and MT-II show limited sequence homology with each other. The MT-I gene product has 62 amino acids including 18 cysteine residues, 14 of which are present in cys-x-cys clusters: MT-II has 16 cyteinyl residues (seven cys-x-cys clusters) in a protein 51 amino acids long. MT-I and MT-II bind 11–12 and 10 moles of Cu(I), respectively.

As with *S. cerevisiae*, the transcription of both MT genes is induced by copper and silver, but not by cadmium, which lowers their transcript levels (72). MT-II mRNA is induced to much higher levels than MT-I mRNA, which correlates with the small amounts of MT-I protein present in copper-grown cells. The observation that there are multiple copies of the MT-II gene, but only one copy of the MT-I gene, in these strains can be excluded as a major difference in their expression (see 48). MT-II, like the Cu-MT of *S. cerevisiae*, may also undergo posttranslational processing, having the first six amino acid residues removed (72).

Analysis of several wild-type isolates of *C. glabrata* has shown that, while the MT-II gene is present in three to nine copies, all contain a single MT-I gene (26). Copper-resistant strains of this species have been selected by stepwise culture, which contain more than 30 copies of the MT-II gene, while the MT-I gene remained as single copy. This selective, tandem amplification of the MT-II gene indicates it has a direct role in copper detoxification. No conclusions can be drawn about MT-I, however, because the fact that MT-II is naturally present in multiple copies greatly increases the likelihood of its amplification by misalignment and unequal crossing-over. The role of MT-I may require MT-II deletion strains to be isolated.

4. Metal-Activated Transcription

The transcription of fungal gene-encoded MTs is specifically directed by copper; silver mimics this effect, because of the ionic similarity between Ag(I) and Cu(I). Nothing is known about this activation in the filamentous fungi, but in *S. cerevisiae* and more recently *C. glabrata*, a protein and its gene have been identified that mediate copper induction.

As described above (Section II.A.2), the *CUP2* and *ACE1* genes of *S. cerevisiae* were identified by the isolation of copper-sensitive mutants (19–21) and are separate isolates of the same gene, whose product is most frequently referred to as the ACE1 protein (68,73). ACE1 is a copper-dependent DNA binding protein, which specifically interacts with the *CUP1* gene control sequences in the presence of copper, but not in its absence. Silver can also activate this DNA binding and transcription, whereas a series of divalent cations cannot (68). Copper activates the amino-terminal domain of ACE1 by causing a change of conformation, stabilizing the structure of this molecule for specific-DNA binding (73).

The ACE1 transcription factor has a novel structure and mechanism: many other transcription factors share common structural motifs, such as zinc fingers (50). The similarity of the Cu-cysteine clusters of the ACE1 and CUP1 proteins has led to the suggestion that they may have evolved from a common ancestral sequence (68,73).

The *AMT1* gene of *C. glabrata* encodes the functional equivalent of *ACE1* of *S. cerevisiae* (74). The gene has been cloned by complementation of *ace1-1* (*cup2*) copper-sensitive mutants of *S. cerevisiae*, transformants being recognized as copper-resistant; clearly, the AMT1 protein can activate *CUP1* gene transcription. Purified AMT1 protein was also shown to bind to both the MT-I and MT-II gene promoters from *C. glabrata* in the presence of Cu(I) and Ag(I) ions. Definitive proof of the role of AMT1 will await gene disruption experiments in *C. glabrata*.

C. Class III MTs; γ-Glutamyl Peptides; Phytochelatins

The third class of metal-binding polypeptides have recently been classified as metallothioneins (52), even though they are normally referred to by a variety of other names: cadystins (75–77), phytochelatins (78), poly(gamma-glutamyl-cysteinyl) glycines (79), gamma-glutamyl peptides (80), metallothiopeptides (81). Detailed reviews of these molecules have recently been published (48,82,83). For brevity in this review, I shall refer to them as PCs to make the contrast to gene-encoded MTs clear, despite the inaccuracy of the prefix *phyto-* in the context of the material discussed.

The initial description of these molecules was as cadmium-binding peptides from the fission yeast *S. pombe* (75–77,84–87). Two structures were described, $[\gamma\text{-glu-cys}]_3\text{gly}$ and $[\gamma\text{-glu-cys}]_2\text{gly}$, as shown in Figure 1, which have molecular masses of approximately 4000 and 1800 Da, respectively. Subsequent analysis showed that this organism could produce higher polymeric forms $[\gamma\text{-glu-cys}]_n\text{gly}$, where $n = 2$–8, and molecules lacking the terminal glycine (called desGly peptides), in which $n = 3$ or 4 (80,88).

glu -γ - cys - gly — Glutathione

glu -γ - cys - glu - γ - cys - gly — PC2

glu -γ - cys - glu - γ - cys - glu - γ - cys - gly — PC3

glu -γ - cys - glu - γ - cys - glu - γ - cys — PC3 desGly

Figure 1 Structure of glutathione and phytochelatins. Unlike proteins in which amino acids are linked via the carboxyl group on the α-carbon, these molecules all have the glu-cys linkage through the carboxyl group of the γ-carbon of glutamic acid: the cys-glu and cys-gly linkages are normal α-peptide bonds.

The structure of these molecules is clearly related to that of glutathione (i.e., γ-glu-cys-gly [Fig. 1]). Glutathione is synthesized by two enzymes, γ-glutamylcysteine synthetase (EC 6.3.2.2.) and glutathione synthetase (EC 6.3.2.3.), as shown in Figure 2. PCs are synthesized by the addition of extra [γ-glu-cys] dipeptides to glutathione, although two pathways may be involved (22,89). The only enzyme characterized thus far is from the plant *Silene cucubalus* (90,91). γ-glutamylcysteine dipeptidyl transferase, trivial name phytochelatin synthase, catalyzes the transfer of γ-glu-cys dipeptides of glutathione to an acceptor glutathione molecule or a growing chain [γ-glu-cys]$_n$gly molecule to produce a PC (Fig. 2). This enzyme has not yet been described from any fungi, but the similarities in the systems are so strong that it seems likely that it will be analogous.

Analysis of PC contents of *S. pombe* cells indicates that, upon addition of cadmium to a logarithmic culture, PC_2 is synthesized immediately (no detectable lag) followed by PC_3 and PC_4 (88,92). The pool of PC molecules always contains a mixture of PCs but the concentration gradient is always $PC_2 > PC_3 > PC_4$, and so forth. This PC synthesis results in a reduction of the glutathione pool.

The rapid synthesis of PCs is indicative of the novel regulatory mechanism of PC synthase. PC synthesis is induced specifically by a wide range of heavy metals in *S. pombe*: cadmium, bismuth, arsenate, copper, lead, zinc, silver (88). In plant cell cultures, induction is also mediated by antimony, nickel, mercury, tin, gold, selenate, tellurium, and tungsten (93). The imme-

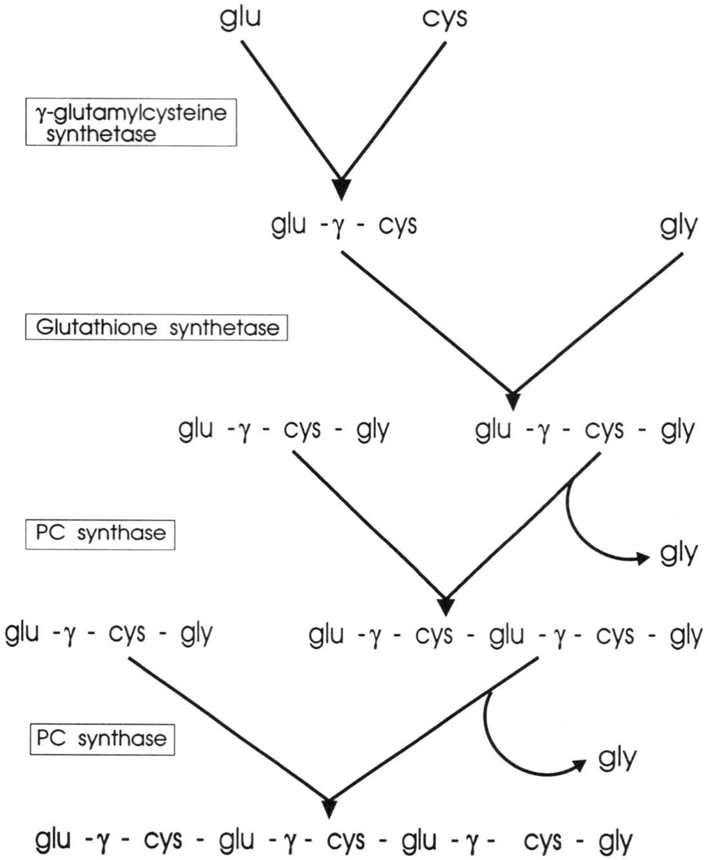

Figure 2 Biosynthesis of phytochelatins. The major pathway of PC synthesis involves (top) the two enzymes of glutathione biosynthesis, γ-glutamylcysteine synthetase and glutathione synthetase, and (bottom) PC synthase. The latter enzyme can transfer glu-γ-cys dipeptides from either glutathione or short-chain PCs, to glutathione or growing chain PCs, releasing one glycine molecule.

diate appearance of PCs on addition of cadmium is because PC synthase is constitutive: metal activation is a posttranslational effect (90,91). Indeed, the enzyme is regulated by the reaction products, because the PCs chelate the metal and hence terminate the reaction by deactivating the enzyme (Fig. 3).

Unlike *S. pombe, C. glabrata* produces both MTs and PCs (71). In this organism, PCs are synthesized in response to cadmium but not copper, and therefore if PC synthase is constitutive, it must differentially respond to these metals during activation.

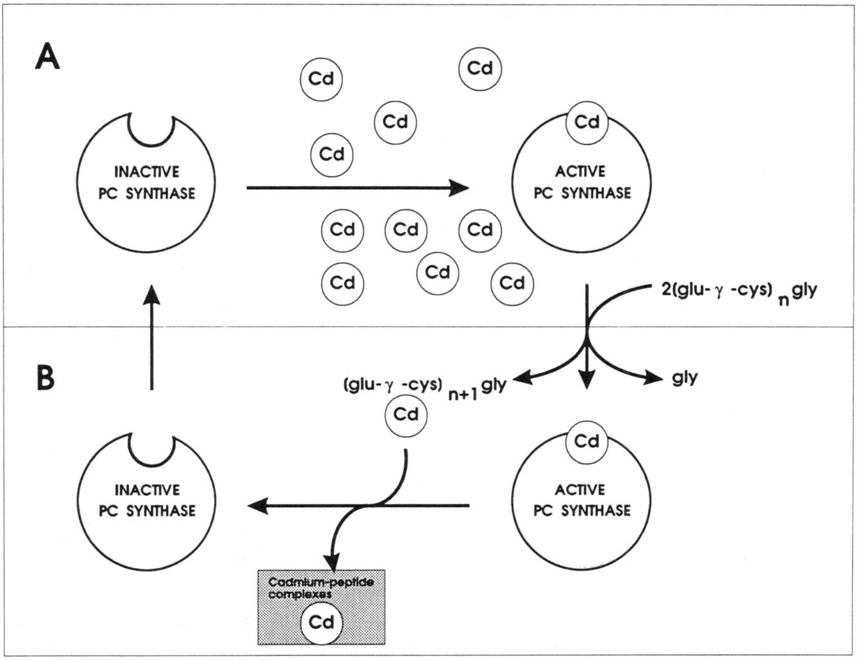

Figure 3 Metal activation of phytochelatin synthase. (A) Inactive PC synthase is activated by the presence of metal ions, e.g., cadmium, and PC synthesis is initiated. (B) PCs and cadmium form cadmium-peptide complexes until no free cadmium ions are available to maintain PC synthase in its active form.

PCs clearly function in metal detoxification, and their capacity to bind metal ions confers resistance. Buthionine sulfoximine is a potent inhibitor of γ-glutamylcysteine synthetase, and its use results in cadmium toxicity because of an inability to synthesize glutathione and PCs (see 12,82). Similarly, cadmium-hypersensitive mutants of *S. pombe* have been isolated which are unable to synthesize PCs; of these, some lack γ-glutamylcysteine synthetase or glutathione synthetase, whereas others have normal levels for these activities (22). In plants, PCs are ubiquitous (94). In fungi, however, only certain species appear to be capable of their synthesis; for example, there is no evidence that *N. crassa, A. nidulans,* or *S. cerevisiae* produce PCs. Because of constitutive synthesis of PCs, there is no evidence that organisms that produce them can be mutated to higher levels of tolerance by increased synthesis, or that this would be an advantage. Theoretically, increased PC synthase gene expression could increase the rate of synthesis of the larger

PC oligomers and hence increase resistance; molecular analysis of PC synthase gene transcription would be required to determine this.

Despite the induction of PC synthesis by a very wide range of metals, binding has only been characterized to Cu, Zn, Pb, and Cd (see 82). These metals form complexes with PCs, which contain both peptides of various lengths, and sulfide. Such complexes will change during the induction period (or with changing cadmium concentrations) as more of the longer PC peptides are synthesized. The larger peptides bond cadmium more firmly than smaller ones, and the complexes containing sulfide bind more metal (92). In fact, the high-sulfide complexes (mostly PC_3) increase with time, whereas the low-sulfide form (mostly PC_2) declines with time. Addition of cadmium to the medium leads to enhanced generation of sulfide ions (95,96), and to stabilization of the PC complexes.

Sulfide is an important component. Cadmium sulfide crystallites have been observed in *S. pombe* after growth in cadmium-containing medium, and similar culture of *C. glabrata* led to the demonstration of PC-coated cadmium sulfide crystallites (97,98). There is a direct correlation between the sulfide levels and the cadmium content of the complexes (71), which has led to the suggestion that sulfide is a key component, and that metal tolerance may result from the extracellular precipitation of CdS. An alternative suggestion has arisen from studies of tobacco: PCs may transport cadmium to the vacuole (99; and see 3).

IV. CONCLUSIONS

Genetic and molecular analyses have led to the definition of specific mechanisms for metal ion homeostasis, detoxification, and tolerance. A gene from *S. cerevisiae*, *ZRC1*, has been cloned that encodes a membrane protein, probably mediating cadmium/zinc efflux.

Genes have been cloned from filamentous fungi and yeast species that encode class I/II copper-specific metallothioneins, which almost certainly have a role in copper homeostasis. In the yeasts, overexpression of these proteins results in copper tolerance; mutants have been recognized with tandem amplifications of these MT genes. A second class of metal-binding polypeptides (PCs) are synthesized by enzymes related to glutathione metabolism, and which may transport metal to the vacuole.

The isolation of copper-resistant mutants had led to the characterization of the MTs. Metal-sensitive mutants have aided the elucidation of PC synthesis and the identification of the novel copper-specific transcription factor from yeasts. There is a wealth of information yet to be gleaned from the analysis of metal-resistant mutants. It is clear that in many cases resistance is metal-specific, indicating a wide range of potential metal tolerance mechanisms.

There is a clear relationship between metal and sulfur metabolism. In addition to MTs and PCs being cysteine-rich polypeptides, sulfide is implicated in metal-binding complexes. The *ZRC1* gene product of *S. cerevisiae* has some sequence similarity to a prokaryotic protein responsible for the regulation of cysteine biosynthesis. Cadmium and cobalt-resistant mutants of *A. nidulans* appear to have a direct involvement in cysteine/methionine biosynthesis, and cadmium stimulates sulfide production. These factors indicate that the future understanding of metal tolerance will require a detailed analysis of sulphur metabolism.

ACKNOWLEDGMENTS

I would like to thank Neil Cooley, Kath Hodges, and Anne Phelan for their hard work on *A. nidulans* to produce the data reported here, and Helen Haslock, Isobel Connell, and Jane Wilkinson for their excellent technical assistance. Special thanks are due to my good friend and co-worker David Thurman, who stimulated my interest in this field in the first place and has kept me on the right track ever since.

REFERENCES

1. Friedland, A. J. (1989). Movement of metals through soils and ecosystems, in *Heavy Metal Tolerance in Plants—Evolutionary Aspects* (A. J. Shaw, ed.), CRC Press, Orlando, FL, pp. 7–19.
2. Nieboer, E., and Richardson, D. H. S. (1990). The replacement of the nondescript term "heavy metals" by a biologically and chemically significant classification of metal ions, *Environ. Pollut., Series B1*: 3–26.
3. Tomsett, A. B., Sewell, A. K., Jones, S. J., de Miranda, J. R., and Thurman, D. A. (1992). Metal-binding proteins and metal-regulated gene expression in higher plants, in *Inducible Plant Proteins* (J. L. Wray, ed.), Cambridge University Press, New York. *Soc. Exp. Biol. Sem. Series, 49*: 1–24.
4. Shaw, A. J., ed. (1989). *Heavy Metal Tolerance in Plants-Evolutionary Aspects*, CRC Press, Orlando, FL, 355 pp.
5. Macnair, M. R. (1989). The genetics of metal tolerance in natural populations, in *Heavy Metal Tolerance in Plants—Evolutionary Aspects* (A. J. Shaw, ed.), CRC Press, Orlando, FL, pp. 235–253.
6. Bradshaw, A. D., McNeilly, T., and Putwain, P. D. (1989). The essential qualities, in *Heavy Metal Tolerance in Plants—Evolutionary Aspects* (A. J. Shaw, ed.), CRC Press, Orlando, FL, pp. 323–334.
7. Brown, M. T., and Hall, I. R. (1989). Metal tolerance in fungi, in *Heavy Metal Tolerance in Plants—Evolutionary Aspects* (A. J. Shaw, ed.), CRC Press, Orlando, FL, pp. 95–104.
8. Denny, H. J., and Wilkins, G. H. (1987). Zinc tolerance in *Betula* spp. III. Variation in response to zinc among ectomycorrizal associates, *New Phytol. 106*: 535.

9. Zibilske, L. M., and Wagner, G. H. (1982). Bacterial growth and fungal genera distribution in soil amended with sewage sludge containing cadmium, chromium, and copper, *Soil Sci. 134*: 364.
10. Arnebrandt, K., Baath, E., and Nordgren, A. (1987). Copper tolerance of microfungi isolated from polluted and unpolluted forest soil, *Mycologia 79*: 890.
11. Gadd, G. M., and White C. (1989). Heavy metal and radionuclide accumulations and toxicity in fungi and yeast, in *Metal-Microbe Interactions*, Vol. 26 IRL Press, Oxford (R. K. Poole and G. Gadd, eds.), Soc. Gen. Microbiol., pp. 19–38.
12. Tomsett, A. B., and Thurman, D. A. (1988). Molecular biology of metal tolerances in plants, *Plant, Cell Environ. 11*: 383–394.
13. Wood, J. M., and Wang, H-K. (1983). Microbial resistance to heavy metals, *Environ. Sci. Technol. 17*: 582–590.
14. Wood, J. M. (1984). Microbiological strategies in resistance to metal ion toxicity, in *Metal Ions in Biological Systems* (H. Sigel, ed.), Marcel Dekker, Inc., New York, Vol. 18, pp. 333-351.
15. Tohoyama, H., Inouhe, M., Joho, M., and Murayama, T. (1990). Resistance to cadmium is under the control of the *CAD2* gene in the yeast *Saccharomyces cerevisiae, Curr. Genet. 18*: 181–185.
16. Garcia-Toledo, A., Babich, H., and Stotsky, G. (1985). Training of *Rhizopus stolonifer* and *Cunninghamella blakesleeana* to copper: cotolerance to cadmium, cobalt, nickel and lead, *Can. J. Microbiol. 31*: 485–492.
17. Phelan, A., Thurman, D. A., and Tomsett, A. B. (1990). The isolation and characterization of copper resistant mutants in *Aspergillus nidulans, Curr. Microbiol. 21*: 255–260.
18. Tomsett, A. B., Hodges, K. E., Cooley, R. N., and Thurman, D. A. (1989). Metal resistance in *Aspergillus nidulans*, in *Metal Ion Homeostasis—Molecular Biology and Chemistry* (D. H. Hamer and D. R. Winge, eds.), Alan R. Liss, New York, pp. 375–384.
19. Thiele, D. J. (1988). *ACE1* regulates expression of the *Saccharomyces cerevisiae* metallothionein gene, *Mol. Cell. Biol. 8*: 2745–2752.
20. Welch, J., Fogel, S., Buchman, C., and Karin, M. (1989). The *CUP2* gene product regulates the expression of the *CUP1* gene, coding for yeast metallothionein, *EMBO J. 8*: 255–260.
21. Buchman, C., Skroch, P., Welch, J., Fogel, S., and Karin, M. (1989). The *CUP1* gene product, regulator of yeast metallothionein expression, is a copper-activated DNA-binding protein, *Mol. Cell. Biol. 9*: 4091–4095.
22. Mutoh, N., and Hayashi, Y. (1988). Isolation of mutants of *Schizosaccharomyces pombe* unable to synthesize cadystin, small cadmium-binding peptides, *Biochem. Biophys. Res. Commun. 151*: 32–39.
23. Glaeser, H., Coblenz, A., Kruczek, R., Ruttke, I., Ebert-Jung, A., and Wolf, K. (1991). Glutathione metabolism and heavy metal detoxification in *Schizosaccharomyces pombe*—isolation and characterisation of glutathione-deficient, cadmium-sensitive mutants, *Curr. Genet. 19*: 207–213.
24. Brenes-Pomales, A., Lindegren, G., and Lindegren, C. C. (1955). Gene control of copper sensitivity in *Saccharomyces, Nature* (*Lond.*) *176*: 841–842.

25. Butt, T. R., and Ecker, D. J. (1987). Yeast metallothionein and applications to biotechnology, *Microbiol. Rev. 51*: 351–364.
26. Mehra, R. K., Garey, J. R., and Winge, D. R. (1990). Selective and tandem amplification of a member of the metallothionein gene family in *Candida glabrata, J. Biol. Chem. 265*: 6369–6375.
27. Munger, K., Germann, U. A., and Lerch, K. (1987). The *Neurospora crassa* metallothionein gene, *J. Biol. Chem. 262*: 7363–7367.
28. Lerch, K. (1980). Copper metallothionein, a copper-binding protein from *Neurospora crassa, Nature* (*Lond.*) *284*: 368–370.
29. Lerch, K., personal communication.
30. Failla, M. L., Benedict, C. D., and Weinberg, E. D. (1976). Accumulation and storage of zinc by *Candida utilis, J. Gen. Microbiol. 94*: 23–36.
31. Norris, P. R., and Kelly, D. P. (1977). Accumulation of cadmium and cobalt by *Saccharomyces cerevisiae, J. Gen. Microbiol. 99*: 317–324.
32. Jacobsen, K. B., and Turner, J. E. (1980). The interaction of cadmium and certain other metal ions with proteins and nucleic acids, *Toxicology 16*: 1–37.
33. Kamizono, A., Nishizawa, M., Teranishi, Y., Murata, K., and Kimura, A. (1989). Identification of a gene conferring resistance to zinc and cadmium ions in *Saccharomyces cerevisiae, Mol. Gen. Genet. 219*: 161–167.
34. Cooley, R. N., Haslock, H. R., and Tomsett, A. B. (1986). Isolation and characterization of cadmium-resistant mutants of *Aspergillus nidulans, Curr. Microbiol. 13*: 265–268.
35. Cooley, R. N. (1986). Metal tolerance and nitrate assimilation in *Aspergillus nidulans*, Ph.D. thesis, University of Liverpool.
36. Stepien, P., Pieniazek, N., Bal, J., and Morzycka, E. (1975). Cysteine biosynthesis in *Aspergillus nidulans, Acta. Microbiol. Polon. A 7*: 201–210.
37. Phelan, A. (1991). Heavy metal tolerance in *Aspergillus nidulans*, Ph.D. thesis, University of Liverpool.
38. Levine, W. B., and Marzluf, G. A. (1989). Isolation and characterization of cadmium-resistant mutants of *Neurospora crassa, Can. J. Microbiol. 35*: 359–365.
39. Healy, W. B., Cheng, S-C., and McElroy, W. D. (1955). Metal toxicity and ion deficiency effects on enzymes in *Neurospora, Arch. Biochem. Biophys. 54*: 206–214.
40. White, C., and Gadd, G. M. (1986). Uptake and cellular distribution of copper, cobalt and cadmium in strains of *Saccharomyces cerevisiae* cultured on elevated concentrations of these metals, *FEMS Microbiol. Ecol. 38*: 277–283.
41. Venkateswerlu, G., and Sastry, K. S. (1973). Interrelationships in trace-element metabolism in metal toxicities in a cobalt-resistant strain *Neurospora crassa, Biochem. J. 132*: 673–680.
42. Mohan, P. M., and Sastry, K. S. (1983). Interrelationships in trace-element metabolism in metal toxicities in nickel-resistant strains of *Neurospora crassa, Biochem. J. 212*: 205–210.
43. Elorza, M. V. (1969). Toxicity of metal ion for *Aspergillus nidulans, Microbiol. Esp. 22*: 131–137.
44. Hodges, K. E. (1990). Cobalt tolerance and sulfur metabolism in *Aspergillus nidulans*, Ph.D. thesis, University of Liverpool.

45. Hodges, K. E., Thurman, D. A., and Tomsett, A. B. unpublished results.
46. Duerre, J. A., and Walker, R. D. (1977). Metabolism of adenosylhomocysteine, in *The Biochemistry of Adenosylmethionine* (F. Salvatore, E. Roek, V. Zappa, W. Ashman, and F. Schlenk, eds.), Columbia University Press, New York, p. 43.
47. Ballou, L., Hitzeman, R. A., Lewis, M. S., and Ballou, C. E. (1991). Vanadate-resistant yeast mutants are defective in protein glycosylation, *Proc. Natl. Acad. Sci. USA 88*: 3209–3212.
48. Mehra, R. K., and Winge, D. R. (1990). Metal ion resistance in fungi—molecular mechanisms and their regulated expression, *J. Cell. Biochem. 45*: 1–11.
49. Fukuda, Y., Saikusa, T., Watanabe, K., Shimosaka, M., Murata, K., and Kimura, A. (1986). Cloning of genes enhancing the resistance of *Saccharomyces cerevisiae* to zinc and cadmium ions, *Agric. Biol. Chem. 50*: 1341–1343.
50. Evans, R. M., and Hollenberg, S. M. (1988). Zinc fingers—gilt by association, *Cell 52*: 1–3.
51. Ostrowski, J., Burdzy, C. J., and Kredich, N. M. (1987). DNA sequence of the *cysB* regions of *Salmonella typhinurium* and *E. coli, J. Biol. Chem. 262*: 5999–6005.
52. Kagi, J. H. R., and Kojima, Y. (1987). Chemistry and biochemistry of metallothionein, in *Metallothionein II* (J. H. R. Kagi and Y. Kojima, eds.), Birkhauser Verlag, Basel, pp. 25–61.
53. Munger, K., and Lerch, K. (1985). Copper-metallothionein from the fungus *Agaricus bisporus*—chemical and spectroscopic properties, *Biochemistry 24*: 6751–6756.
54. Nielson, K. B., Atkin, C. L., and Winge, D. R. (1985). Distinct metal-binding configurations in metallothionein, *J. Biol. Chem. 260*: 5342–50.
55. Munger, K., Germann, U. A., and Lerch, K. (1985). Isolation and structural organization of the *Neurospora crassa* copper metallothionein gene, *EMBO J. 4*: 2665–2668.
56. Hamer, D. H. (1986). Metallothionein, *Annu. Rev. Biochem. 55*: 913–51.
57. Premakumar, R., Winge, D. R., Wiley, R. D., and Rajagopalan, K. V. (1975). Copper chelatin—isolation from various eucaryotic sources, *Arch. Biochem. Biophys. 170*: 278–288.
58. Prinz, R., and Weser, U. (1975). Naturally occurring copper-thionein in *Saccharomyces cerevisiae, J. Physiol. Chem. 356*: 767–776.
59. Fogel, S., and Welch, J. W. (1982). Tandem gene amplification mediates copper resistance in yeast, *Proc. Natl. Acad. Sci. USA 79*: 5342–5346.
60. Butt, T. R., Sternberg, A. J., Gorman, J. A., Clark, P., Hamer, D., Rosenberg, M., and Crooke, S. T. (1984). Copper metallothionein of yeast, structure of the gene, and regulation of expression, *Proc. Natl. Acad. Sci. USA 81*: 3332–3336.
61. Butt, T. R., Sternberg, E. J., Herd, J., and Crooke, S. T. (1984). Cloning and expression of a yeast copper metallothionein gene, *Gene 27*: 23–33.
62. Karin, M., Najarian, R., Haslinger, A., Valenzuela, P., Welch, J., and Fogel, S. (1984). Primary structure and transcription of an amplified genetic locus—the *CUP1* locus of yeast, *Proc. Natl. Acad. Sci. USA 81*: 337–341.

63. Winge, D. R., Nielson, K. B., Gray, W. R., and Hamer, D. H. (1985). Yeast metallothionein—sequence and metal-binding properties, *J. Biol. Chem. 260*: 14464–14470.
64. Ecker, D. J., Butt, T. R., Sternberg, E. J., Neeper, M. P., Debouck, C., Gorman, J. A., and Crooke, S. T. (1986). Yeast metallothionein function in metal ion detoxification, *J. Biol. Chem. 261*: 16895–16900.
65. Hartmann, H-J., and Weser, U. (1977). Copper-thionein from fetal bovine liver, *Biochim. Biophys. Acta 491*: 211–227.
66. Henderson, R. C. A., Cox, B. S., and Tubb, R. (1985). The transformation of brewing yeasts with a plasmid containing the gene for copper resistance, *Curr. Genet. 9*: 133–138.
67. Fogel, S., Welch, J. W., Cathala, G., and Karin, M. (1983). Gene amplification in yeast-CUP1 copy number regulates copper resistance, *Curr. Genet. 7*: 347–355.
68. Furst, F. S., Hu, S., Hackett, R., and Hamer, D. (1988). Copper activates metallothionein gene transcription by altering the conformation of a specific DNA-binding protein, *Cell 55*: 705–717.
69. Jeyaprakash, A., Welch, J. W., and Fogel, S. (1991). Multicopy *CUP1* plasmids enhance cadmium and copper resistance levels in yeast, *Mol. Gen. Genet. 225*: 363–368.
70. Cano-Canchola, C., Sosa, L., Fonzi, W., Sypherd, P., and Ruiz-Herrera, J. (1992). Developmental regulation of CUP gene expression through DNA methylation in *Mucor* spp., *J. Bacteriol. 174*: 362–366.
71. Mehra, R. K., Tarbet, E. B., Gray, W. R., and Winge, D. R. (1988). Metal-specific synthesis of two metallothioneins and gamma-glutamyl peptides in *Candida glabrata, Proc. Natl. Acad. Sci. USA 85*: 8815–19.
72. Mehra, R. K., Garey, J. R., Butt, T. R., Gray, W. R., and Winge, D. R. (1989). *Candida glabrata* metallothioneins, cloning and sequence of the genes and characterization of proteins, *J. Biol. Chem. 264*: 19747–53.
73. Dameron, C. T., Winge, D. R., George, G. N., Sansone, M., Hu, S., and Hamer, D. (1991). A copper-thiolate polynuclear cluster in the ACE1 transcription factor, *Proc. Natl. Acad. Sci. USA 88*: 6127–6131.
74. Zhou, P., and Thiele, D. J. (1991). Isolation of a metal-activated transcription factor gene from *Candida glabrata* by complementation in *Saccharomyces cerevisiae, Proc. Natl. Acad. Sci. USA 88*: 6112–6116.
75. Kondo, N., Isobe, M., Imai, K., Goto, T., Marusugi, A., and Hayahsi, Y. (1983). Structure of cadystin, the unit peptide of cadmium-binding peptides induced in a fission yeast, *Schizosaccharomyces pombe, Tetrahedron Lett. 24*: 925–928.
76. Kondo, N., Imai, K., Isobe, M., Goto, T., Murasugi, A., Wada-Nakagawa, C., and Hayashi, Y. (1984). Cadystin A and B major unit peptides comprising cadmium-binding peptides induced in a fission yeast—separation, revision of structures and synthesis, *Tetrahedron Lett. 25*: 3869–3872.
77. Kondo, N., Isobe, M., Imai, K., and Goto, T. (1985). Synthesis of metallothionein-like peptides cadystin A and B occurring in a fission yeast, and their isomers, *Agric. Biol. Chem. 49*: 71–83.

78. Grill, E., Winnacker, E-L., and Zenk, M. H. (1985). Phytochelatins: the principle heavy metal complexing peptides in higher plants, *Science 230*: 674-76.
79. Jackson, P. J., Unkefer, C. J., Doolen, J. A., Watts, K., and Robinson, N. J. (1987). Poly(gamma-glutamylcysteinyl) glycine: its role in cadmium resistance in plant cells, *Proc. Natl. Acad. Sci. USA 84*: 6619–23.
80. Mehra, R. K., and Winge, D. R. (1988). Cu(I) binding to the *Schizosaccharomyces pombe* γ-glutamyl peptide varying in chain lengths, *Arch. Biochem. Biophys. 265*: 381–389.
81. Verkleij, J. A. C., Koevoets, P. L. M., v't Riet, J., de Knecht, J. A., and Ernst, W. H. O. (1989). The role of metal-binding compounds (metallothiopeptides) in the cadmium tolerance mechanism of bladder campion, *Silene cucubalus*, in *Sulfur Metabolism* (C. Brunold, L. J. de Kok, P. C. Kunper, H. Rennenberg, and I. Stulen, eds.), SPB Academic Publishing, The Hague.
82. Steffens, J. C. (1990). The heavy metal-binding peptides of plants, *Annu. Rev. Plant Physiol. 41*: 553–75.
83. Rauser, W. E. (1990). Phytochelatins, *Annu. Rev. Biochem. 59*: 61–86.
84. Murasugi, A., Wada, C., and Hayashi, Y. (1981). Cadmium-binding peptide induced in fission yeast, *Schizosaccharomyces pombe, J. Biochem.* (*Tokyo*) *90*: 1561–64.
85. Murasugi, A., Wada, C., and Hayashi, Y. (1981). Purification and unique properties in UV and CD spectra of cadmium-binding peptide 1 from *Schizosaccharomyces pombe, Biochem. Biophys. Res. Commun. 103*: 1021–1028.
86. Murasugi, A., Wada, C., and Hayashi, Y. (1983). Occurrence of acid-labile sulfide in cadmium-binding peptide 1 from fission yeast, *J. Biochem. 93*: 661–664.
87. Murasugi, A., Wada-Nakagawa, C., and Hayashi, Y. (1984). Formation of cadmium-binding peptide allomorphs in fission yeast, *J. Biochem. 96*: 1375–1379.
88. Grill, E., Winnacker, E-L., and Zenk, M. H. (1986). Synthesis of seven different homologous phytochelatins in metal-exposed *Schizosaccharomyces pombe* cells, *FEBS Lett. 197*: 115–20.
89. Hayashi, Y., Nakagawa, C. W., and Mutoh, N. (1991). Two pathways in the *biosynthesis of cadystins* $(\gamma EC)_nG$ in the cell-free system of the fission yeast, *Biochem. Cell. Biol. 69*: 115–121.
90. Grill, E., Loeffler, S., Winnacker, E-L., and Zenk, M. H. (1989). Phytochelatins, the heavy metal-binding peptides of plants are synthesized from glutathione by a specific gamma-glutamyl cysteine dipeptidyl transpeptidase (phytochelatin synthase), *Proc. Natl. Acad. Sci. USA 86*: 6838–42.
91. Loeffler, S., Hochberger, A., Grill, E., Winnacker, E-L., and Zenk, M. H. (1989). Termination of the phytochelatin synthase reaction through sequestration of heavy metals by the reaction product, *FEBS Lett. 258*: 42–46.
92. Hayashi, Y., Nakagawa, C. W., Uyakul, D., Imai, K., Isobe, M., and Goto, T. (1988). The change of cadystin components in cadmium-binding peptides from the fission yeast during their induction by cadmium, *Biochem. Cell. Biol. 66*: 288–295.

93. Grill, E., Winnacker, E-L., and Zenk, M. H. (1987). Phytochelatins, a class of heavy metal binding peptides from plants are functionally analogous to metallothioneins, *Proc. Natl. Acad. Sci. USA 84*: 439–443.
94. Gekeler, W., Grill, E., Winnacker, E-L., and Zenk, M. H. (1989). Survey of the plant kingdom for the ability to bind heavy metals through phytochelatins, *Z. Naturforsch. 44c*: 361–69.
95. Reese, R. N., Mehra, R. K., Tarbet, E. B., and Winge, D. R. (1988). Studies on the γ-glutamyl copper-binding peptide from *Schizosaccharomyces pombe, J. Biol. Chem. 263*: 4186–4192.
96. Reese, R. N., and Winge, D. R. (1988). Sulfide stabilization of the cadmium-γ-glutamyl peptide complex of *Schizosaccharomyces pombe, J. Biol. Chem. 263*: 12832–12835.
97. Dameron, C. T., Reese, R. N., Mehra, R. K., Korlan, P. J., Carroll, M. L., Steigerwalg, L., Brus, L. E., and Winge, D. R. (1989). Biosynthesis of cadmium sulphide quantum semiconductor crystallites, *Nature* (*Lond.*) *338*: 596–597.
98. Dameron, C. T., and Winge, D. R. (1990). Peptide-mediated formation of quantum semiconductors, *Trends Biotech. 8*: 3–6.
99. Vogeli-Lange, R., and Wagner, G. J. (1990). Subcellular localization of cadmium and cadmium-binding peptides in tobacco leaves, *Plant Physiol. 92*: 1086–1093.

4

Membrane Transport Processes and Response to Exposure to Heavy Metals

I. S. Ross

University of Keele
Keele, Staffordshire, England

I. INTRODUCTION

In recent years there has been increasing concern about heavy metal contamination of both aqueous and terrestrial environments, and a number of reviews have considered the effects of such contamination on fungi (1,2). This discussion will consider the ways in which fungal cells respond to stress imposed by elevated heavy metal ion concentrations with respect to regulation of membrane transport, efflux, and compartmentalization of metal ions into vacuoles. The role of the cell wall in binding metals and the direct effects of heavy metal ions on membrane integrity will also be discussed. There is no precise definition of "heavy metal" in the literature (1), and this discussion will consider those metals considered essential for growth but toxic at high concentrations, including copper, zinc, cobalt, manganese, and nickel, and those with no known biological function, such as cadmium, mercury, silver, and lead, which may be very toxic to fungi (3). Calcium has also been included, though it is not regarded as a heavy metal, because calcium homeostasis in fungi has received considerable attention and may provide a background for a better understanding of the regulation of cellular concentrations of other metals, especially the role of the vacuole and possibly efflux from cells.

II. UPTAKE OF METALS BY CELL AND HYPHAL WALLS

In both yeasts and filamentous fungi, the first potential barrier to penetration and possible damage to the organisms by metals is the wall. There is much literature on metal uptake by fungal cell walls with emphasis on use as biosorbents, with Langmuir or Freundlich isotherms having been used to describe binding. This work together with data on the chemical nature of metal binding has recently been reviewed by Gadd (4). However, not all of this work has taken account of fungal viability or the relevance of the metal concentrations used to those likely to be tolerated by growing fungi. Also, the application of the Langmuir or Freundlich isotherm needs to be done with caution, as both assume that binding is to a uniform unimolecular layer with each ion adsorbed to an adsorption site, a situation unlikely to be found in the case of fungal cell walls where binding is likely to be of more than one type and may well be complex (1,4). This discussion will concentrate on data obtained from metal uptake studies aimed at understanding the transport of metal ions into cells, when more attention has generally been paid to cell viability.

A. Yeast

There have been many studies of uptake of metals by yeasts, chiefly *Saccharomyces cerevisiae*, and these have shown that uptake is generally biphasic with an initial rapid phase that is independent of metabolic uptake followed by a slower, energy-dependent phase. The initial phase is generally regarded as due to binding to the cell wall, and the second phase is regarded as the result of membrane transport of metal ions. These studies have largely been based on time course experiments on nongrowing suspensions in the presence and absence of energy sources and/or metabolic inhibitors. The initial surface binding has been recognized as the amount of metal bound within a few minutes in the absence of energy sources, at low temperature (usually about 4 °C) or in the presence of metabolic inhibitors. Such methods do not distinguish between the metal ions bound by the cell wall and those bound to the outer surface of the cytoplasmic membrane. A typical example is the uptake of copper by the yeast *Candida utilis* in the presence and absence of an energy source (glucose), a nonmetabolized glucose analog (2-deoxyglucose), and at 4 °C (5). In this case about 2 nmol Cu (mg dry wt)$^{-1}$ was bound within 1 min from an external copper concentration of 10 μM, and this did not increase over a 20-min incubation period in the absence of glucose, the presence of 2-deoxyglucose, or at low temperature, whereas when glucose was present uptake reached about 7 nmol (mg dry wt)$^{-1}$ at the end of the incubation period (5). This pattern of uptake has been seen in a number of studies using various strains of *S. cerevisiae* and a range of metal ions, including copper (6), cobalt, (7,8), cadmium (8), manganese (9) nickel (7), and zinc (7,10), and in studies of uptake of zinc (11) and manganese in *C. utilis* (12,13).

It is generally not particularly useful to compare published data on amounts of various metal ions bound by yeast or filamentous fungi (see below) as the actual figures will depend on the particular experimental conditions used, especially the composition and pH of the uptake medium (usually a buffer solution). Many metal ions form complexes with various medium or buffer components or may be precipitated by, for example, phosphates, especially at pH near neutrality or higher, and this may result in the concentration of *available* metal ions being significantly less than the added metal concentration. The choice of pH also influences binding, which generally decreases as the pH falls (1), probably owing to competion for binding sites by hydrogen ions. It should be noted however, that uptake at high pH may be elevated by precipitation of metal hydroxides at the cell surface as has been reported for manganese uptake by *Can. utilis* at pH 7 or above (13). Further, the cell density used (usually expressed as dry weight but sometimes expressed as cells ml^{-1}) varies between reports and this may influence metal ion binding when the external metal ion concentration is the same. However, there is some unequivocal data that suggest that the capacity of yeast cell walls to bind metals varies between species. For example, the amount of zinc bound externally by *Sporobolomyces roseus* from a solution containing 100 μM Zn^{2+} was 28.9 nmol (mg dry wt)$^{-1}$ while under similar conditions that bound by *S. cerevisiae* was 3.8 nmol (mg dry wt)$^{-1}$ (10). In this case differences in cell wall composition may have been responsible, since the wall of *S. roseus* consists largely of chitin and mannan whereas that of *S. cerevisiae* is mainly glucan and mannan with traces of chitin (10). Other workers have compared the ability of various yeast species to bind metal ions under standard conditions (14,15), but it is not possible to readily identify differences in external binding as opposed to combined external and intracellular uptake from the data presented.

Alternative approaches to investigation of metal binding by cell walls of yeast have included the use of isolated cell walls and electron microscopy. Electron microscope autoradiography showed that most of the mercury taken up by cells of *S. cerevisiae* grown in the presence of ^{203}Hg was located in the cell wall, and studies of uptake by isolated cell walls showed that they were capable of taking up their own weight of mercury (16). Electron microscope studies have also revealed that silver is deposited in the cell walls of *Cryptococcus albidus* grown in medium containing silver nitrate, and X-ray microanalysis indicated that the deposits were elemental silver (17). Similarly, elemental silver was found in the walls of *S. cerevisiae* and several other yeast species after exposure to silver ions (18). Uptake of thorium by *S. cerevisiae* was shown to be biphasic as reported for many other metal ions, but electron microscopy revealed substantial deposits of thorium in the walls of the cells only when stationary-phase cells were exposed to the metal (19).

It is clear from the above that yeast cell walls are able to bind considerable amounts of metal ions, though with the exception of uranium and possibly silver and thorium, the amount bound is much less than overall uptake due to transport across the plasma membrane.

B. Filamentous Fungi

About 30–40% of cobalt taken up by the mycelium of *Neurospora crassa* was bound to the hyphal surface by a process independent of metabolic energy (20), while studies of zinc uptake in *Neocosmospora vasinfecta* (21) showed that about 84% of the zinc taken up after 30 min was desorbed into phthalate buffer, indicating binding to the exterior of the hyphae. Further investigation showed that this binding was unaffected by anaerobiosis, sodium azide, or low temperature and was complete within 2 min. Paton and Budd (21) also observed that incubation of mycelium for 3 hr in anaerobic conditions or in the presence of sodium azide prior to addition of zinc resulted in a reduction in binding of 25–30%. They suggested that this indicated that some binding sites were subject to metabolic turnover, and that these might be located in the cytoplasmic membrane. If this is the case, some 25–30% of overall surface binding of zinc represents binding to the plasma membrane in this organism.

Townsley and Ross (22,23) found that large amounts of copper were bound by nongrowing suspensions of mycelium of *Penicillium spinulosum* and *Aspergillus niger* from an external copper concentration of 40 μM that was not toxic to the cells. This binding was rapid, independent of metabolic energy, and largely masked any intracellular uptake. When copper was added to growth media, the uptake of copper by growing mycelium appeared maximal in lag phase and fell rapidly on a cell dry weight basis during subsequent growth. Mycelial copper content of rapidly growing mycelium showing a linear increase in dry weight and in stationary phase cells was much lower than that seen in the nongrowing mycelium. Similar results were obtained for *P. spinulosum* and *A. niger* with zinc and cadmium (24). Although there is evidence for complexing of copper with extracellular material produced during growth (23), it is possible that high levels of surface binding seen in nongrowing mycelium are not akin to the situation in growing cultures.

Uptake of cadmium, lead, and zinc by the aquatic fungi *Pythium* sp., *Scytalidium lignicola*, and *Dictyuchus sterile* during growth in medium containing these metals has been reported to be largely due to adsorption to external surfaces, since good fit was obtained to either the Langmuir or the Freundlich adsorption isotherm (25). Uptake of copper by wild-type and a wall-less mutant ("slime") strain of *N. crassa* has been compared during active growth, and much less copper (assessed as Cu removed from the medium) was taken up by the wall-less mutant, suggesting that in this case a significant component of copper uptake by the wild-type was due to binding to the wall (26). It was observed that uptake was maximal during exponential growth, which is in contrast to copper uptake in *P. spinulosum* (22) and *A. niger* (23).

III. ROLE OF WALLS IN RESPONSE TO METAL STRESS

It will be clear from the above discussion that walls of both yeasts and filamentous fungi are capable of binding significant amounts of metal ions. Some examples are shown in Table 1. It should be noted that the experimental conditions used vary between reports and direct comparisons of data from different authors are not possible. The question arises as to whether this binding plays a part in the response of fungi to exposure to metal ions. There is now little reason to believe that binding of metals to the cell wall is an essential prerequisite to intracellular uptake as this has been demonstrated in protoplasts of a number of fungi (see below) but there is some evidence that cell

Table 1 Binding of Metal Ions by Fungal Walls

Fungus	Metal concentration (mM)		Metal bound (nmol mg^{-1} dry wt)[a]	Reference
Saccharomyces cerevisiae	Mn	3.0	5.1	41
S. carlsbergensis	Mn	3.0	4.8	41
Candida utilis	Mn	0.05	0.5	12
S. cerevisiae	Cu	0.01	12.0	6
C. utilis	Cu	0.01	2.0	12
Penicillium spinulosum	Cu	0.04	15.7	22
Aspergillus niger	Cu	0.04	15.0	23
S. cerevisiae	Zn	0.1	3.8	10
Sporobolomyces roseus	Zn	0.1	28.9	10
P. spinulosum	Zn	0.04	7.6	24
Neurospora crassa	Co	3.4	5.6–8.8	20
Aureobasidium pullulans				
Yeast-like cells	Cd	0.1	10.0	33
Mycelium	Cd	0.1	8.5	33
Chlamydospores	Cd	0.1	24.6	33

[a]Except Ref. 41 where data are given as wet wt.

walls may have a protective role. In the copper-tolerant fungus *Penicillium ochro-chloron* growing in copper-containing medium, the copper was largely located at the cell surface (27), and protoplasts of this fungus were more sensitive to copper toxicity as assessed by reduction in respiratory activity (28). Comparison of copper uptake in whole cells and protoplasts of a copper-sensitive and a copper-resistant strain of *S. cerevisiae* showed that intact cells of the resistant strain took up less copper than cells of the sensitive strain, and that protoplasts of this strain took up less copper than intact cells but showed similar uptake kinetics and copper resistance to intact cells (29). This suggests that the cell walls play no significant part in the enhanced resistance of this strain. However, while intact cells of the sensitive strain took up copper more rapidly and to a greater extent than intact cells of the resistant strain, protoplasts of the sensitive strain showed extensive lysis under the uptake conditions (29). It therefore seems probable, that, while the enhanced tolerance of copper of the resistant strain of *S. cerevisiae* was not due to changes in the cell walls leading to reduced uptake of copper across the plasma membrane into the cells, the walls do play some part in protecting the plasma membranes of the cells of the sensitive strain from lysis under copper stress, though without presenting a significant barrier to the intracellular transport of copper.

It should be remembered that a number of metal ions, including copper, zinc, manganese, and probably cobalt, are essential trace elements for fungi (3), and therefore the wall cannot act as a barrier to entry of such metal ions to the cells under normal conditions (i.e., when these metals are present at trace concentrations). This would not apply in the case of metal ions of no known biological function. Indeed, there is evidence that the cell wall is involved in mercury resistance of *S. cerevisiae*. Two mercury-resistant mutants of this yeast, designated *HGR1-1* and *HGR2-1*, which grew faster in the presence of sublethal mercury concentrations than the parent cells were found to accumulate similar amounts of mercury to the parent cells (30). However, when the cell walls were removed by enzymatic digestion, the spheroplasts of parental cells contained 15% of the total cell-associated mercury, while those of the resistant mutants contained 5% of the total cell-associated mercury. Therefore, most of the mercury in both parental and resistant cells was bound to the cell wall, and in resistant cells more of the total cell mercury was bound by the cell walls than in the sensitive cells. It was therefore suggested that mercury resistance conferred by the mutations *HGR1-1* and *HGR2-1* was due to increased binding in the cell walls of these mutants, with reduced penetration to the plasma membrane.

In the polymorphic fungus *Aureobasidium pullulans*, an organism commonly found in heavy metal–polluted habitats (31,32), it has been observed that thick-walled, melanized chlamydospores bind much greater amounts of heavy metals than yeast-like cells and mycelium (32,33), and further investigation using intact cells and protoplasts has shown that the walls of chlamydospores play a major role in tolerance by preventing entry of metal ions into the cells (34). Also in yeast-like cells, uptake of zinc and cobalt at nontoxic concentrations was greater in protoplasts than in whole cells, and with metals of higher toxicity such as copper and cadmium, protoplasts of yeast-like cells were more liable to toxic effects as observed by viability loss and release of potassium ions (34). At copper concentrations at which toxicity is apparent, protoplasts have been shown to suffer lysis, as previously seen in *S. cerevisiae* (29), while in the presence of cadmium, increased uptake accompanied by potassium loss was seen at toxic concentrations (34). These results suggest that as with *S. cerevisiae*, the cell walls of yeast-like cells of *Aur. pullulans* play some role in protection of the cytoplasmic membrane from damage by heavy metals. The greater protective effect of the chlamydospore walls discussed above may be due to the presence of melanin (34), which has been shown to effectively bind copper (35). It is also relevant that chlamydospores are resting structures so that in this case a cell wall that effectively prevents entry of heavy metals would not impair growth by deprivation of essential trace metals, as could potentially be the case in actively growing yeast-like or mycelial forms.

IV. UPTAKE OF METALS ACROSS THE PLASMA MEMBRANE

A. Yeasts

Uptake of divalent metal ions across the plasma membrane in yeasts occurs more slowly than external binding and shows a marked dependence on metabolic energy, being inhibited by uncouplers and other metabolic inhibitors, absence of an energy source, and low temperature (5,10,13,33,36). In *S. cerevisiae* a number of divalent metal ions were taken up by an energy-dependent process with an affinity series of magnesium, cobalt, and zinc > manganese > nickel > calcium > strontium (7). This series is similar to that for the activation of the plasma membrane ATPase and led to the suggestion that this enzyme was directly involved in transport of these ions (37). However, manganese ions, which strongly stimulate ATPase activity, were taken up with an initial rate similar to initial uptake rates for calcium and strontium ions (38,39) and these ions do not activate ATPase (40). This rules out a direct role of the plasma membrane ATPase in divalent cation uptake. Exchange of cobalt (7,8), zinc (7), nickel (7), manganese (41), and copper (6) ions for cellular potassium in the ratio 1:2 during uptake has been observed, and an association between potassium efflux and uptake of such metals has been suggested. However, release of potassium was not always correlated with metal uptake in this way (10), and there is evidence that potassium may be released from cells as a result of damage or death caused by the toxic action of certain heavy metal ions, especially cadmium ions (8,42). Kinetic studies have yielded widely varying data for apparent Michaelis constants (K_m) (Table 2), and it is possible that this is due to the wide variety of uptake conditions and yeast strains used by different workers, so although it is clear that the uptake of a number of divalent metal ions by *S. cerevisiae* can be described by the Michaelis-Menten equation, comparison of data from different studies even for the same metal ion is difficult. Most of the data for *S. cerevisiae* probably relate to a relatively nonspecific uptake system as competitive inhibition of uptake of a particular divalent metal ion by others is frequently seen (6,14,41).

In the yeast *Can. utilis*, studies of zinc uptake demonstrated a relatively specific uptake system inhibited only by cadmium (11). Similarly specific systems for uptake of manganese (13) and copper (5) by *Can. utilis* have been observed. In the case of zinc and manganese, relatively low concentrations of the metals (1 μM for zinc, 10 nM for manganese) were used. When higher manganese concentrations of the order of 50 μM were used, there was a nonspecific uptake of manganese, which was strongly inhibited by magnesium (12). It therefore appears that at least in the case of manganese, a specific transport system is active at low external manganese concentrations, whereas at higher concentrations a nonspecific system operates. Kinetic studies gave a K_m of 16.4 nM for the specific system (13) and 65 μM for the nonspecific system

Table 2 Half-Saturation Constants (K_m) for Uptake of a Selection of Divalent Metal Ions by *Saccharomyces cerevisiae*

Metal ion	K_m (μM)	Reference
Zinc	3.7	42
	<10	7
	1150	63
	5000	10
Cobalt	<10	7
	77	8
	100	14
	100	43
	800	43
Manganese	<10	7
	100	14
	860	79

(12), indicating that the specific system has a high affinity for manganese. The specific zinc uptake system in *Can. utilis* described by Failla et al. (11) also has a high affinity with a K_m of 1.3 μM, as does that of *S. cerevisiae* with a K_m of 3.7 μM (42), but the latter is less specific, being inhibited by magnesium and manganese. In the case of copper, uptake by *Can. utilis* was of high affinity, with a K_m of 3.1 μM, and was found not to be inhibited by magnesium, manganese, zinc or cobalt (5). Again, the copper transport system in *S. cerevisiae* showed a similar high affinity, with a K_m of 1.13 μM, but was less specific, being inhibited by magnesium, manganese, zinc, and calcium ions (6).

It appears from the available data that although both *Can. utilis* and *S. cerevisiae* possess high-affinity transport systems for at least some divalent cations, those of *S. cerevisiae* are less discriminatory against other divalent metal ions. In the case of *Can. utilis*, there is evidence of nonspecific transport of manganese when external concentrations are high, and it is perhaps significant that many of the studies of divalent metal ion transport in *S. cerevisiae* discussed above used metal ion concentrations approaching 0.1–0.2 mM and in some cases as high as 2–3 mM. It is therefore possible that the relatively nonspecific transport seen in these studies is due to a system similar to the low-affinity transport system for manganese in *Can. utilis*. However, where lower concentrations of metal ions have been used to study relatively high-affinity transport systems in *S. cerevisiae*, these systems do not have the high specificity of similar systems in *Can. utilis*. It should also be noted that in contrast to the situation in *S. cerevisiae*, no efflux of potassium ions was observed during uptake of manganese (13) or copper (5) in *Can. utilis*, though high external concentrations of potassium ions depressed copper uptake, possibly owing to depolarization of the cytoplasmic membrane. There are, however, too few reports of divalent metal ion uptake by *Can. utilis* which have included determinations of potassium efflux to confirm the absence of potassium exchange during divalent metal ion uptake.

Most of the work discussed above has concentrated on those divalent metal ions that are generally regarded as essential for the growth of yeasts (with the exception of cadmium) and for which transport systems may reasonably be expected. In the context of this discussion it is pertinent to consider whether other toxic metal ions, such as silver, lead, and mercury, can enter yeast cells. There is, in fact, little information on uptake of these metals, though lead has been reported to enter the cells of *S. cerevisiae* by diffusion (43). Electron micrographs have revealed the presence of silver within cells of the yeast *Cryptococcus albidus* (17) but apparently not in cells of *S. cerevisiae* (18). There does not appear to be any other information. In the case of mercury, there is a similar lack of data, though autoradiography indicated penetration into the cytoplasm (16). Mercury and silver are both highly toxic to yeasts, and it is likely that transport across the plasma membrane does occur, though unambiguous evidence is lacking.

B. Filamentous Fungi

In general, divalent metal ion transport has been less well studied in filamentous fungi than in yeasts, but this situation is being rectified. Uptake of a number of divalent metal ions, including cobalt, nickel, zinc, manganese, magnesium, and copper, has been shown to be dependent on metabolic activity by the use of inhibitors (20,21,28,44–47). Unlike yeasts, a direct dependence of metal ion uptake on the supply of glucose in the uptake medium was not demonstrated in a number of studies with unstarved mycelium (21, 48–50) but was apparent with mycelium starved for 18 hr (49,50). This is not always the case, however, as uptake of manganese ions by *A. niger* was reduced in the absence of an energy source (sucrose) in the medium (46) and copper uptake by protoplasts of *P. ochro-chloron* was dependent on the presence of glucose in the uptake medium (28).

Cobalt and nickel uptake by cobalt- or nickel-resistant strains of *N. crassa* was found to be inhibited strongly by magnesium (44,45) though the effects of other metal ions were more variable, and strain-dependent. Copper ions did not inhibit cobalt uptake, while strongly inhibiting that of nickel, while zinc ions inhibited uptake of both cobalt and nickel. Manganese ions inhibited both cobalt and nickel uptake, but not in all strains tested (44,45). Uptake of zinc by *N. vasinfecta* was competitively inhibited by manganese ions, while magnesium ions inhibited uptake by increasing the apparent K_m for zinc, while decreasing the rate of uptake (21). Copper, nickel, and cobalt ions were less effective in inhibiting zinc uptake. These data are rather confusing overall, but do suggest that uptake of cobalt, nickel, and zinc in these experiments was by a relatively nonspecific transport system.

More recently, a number of reports have suggested that at least for zinc and manganese, both low-affinity, relatively nonspecific and high-affinity, relatively specific transport systems can exist. Budd (48) described two transporters of zinc ions, one of which had an apparent K_m of 1.1 μm and a second with an apparent K_m of 770 μM. The high-affinity system was relatively insensitive to inhibition by calcium, magnesium, manganese, cadmium, cobalt, or nickel ions at equimolar concentration with zinc ions, indicating considerable selectivity for zinc. No information was given on the specificity of the second, low-affinity transport system. Recent work on zinc transport in *Penicillium notatum* at low concentrations (5–100 nM) (50) has demonstrated the existence of a high-affinity, highly specific transporter with an apparent K_m of 6.66 nM, which was inhibited competitively by cadmium and noncompetitively by copper ions. None of the other metal ions tested, including magnesium, had any effect. Further evidence for the existence of high-affinity, specific zinc transport in filamentous fungi has been provided by studies on zinc uptake by *Sclerotium rolfsii* (51), which demonstrated the presence of a zinc transport system with an apparent K_m of 3.74 μM and which was not subject to inhibition by calcium, manganese, cobalt, nickel, cadmium, iron, or copper ions at 1×10^6 fold excess concentration relative to zinc or by magnesium ions at 1×10^7 fold excess. This system seems to be the most selective zinc transport system so far described in fungi.

Starling and Ross (50) also detected a second transport system active at higher zinc concentrations in the range of 10–100 μM, which had an apparent K_m of 5.9 μM and was strongly inhibited by magnesium, manganese, nickel, copper, cobalt, and cadmium ions. The activity of this relatively nonspecific uptake system, unlike the high-affinity system, was stimulated by growth of the mycelium in a medium with a low magnesium content. Some broadly similar data on manganese uptake have been obtained for *A. niger* (46) and *P. notatum* (49). In *A. niger* a relatively specific high-affinity uptake system, with an apparent K_m of 3.04 μM in growing cells, was found to be insensitive to inhibition by a 5000-fold concentration of magnesium or calcium ions, though strong inhibition by zinc, copper, and cadmium was observed at this concentration (46). In *P. notatum*, uptake by resting mycelium from concentrations of manganese between 5 and 30 nM occurred by a system with an apparent K_m of 4.37 nM, and which was insensitive to competition by magnesium, zinc, nickel, cobalt, or copper ions at concentrations 1000-fold greater than the manganese concentration (49). This system was, however, strongly inhibited by cadmium and this inhibition was found to be competitive (52). A second uptake system with lower affinity for manganese and active at concentrations of 10–100 μM Mn^{2+} was also detected in *P. notatum* (52). This system had an apparent K_m of 22.2 μM and was inhibited, apparently competitively, by a range of metal ions in the order magnesium > nickel > cadmium > zinc > copper with a fivefold excess of competing cation. These data are similar to those obtained for manganese uptake in the yeast *C. utilis* (12,13) discussed above.

Despite the fact that copper is essential to the growth of fungi, but toxic in excess (hence its use in fungicides), uptake into the cells of filamentous fungi has received relatively little attention. In *P. spinulosum* it has been suggested that copper uptake may be by diffusion, as energy-dependent uptake could not be demonstrated (22). However, much copper was bound to external parts of the cells in these experiments, a situation also found in *A. niger* (23) and *Trichoderma viride* (24), and an inability to demonstrate energy-dependent uptake may have been due to masking of intracellular uptake by the external binding. Some experiments with protoplasts from the copper-tolerant fungus *P. ochro-chloron* (28) demonstrated energy-dependent copper uptake with an apparent K_m of 390 μM at pH 3. At pH of 6 or above, saturation kinetics were not observed. There do not seem to be any other data on copper uptake processes in filamentous fungi.

There is some evidence that uptake of certain metal ions in filamentous fungi is regulated by the metal ion content of the mycelium. The rate of zinc transport by the high-affinity system of *Neo. vasinfecta* has been shown to be stimulated by pregrowth of mycelium in medium with a low zinc content (48), and a similar increase in zinc uptake associated with low medium zinc content was seen in *Aspergillus parasiticus* (53). In *P. notatum*, zinc uptake by the high-affinity transport system was low in mycelium grown in medium with a normal zinc content (26.8 μM), much higher in medium to which no zinc had been added (but which was not necessarily zinc deficient), and suppressed almost totally in medium containing 1 mM zinc (50). Budd (48) also reported that growth of mycelium of *Neo. vasinfecta* in medium with a low magnesium content stimulated zinc uptake. In *P. notatum*, low magnesium had no effect on the uptake of zinc via the high-affinity system, though such growth conditions did stimulate uptake of zinc by the low-affinity, relatively nonspecific system (50). Uptake of manganese by *P. notatum* could only be demonstrated in mycelium grown in low-manganese medium (49), which also suggests that manganese uptake is regulated by mycelial manganese content. Experiments with *A. niger* (46) have also been carried out on manganese uptake by mycelium grown in medium with a low manganese content, but it is not clear from this work whether manganese uptake could be seen when the mycelium was grown in medium with a normal manganese content.

The uptake of metal ions by the polymorphic fungus *Aureobasidium pullulans* has been extensively studied. Cadmium, copper, and zinc ions were all taken up by energy-dependent systems that followed saturation kinetics in both yeast-like and mycelial cells, though no such uptake was detected in chlamydospores (32–34). Cadmium uptake was similar in both yeast-like and mycelial cell types with an apparent K_m of 0.11 mM and 0.10 mM, respectively (33). As only calcium ions caused a significant inhibition of cadmium uptake, it was suggested that cadmium is taken up by a system that, usually transports calcium.

Copper uptake was also similar in both yeast-like and mycelial cells with an apparent K_m of 0.22 mM and 0.20 mM, respectively (32). Copper uptake was associated with release of intracellular potassium in the ratio $2K^+:1Cu^{2+}$. Uptake of copper was inhibited by zinc, cadmium, manganese, and cobalt but not by magnesium when these ions were supplied at equimolar concentration with copper (0.04 mM). Calcium ions reduced surface binding of copper but only inhibited intracellular uptake at a concentration of 1 mM, while copper reduced calcium binding but not intracellular uptake. The uptake of copper was reduced by metabolic inhibitors and high potassium concentrations that caused depolarization of the cell membrane, and it was suggested that uptake was driven by the membrane potential (32).

Studies with intact cells and protoplasts of yeast-like and mycelial cells of *Aur. pullulans* showed that for zinc, uptake in both cell types was similar, with an apparent K_m of around 11 μM for both intact cells and protoplasts (34). With other metal ions including cadmium, copper, and cobalt, protoplasts showed reduced surface binding when compared to intact cells (as might be expected since binding to cell walls is eliminated), but uptake across the plasma membrane was increased, especially at higher concentrations of 100 μM cadmium and 500 μM cobalt or zinc, though at 100 μM copper uptake was reduced. However, there was extensive lysis of protoplasts at this copper concentration. It is probable that some damage to cell membranes was caused by high concentrations of metal ions other than copper in protoplasts, leading to increased permeability and therefore an apparent increase in metal ion uptake. Protoplasts of chlamydospores showed a marked difference in metal ion uptake compared to whole cells (34). No intracellular uptake could be detected in intact chlamydospores though there was extensive binding to the exterior of the cells. Protoplasts of these cells, however, had much lower surface binding, and energy-dependent uptake of zinc, cobalt, cadmium, and copper ions could be detected, though at a lower rate than in yeast-like cells. Protoplasts of chlamydospores were more sensitive to toxic effects of copper and cadmium than whole cells as evidenced by loss of viability and release of intracellular potassium.

C. Calcium Uptake

Calcium uptake in yeasts has received considerable attention, though it is not entirely clear how it relates to uptake of other divalent metal ions as few studies have included competition experiments with other divalent metal ions. Calcium uptake is clearly dependent on metabolic energy (38,54), and there is evidence that at least in *S. cerevisiae*, uptake is driven by the membrane potential (55,56), though surface potential and cell pH may also be involved (38). More recently, Eilam and Chernichovsky (57) suggested that calcium uptake was mediated by gated channels open only when the membrane potential fell below -70 mV, and in the absence of an energy source, the membrane potential may be above this threshold and most of the channels remain closed. Further work by Eilam and Othman (58) suggests that additional mechanism(s) other than the membrane potential are involved though it is unclear what these may be. No simple correlation between calcium and cellular ATP levels has been observed, and ATP levels did not affect initial rates of calcium transport into *S. cerevisiae*.

Calcium uptake in *Schizosaccharomyces pombe* (54) exhibited saturation kinetics with an apparent K_m of 45 μM at pH 4.5. Uptake was inhibited by several cations at a concentration of 400 μM in the order strontium > manganese > cobalt > magnesium. Uptake was faster at pH 8.5 than at pH 4.5 by a factor of 10. In *S. cerevisiae* calcium uptake was found not to follow Michaelis-Menten kinetics (38), and the apparent K_m and also V_{max} appeared to increase with increasing calcium concentration. This was thought to be related to changes in surface potential caused by increasing calcium concentrations at least in the case of the apparent K_m. Other data for *S. cerevisiae* gave an apparent K_m of 1.9 μM when the calcium concentrations used to determine the kinetics were corrected for calcium bound to the exterior of the cells (59). If this was not done, an apparent K_m of 75 μM was obtained. There does not seem to be other information on kinetics and specificity of calcium transport across the plasma membrane in yeast.

There are apparently few studies of transport of calcium into filamentous fungi. Uptake was shown to be biphasic in *Penicillium cyclopium*, with an initial energy-independent binding followed by energy-dependent intracellular uptake with an apparent K_m of 1.88 mM (60). No data on specificity were given. Uptake of calcium by two strains of *P. notatum*, IMI 15378, which sporulated readily in response to 10 mM calcium in the medium, and IMI 832, which was unresponsive to calcium, has been shown to be energy dependent, with an apparent K_m for one strain of 0.28 mM and 0.42 mM for the second (61). As with *P. cyclopium*, there was an initial energy-independent binding of calcium to the exterior of the mycelium. Protoplasts of strain IMI 15378 showed energy-dependent uptake with an apparent K_m of 7.45 mM, while protoplasts of strain IMI 832 showed no detectable uptake. It was suggested that, though there was uptake by protoplasts of the responsive strain, the K_m was much higher than in intact cells, and that this and the failure of protoplasts of the unresponsive strain to take up calcium into the cells may reflect damage to the cytoplasmic membrane during the protoplasting procedure. It is unlikely that the failure of protoplasts of strain IMI 832 to take up calcium was a feature of this strain, since whole cells took up calcium. The only other report on calcium transport into filamentous fungi appears to be that of LeJohn et al. (62) on *Achyla* sp. This fungus appeared to transport calcium into the cells by an energy-independent process involving facilitated diffusion.

V. EFFLUX OF ACCUMULATED METAL IONS

Despite the considerable body of information on intracellular uptake of metal ions by yeasts and filamentous fungi, data on efflux of metal ions from fungal cells is relatively scarce, except for calcium, which will be dealt with later. In the relatively few investigations of zinc transport that have included data on efflux, it has not been detected (11,42,51,63,64). Similarly, efflux of cobalt ions has not been detected in the one report available (20). Copper efflux from copper-loaded yeast-like cells of *Aur. pullulans* was reported to occur in two phases, one rapid with a half-life of 4 min and accounting for 36.4% of loaded copper, and a second much slower (half-life 484 min), which accounted for 63.6% of loaded copper (32). Neither phase was energy-dependent, and it was not clear whether the two phases represented bound copper and intracellular copper or whether there was internal compartmentalization of copper. In the yeast *Can. utilis* no copper efflux could be detected from preloaded cells (5). In *S. cerevisiae* efflux of manganese and strontium has been observed, the rate of strontium efflux being three times that of manganese efflux (39). However, this difference may have resulted from a lower concentration of manganese in the cytoplasm due to increased compartmentalization of manganese in the vacuole (see Section VII). Energy-dependent manganese efflux was detected in *Can. utilis* (13) but was relatively slow, with about 25% of preloaded $^{54}Mn^{2+}$ exchanged with nonradioactive manganese in 30 min. An energy-dependent efflux of cadmium has been reported in yeast-like cells of *Aur. pullulans* (33), and Budd (47) detected a very slow efflux of magnesium ions from *Neo. vasinfecta*, but this did not appear to be related to the magnesium status of the mycelium.

Efflux of calcium ions from both yeasts and filamentous fungi has received rather more attention than other divalent cations. An energy-dependent efflux of $^{45}Ca^{2+}$ has been shown to occur from cells of *Schizosaccharomyces pombe*, which was not affected by addition of unlabeled calcium (54), and similar calcium efflux from *S. cerevisiae* has been reported by Eilam (65). However, only 20% of cell calcium was lost over a 2-hr period, and further loss did not occur over a 24-hr period unless potassium, sodium, calcium, or magnesium ions were added when efflux continued throughout the experiment (65,66). The efflux of calcium was accompanied by uptake of those monovalent or divalent cations used to induce it. The initial efflux of calcium was probably due to transport across the plasma membrane via Ca^{2+}/H^+ antiport, while the second phase of efflux stimulated by divalent cations represented release of calcium sequestered in cellular organelles, probably vacuoles (66). A further study (67) of calcium efflux induced by potassium ions, using ionophores to distinguish between a Ca^{2+}/K^+ antiport and an electrical coupling of calcium efflux to potassium influx, but mediated by different carriers, suggested that, as calcium efflux was not stimulated by valinomycin or nigericin in the absence of glucose, calcium efflux must be mediated by antiport carriers that exchange calcium and potassium ions. In proposing Ca^{2+}/K^+ antiport, Eilam (67) makes no reference to his earlier suggestion that Ca^{2+}/H^+ antiport occurs (66), so it is not clear whether it is considered that both systems may operate under different conditions. Presumably, Ca^{2+}/K^+ antiport could not operate under the conditions in which Ca^{2+}/H^+ antiport was demonstrated (20 mM HEPES buffer plus 100 mM glucose) since potassium was not added to the medium. However, it is not explicitly stated that the HEPES buffer was potassium-free.

In the filamentous fungus *Neurospora* sp., calcium efflux is apparently mediated by a Ca^{2+}/H^+ antiporter energized by a transmembrane electrical potential and pH gradient that is generated by a proton-translocating ATPase (68). There is as yet little evidence for a Ca^{2+}-ATPase involved in calcium efflux in fungi, though a gene encoding for a Ca^{2+}-ATPase apparently located in intracellular membranes has been identified in *Sch. pombe* (69). It is therefore possible that Ca^{2+}-ATPase does play a role in mediating cellular calcium levels, but more evidence is required to confirm this.

It will be clear from the above that there is relatively little detailed information on the efflux of divalent metal ions other than calcium and that the possibility that efflux may play a part in tolerance of toxic metal ions has not been directly addressed. However, if the data presented are representative, it is unlikely that efflux of heavy metal ions from cells is a significant response to metal ion stress.

VI. REDUCED UPTAKE OF METALS AS A RESISTANCE MECHANISM

There are a number of reports of reduced uptake of divalent metal ions by strains of fungi resistant to the toxic effects of such metals, and these fall into two categories. The first consists of measurement of total cellular accumulation after a given period, often in actively growing cultures. Joho et al. (70) found that cadmium-resistant cells of *S. cerevisiae* took up less cadmium during 3 hr of exposure in growth medium than sensitive cells, and this result was confirmed in a second study in which resistant cells were found to take up about 33% less cadmium than sensitive cells (71). Uptake of copper was significantly less in growing cultures of a copper-resistant strain of *S. cerevisiae* (72), while nickel-resistant cells of *S. cerevisiae* accumulated similar amounts of nickel to sensitive cells (73). However, in this case the data presented do suggest that at 4 mM nickel, which causes about 50% growth inhibition of nickel-resistant stains and 63% inhibition of sensitive strains, uptake was consistently less in resistant strains by about 25%. As no error bars or other statistical information was presented it is not possible to confirm the authors' view of the significance of this difference.

A more recent study of a strain of *S. cerevisiae* resistant to both cobalt and nickel (74) showed that this strain had a sharply reduced uptake of magnesium, and cobalt and nickel uptake assessed after 4.5 hr of incubation in growth medium containing the appropriate metal was reduced by 20% and 37%, respectively. Again, no error bars or other statistical information was provided to indicate the significance of the difference between resistant and sensitive strains, but in view of the marked difference in magnesium uptake between the wild-type and metal-resistant strains, the suggestion (74) that nickel and cobalt resistance in this mutant might be associated with reduced magnesium transport is plausible. It should also be noted that in experiments of the type discussed here it is not possible to distinguish between metal ions bound to the wall and intracellular metal uptake, so it is not generally clear whether reduced uptake in metal-resistant strains is due to changes in wall binding, as has been reported in mercury-resistant *S. cerevisiae* (30), or to altered metal ion transport systems.

The second approach to uptake by metal-resistant and metal-sensitive strains involves more detailed investigation of the transport properties of the plasma membranes of the cells. Cobalt uptake in a cobalt-resistant strain of *N. crassa* has been shown to be a passive process independent of metabolic energy, apparently diffusion, whereas the cobalt-sensitive parent strain had an energy-dependent uptake system for cobalt (75). One of four nickel-resistant strains of *N. crassa* showed a reduced nickel uptake by comparison with the parent strain (44), and this was apparently the result of a lower rate of nickel transport with a V_{max} of 24 μg Ni 100 mg·dry $wt^{-1}hr^{-1}$ compared to 91 μg Ni 100 mg·dry $wt^{-1}hr^{-1}$ in the sensitive parent strain. However, two other nickel-resistant strains had a higher rate of nickel uptake than the parent, with a V_{max} of 143 μg Ni 100 mg·dry $wt^{-1}hr^{-1}$. All three nickel-resistant isolates had an apparent K_m for nickel of 2 mM, while that for the parent strain was 0.29 mM.

When cobalt uptake was examined in the nickel-resistant strains of *N. crassa* described above (45), the two strains that did not show reduced nickel uptake were found to have two uptake systems for cobalt, one of which was sensitive to inhibition by magnesium and was functional for the first 40 hr, while the second was insensitive to magnesium and active beyond 40 hr of growth. The third nickel-resistant strain that showed reduced nickel uptake had only one uptake system for cobalt, which was inhibited by magnesium and took up less cobalt than the other two strains. There were also differences in specificity of the cobalt uptake system with respect to zinc and manganese between the first two nickel-resistant strains and the third (45). These results suggest that in some metal-resistant strains of *N. crassa*, resistance may be associated with reduced transport of the metal across the plasma membrane, but not in others, and that the interrelationships between uptake of different metals in metal-resistant strains may be rather complex. There is a need for more kinetic studies of intracellular metal transport by metal-resistant strains of *N. crassa* to clarify these relationships. Such studies might be valuable in probing further divalent metal ion transport in general as well as the role of altered transport in metal ion resistance.

The only other clearly documented example of a relationship between metal uptake and resistance in a filamentous fungus is the response of *P. ochro-chloron* to copper. This fungus is well known to exhibit high resistance to copper at low pH (76). Over the pH range 3.0–5.5 copper uptake by and toxicity to protoplasts of *P. ochro-chloron* was found to be roughly constant, and uptake followed saturation kinetics (28), but uptake and toxicity rose sharply at pH 6 with a deviation from saturation kinetics. At low pH (3.0–5.5) copper uptake with 1 mM Cu^{2+} was low, with a rate of about 7 $nmol \cdot hr^{-1}$ 10^7 $cells^{-1}$, while at pH 6, the copper uptake rate increased to 38.6 $nmol \cdot hr^{-1}$ 10^7 $cells^{-1}$. These data suggest that the high copper tolerance of *P. ochro-chloron* at low pH is due to a low rate of copper influx even with very high external concentrations. Although there is ample evidence that divalent metal ion transport is inhibited by low pH (1), copper at high external concentrations can induce serious membrane damage leading to increased permeability (see Section VIII) so that at low pH the membrane of *P. ochro-chloron* must be protected from this effect, possibly by competion for binding sites between copper and hydrogen ions. The increased uptake at pH 6 that does not follow saturation kinetics may arise from damage to the membrane resulting in increased permeability (28). This needs further examination.

Comparison of copper uptake by a wild-type and a copper-resistant strain of the yeast *S. cerevisiae* (29) showed that copper accumulation and initial uptake rate was significantly reduced in cells of the copper-resistant strain. Protoplasts prepared from the wild-type strain took up very little copper and uptake reached a maximum very rapidly with no further uptake occurring. This was apparently due to extensive lysis of the protoplasts, the measured uptake being due to binding of copper to cell debris. Protoplasts of the copper-resistant mutant were tolerant to copper, and the uptake kinetics were similar to those of whole cells (29). This suggests that the reduced uptake of copper by the resistant mutant was due to altered membrane transport, possibly associated with a change in membrane structure as the protoplasts of the mutant were resistant to lysis by copper.

A series of metal-resistant strains of *S. cerevisiae* obtained by repeated subculture on elevated concentrations of copper, cobalt, or cadmium (a procedure often known as "training") all showed reduced uptake of the metal ion to which resistance was obtained based on initial rates of membrane transport over a 15-min period, and this was retained after repeated subculture on metal-free media ("detraining") (77). Also, Belde et al. (78) observed reduced cadmium uptake in a cadmium-resistant strain of *S. cerevisiae*, and Gadd et al. (34) found that protoplasts of chlamydospores of the polymorphic fungus *Aur. pullulans* were less sensitive to the toxic effects of metal ions such as copper, cadmium, cobalt, and zinc than protoplasts of yeast-like cells and had a lower rate of uptake of the metal ions. This is further evidence that reduced plasma membrane transport is a feature of some metal-resistant fungi. However, this is not universal, since a manganese-resistant strain of *S. cerevisiae* accumulated much more manganese than a sensitive strain (79), and in this case the K_m for manganese was similar in both strains but V_{max} for the resistant strain was twice that of the sensitive strain. In this case it is possible that manganese taken up by the resistant strain was compartmentalized into the cell vacuoles (see Section VII).

VII. ROLE OF THE VACUOLE IN REGULATION OF INTRACELLULAR METAL CONTENT

Data on the intracellular distribution of several heavy metals, magnesium and calcium, within fungal cells points to a role for the vacuole in storage and detoxification of these metals. Differential extraction methods have shown that in the yeast *Saccharomyces carlsbergensis*, magnesium and manganese ions accumulate in the vacuoles to concentrations significantly higher than in the cytosol (9,80) possibly with binding to polyphosphate. Further studies have shown that during uptake of magnesium or manganese the vacuolar magnesium or manganese concentration can rise markedly but the cytosolic concentration is not greatly increased, suggesting that the vacuole plays a part in regulation of metal ion concentration in the cytoplasm (81). The ability of the vacuole to accumulate metal ions depends on transport across the vacuolar membrane, and investigations of pH and membrane potential using fluorescence quenching techniques have indicated the presence of $Ca^{2+}/2H^+$, $Mn^{2+}/2H^+$, $Zn^{2+}/2H^+$, $Mg^{2+}/2H^+$, and K^+/H^+ antiport systems in the vacuolar membrane of *S. carlsbergensis* (82). In *S. cerevisiae* about 56% of accumulated zinc can be present in the vacuolar fraction, and inhibitor studies suggested Zn^{2+}/H^+ antiport across the vacuolar membrane (42). Other studies using *S. cerevisiae* have indicated that most of the cobalt accumulated by this yeast is located in the vacuoles (77). When the yeast was trained for cobalt resistance by repeated subculture on medium containing elevated cobalt concentrations, the accumulation of cobalt in the vacuoles was increased, suggesting a role in metal resistance. Accumulation of cadmium and copper in vacuoles of *S. cerevisiae* was much less than that of cobalt and was not altered in cells trained to metal resistance (77). These results indicate that the vacuole may not necessarily be a major site of metal ion accumulation, and this is supported by the earlier observation that the difference between the rates of manganese and strontium uptake in *S. cerevisiae* resulted from greater compartmentalization of manganese in the vacuoles (39).

The above suggests that the vacuole may play a role in the homeostasis of some heavy metal ions in yeast cells; data on filamentous fungi are lacking. However, the role of the vacuole in maintaining cytosolic concentrations stable is rather better documented for calcium. Vacuoles have been shown to be a major storage site of calcium in yeast (83,84), and there is evidence for a Ca^{2+}/nH^{+} antiport system (85) driven by a proton motive force generated by the H^{+}-ATPase in the vacuolar membrane (86,87). This system apparently maintains calcium homeostasis in yeast cells under a wide range of calcium concentrations (87). Recently, calcium-sensitive mutants of *S. cerevisiae* have been obtained that show defects in vacuolar membrane H^{+}-ATPase activity, and vacuoles isolated from these mutants are defective in calcium uptake. Free calcium ion concentrations in the cytosol of these mutants was some six times greater than in wild-type cells, suggesting a loss of calcium homeostasis due to defective vacuolar membrane H^{+}-ATPase (88).

There is even less information on calcium homeostasis in filamentous fungi but some data are available for *N. crassa*. Calcium-sensitive mutants of *N. crassa* were found to have a significantly lower uptake of calcium into vacuoles than wild-type cells, but transport of calcium into the cells was not affected, suggesting a role for vacuoles in regulating cytosolic calcium concentration, and there was evidence of involvement of a vacuolar H^{+}-ATPase (89). More recently, it has been suggested (90) that the vacuole of *N. crassa* may be an effective buffer against short-term changes in cytosolic calcium concentrations.

The above discussion suggests that uptake into the vacuoles of fungal cells has a clear role in the regulation of calcium concentrations in the cytosol, and it is reasonable to suppose that the vacuole may play a similar role for other divalent metals, such as manganese, zinc, and cobalt, which have been shown to accumulate in vacuoles. There are, however, not sufficient data for these metals to confirm this unequivocally, and more information on vacuolar uptake, especially comparative data for wild-type and metal-resistant strains under conditions of metal ion stress, is needed.

VIII. EFFECTS OF HEAVY METALS ON MEMBRANE INTEGRITY

Several reports suggest that some heavy metals damage the cytoplasmic membranes of fungal cells. Passow and Rothstein (91) found that mercury caused a marked loss of potassium from the cells of *S. cerevisiae* and this was attributed to an "all or nothing" response of individual cells to mercury with disruption of the cytoplasmic membrane. A more detailed examination of this phenomenon (92) showed that, at low concentrations of mercuric chloride, there was a slow release of cellular potassium, which increased with increasing mercury concentration until virtually all intracellular potassium was lost with a mercury concentration of 0.4 mM. Electron probe X-ray microanalysis demonstrated that at mercury concentrations of 0.1–2 mM some cells showed partial loss of potassium while others showed a total loss. With mercury concentrations of 0.22 mM or above, virtually all intracellular potassium was lost. Copper ions also caused potassium loss, with a copper concentration of 0.06 mM causing 90% loss of intracellular potassium (92). Damage to yeast cell membranes by copper was also demonstrated by Ohsumi et al. (83), who found that copper at concentrations as low as 30 μM caused microscopically visible damage to cells suspended in buffer, and this effect was more severe if cells were suspended in distilled water. Exposure to 100 μM cupric chloride led to loss of nucleotides and an immediate loss of 70% of cell potassium. Leakage of amino acids from cells treated with copper also occurred, and it was noted that arginine, normally found in vacuoles, was not released, while glutamate, normally present in the cytoplasm, was rapidly released (83). This was interpreted as indicating that damage was limited to the cytoplasmic membrane while the vacuolar membrane remained intact. Other ions, including mercury, had no effect. The lack of any marked effect of mercury may be due to the concentration used (0.1 mM), which caused only slow potassium release in *S. cerevisiae* (92). Ohsumi et al. (83) also suggested that copper only affects energized membranes as low temperature and metabolic inhibitors prevented membrane damage. This is in agreement with observations of Ross (93) that the viability of cell suspensions of *S. cerevisiae* exposed to copper fell much more rapidly when an energy source such as glucose was supplied.

A number of workers have reported efflux of potassium during uptake of various heavy metal ions from yeast cells. In some cases, as for example in the uptake of cobalt (8), manganese (41), or copper (6), potassium was released in the ratio of 2 moles of K^+ to 1 mole Co^{2+}, Mn^{2+}, or Cu^{2+}, leading to the suggestion that K^+ was exchanged with the divalent metal to maintain charge balance. However, with zinc (42) and cadmium (8,94,95) the efflux of potassium was not directly related to divalent cation uptake in this way and was probably due to other toxic effects of these metals (8,94). Potassium efflux induced by cadmium was increased in the presence of an energy source in *S. cerevisiae* (8,94) at least at lower concentrations. However at 0.5 mM cadmium, 90% of cellular potassium was lost in 5 min irrespective of the presence or absence of an energy source and this was accompanied by a drop of 90% in cell viability (94). Cadmium concentrations that provoked significant potassium efflux also resulted in increased $^{45}Ca^{2+}$, $^{109}Cd^{2+}$, and [^{14}C]-tetraphenylphosphonium uptake, which suggests that exposure to cadmium had increased the permeability of the cytoplasmic membrane of *S. cerevisiae* to cations (96). A cadmium-resistant strain of *S. cerevisiae* lost less potassium than the wild-type parent strain when incubated in buffer containing cadmium at 0.5 mM in the presence of glucose (95) and also lost less potassium on exposure under the same conditions to mercury and copper than the parent strain. It appears, therefore, that this cadmium-resistant strain of *S. cerevisiae* had a modified plasma membrane structure, which was less sensitive to damage by heavy metals.

There is little information on efflux of potassium induced by divalent metal ion uptake in fungi other than *S. cerevisiae* though an exchange of $2K^+$ for 1 Cd^{2+} and 1 Cu^{2+} during uptake by *Aur. pullulans* has been reported (32, 33). However, in *Can. utilis* no exchange of potassium ions for copper ions was detected (5). There does not appear to be any data of this type for filamentous fungi.

Overall, the role of potassium release during metal ion uptake is somewhat unclear at the present time, some studies suggesting that it is a symptom of cellular, probably cytoplasmic membrane, damage by heavy metals while in other cases there appears to be a regulated exchange for divalent metal ions taken up. It is clear that certain heavy metals, notably copper, cadmium, and mercury, do act to impair membrane function directly, and that this damage generally occurs as a result of their energy-dependent transport into the cells, though at elevated concentrations direct energy-independent damage can result. Whether the exchange of potassium ions for heavy metal ions during the uptake of the latter is a universal property of the transport systems involved is difficult to establish from currently available data, but may become clear if the molecular mechanism of the transport of metals such as copper, manganese, and zinc is established. Transport of metals such as cadmium and mercury with no known biological function presumably occurs via systems primarily concerned with the transport of essential metals, and it might be difficult to distinguish reliably between potassium exchange as part of the transport process and potassium released due to membrane damage by these metals.

IX. CONCLUSIONS

The evidence now points to two mechanisms by which heavy metals enter fungal cells, one of which is relatively specific for individual metals and is active primarily at low external metal concentrations and a second which is relatively nonspecific and active at higher metal concentrations. The former mechanism is presumably responsible for uptake of essential metals when these are present at trace concentrations, while the latter primarily transports magnesium but will transport other metals with lower affinity. In conditions of metal ion stress, the nonspecific system is probably the main route by which heavy metals enter fungi to reach toxic concentrations. The concentrations of metals such as zinc, cobalt, manganese, and nickel likely to saturate the specific high-affinity systems so far described are not toxic to the relevant fungi. However, some metals including cadmium, silver, and mercury are toxic at relatively low concentrations and may enter cells by systems primarily responsible for uptake of essential trace elements, though direct evidence is lacking. It should also be noted that the role of the calcium uptake system in facilitating entry of toxic metals is still uncertain (see Section IV).

There is relatively little information on the regulation of uptake of heavy metal ions in fungi, though uptake of manganese and zinc is apparently dependent on cellular manganese or zinc content in a number of instances (46, 49,50,64,96). However, this was not true for copper in *Can. utilis* (5). Some heavy metal–resistant strains of fungi exhibit reduced uptake of the relevant metal, but there is little information on the underlying mechanism involved. It is clear that in the case of calcium, efflux from the cells is important in maintaining cytoplasmic calcium concentrations, and that the vacuole also plays a role in calcium homeostasis. To what extent calcium provides a possible model for the homeostasis of other metal ions is not yet apparent. Efflux of heavy metals has not been extensively studied, and more work is needed, especially with filamentous fungi, but data currently available suggest that efflux is not a major resistance mechanism in fungi. A number of metals are taken up and apparently immobilized in the vacuoles of fungal cells, and this may be a mechanism for regulating cytoplasmic metal concentration, though again more work is needed to demonstrate the significance of this.

Exclusion of metal ions from fungal cells by binding to cell walls plays little part in tolerance of metals in those fungi which have been investigated, with the notable exception of chlamydospores of *Aur. pullulans.* On the other hand, the wall may provide some protection for the cytoplasmic membrane, which can be directly damaged by heavy metals. Presumably, fungi that develop high resistance to heavy metals, especially copper and cadmium, must have cytoplasmic membranes that are less readily damaged by these metals or are protected from their action in some way. Further insight into these matters may be achieved by comparative studies of transport of heavy metals across the plasma membrane and the vacuolar membrane of wild-type and metal-resistant strains of fungi.

REFERENCES

1. Gadd, G. M. (1986). Fungal responses towards heavy metals, in *Microbes in Extreme Environments* (R. A. Herbert and G. A. Codd, eds.), Academic Press, London, p. 85.

2. Gadd, G. M. (1989). Metal tolerance, in *Microbiology of Extreme Environments* (C. Edwards, ed.), Open University Press, Milton Keynes, p. 178.
3. Bowen, H. G. M. (1966). *Trace Elements in Biochemistry*, Academic Press, London.
4. Gadd, G. M. (1990). Fungi and yeasts for metal accumulation, in *Microbial Mineral Recovery* (H. L. Ehrlich and C. L. Brierley, eds.), McGraw-Hill, New York, p. 249.
5. Ross, I. S., and Parkin, M. J. (1989). Uptake of copper by *Candida utilis*, *Mycol. Res. 93*: 33.
6. de Rome, L., and Gadd, G. M. (1987). Measurement of copper uptake in *Saccharomyces cerevisiae* using a Cu^{2+}-selective electrode, *FEMS Microbiol. Lett. 43*: 283.
7. Fuhrmann, G., and Rothstein, A. (1968). The transport of Zn^{2+}, Co^{2+} and Ni^{2+} into yeast cells, *Biochim. Biophys. Acta 163*: 325.
8. Norris, P. R., and Kelly, D. P. (1977). Accumulation of cadmium and cobalt by *Saccharomyces cerevisiae*, *J. Gen. Microbiol. 99*: 317.
9. Okorokov, L. A., Lichko, L. P., Kadomtseva, V. M., Kholodenko, V. P., Titovsky, V. T., and Kulaev, I. S. (1977). Energy-dependent transport of manganese into yeast cells and distribution of accumulated ions, *Eur. J. Biochem. 75*: 373.
10. Mowll, J. L., and Gadd, G. M. (1983). Zinc uptake and toxicity in the yeasts *Sporobolomyces roseus* and *Saccharomyces cerevisiae*, *J. Gen. Microbiol. 129*: 3421.
11. Failla, M. L., Benedict, C. D., and Weinberg, E. D. (1976). Accumulation and storage of Zn^{2+} by *Candida utilis*, *J. Gen. Microbiol. 94*: 23.
12. Parkin, M. J., and Ross, I. S. (1985). Uptake of copper and manganese by the yeast *Candida utilis*, *Microb. Lett. 29*: 115.
13. Parkin, M. J., and Ross, I. S. (1986). The specific uptake of manganese in the yeast *Candida utilis*, *J. Gen. Microbiol. 132*: 2155.
14. Norris, P. R., and Kelly, D. P. (1979). Accumulation of metals by bacteria and yeasts, *Dev. Ind. Microbiol. 20*: 299.
15. Horikoshi, T., Nakajima, A., and Sakaguchi, T. (1981). Studies on the accumulation of heavy metal elements in biological systems. XIX. Accumulation of uranium by microorganisms, *Eur. J. Appl. Microbiol. Biotechnol. 12*: 90.
16. Murray, A. D., and Kidby, D. K. (1975). Sub-cellular location of mercury in yeast grown in the presence of mercuric chloride, *J. Gen. Microbiol. 86*: 66.
17. Brown, T. A., and Smith, D. G. (1976). The effects of silver nitrate on the growth and ultrastructure of the yeast *Cryptococcus albidus*, *Microb. Lett. 3*: 155.
18. Kierans, M., Staines, A. M., Bennett, H., and Gadd, G. M. (1991). Silver tolerance and accumulation in yeasts, *Biol. Metals 4*: 100.
19. Gadd, G. M., and White, C. (1989). Uptake and intracellular compartmentation of thorium in *Saccharomyces cerevisiae*, *Environ. Pollut. 61*: 187.
20. Venkateswerlu, G., and Sastry, K. S. (1970). The mechanism of uptake of cobalt ions by *Neurospora crassa*, *Biochem. J. 118*: 497.
21. Paton, W. H. N., and Budd, K. (1972). Zinc uptake in *Neocosmospora vasinfecta*, *J. Gen. Microbiol. 72*: 173.

22. Townsley, C. C., and Ross, I. S. (1985). Copper uptake by *Penicillium spinulosum*, *Microbios 44*: 125.
23. Townsley, C. C., and Ross, I. S. (1986). Copper uptake in *Aspergillus niger* during batch growth and in nongrowing mycelial suspensions, *Exp. Mycol. 10*: 281.
24. Townsley, C. C. (1985). Heavy metal accumulation in filamentous fungi, Ph.D. thesis, University of Keele.
25. Duddridge, J. E., and Wainwright, M. (1980). Heavy metal accumulation by aquatic fungi and reduction in viability of *Gammarus pulex* fed Cd^{2+} contaminated mycelium, *Water Res. 14*: 1605.
26. Germann, U. A., and Lerch, K. (1987). Copper accumulation in the cell-wall-deficient slime variant of *Neurospora crassa*, *Biochem. J. 245*: 479.
27. Fukami, M., Yamazaki, S., and Toda, S. (1983). Distribution of copper in the cells of heavy metal tolerant fungus, *Penicillium ochro-chloron*, *Agric. Biol. Chem. 47*: 1367.
28. Gadd, G. M., and White, C. (1985). Copper uptake by *Penicillium ochrochloron*: influence of pH on toxicity and demonstration of energy-dependent copper influx using protoplasts, *J. Gen. Microbiol. 131*: 1875.
29. Gadd, G. M., Stewart, A., White, C., and Mowll, J. L. (1984). Copper uptake by whole cells and protoplasts of a wild-type and copper-resistant strain of *Saccharomyces cerevisiae*, *FEMS Microbiol. Lett. 24*: 231.
30. Ono, B., Ohue, H., and Ishihara, F. (1988). Role of cell wall in *Saccharomyces cerevisiae* mutants resistant to Hg^{2+}, *J. Bacteriol. 170*: 5877.
31. Bewley, R. J. F., and Campbell, R. (1980). Influence of zinc, lead and cadmium pollutants on the microflora of hawthorn leaves, *Microbial Ecol. 6*: 227.
32. Gadd, G. M., and Mowll, J. L. (1985). Copper uptake by yeast-like cells, hyphae and chlamydospores of *Aureobasidium pullulans*, *Exp. Mycol. 9*: 230.
33. Mowll, J. L., and Gadd, G. M. (1984). Cadmium uptake by *Aureobasidium pullulans*, *J. Gen. Microbiol. 130*: 279.
34. Gadd, G. M., White, C., and Mowll, J. L. (1987). Heavy metal uptake by intact cells and protoplasts of *Aureobasidium pullulans*, *FEMS Microbiol. Ecol. 45*: 261.
35. Gadd, G. M., and de Rome, L. (1988). Biosorption of copper by fungal melanin, *Appl. Microbiol. Biotechnol. 29*: 610.
36. Borst-Pauwels, G. W. F. H. (1981). Ion transport in yeast, *Biochim. Biophys. Acta 650*: 88.
37. Fuhrmann, G.F. (1974). Relation between divalent cation transport and divalent cation activated ATPase in yeast plasma membranes, *Experientia 30*: 686.
38. Roomans, G. M., Theuvenet, A. P. R., Van den Berg, T. P. R., and Borst-Pauwels, G. W. F. H. (1979). Kinetics of Ca^{2+} and Sr^{2+} uptake by yeast. Effect of pH, cations and phosphate, *Biochim. Biophys. Acta 551*: 187.
39. Nieuwenhuis, B. J. W. M., Weijers, C. A. G. M., and Borst-Pauwels, G. W. F. H. (1981). Uptake and accumulation of Mn^{2+} and Sr^{2+} in *Saccharomyces cerevisiae*, *Biochim. Biophys. Acta 649*: 83.
40. Peters, P. H. J., and Borst-Pauwels, G. W. F. H. (1979). Properties of plasmamembrane ATPase and mitochondrial ATPase of *Saccharomyces cerevisiae*, *Physiol. Plant. 46*: 330.

41. Okorokov, L. A., Kadomtseva, V. M., and Titovskii, B. I. (1979). Transport of manganese into *Saccharomyces cerevisiae*, *Folia Microbiol. 24*: 240.
42. White, C., and Gadd, G. M. (1987). The uptake and cellular distribution of zinc in *Saccharomyces cerevisiae*, *J. Gen. Microbiol. 133*: 727.
43. Heldwein, R., Tromballa, H. W., and Broda, E. (1977). Aufnahme von Cobalt, Blei und cadmium durch Bäckerhefe, *Z. Allg. Mikrobiol. 17*: 309.
44. Mohan, P. M., and Sastry, K. S. (1984). Cobalt transport in nickel-resistant strains of *Neurospora crassa, Curr. Microbiol. 11*: 137.
45. Mohan, P. M., Rudra, M. P. P., and Sastry, K. S. (1984). Nickel transport in nickel-resistant strains of *Neurospora crassa*, *Curr. Microbiol. 10*: 125.
46. Hockertz, S., Schmid, J., and Auling, G. (1987). A specific transport system for manganese in the filamentous fungus *Aspergillus niger*, *J. Gen. Microbiol. 133*: 3513.
47. Budd, K. (1979). Magnesium uptake in *Neocosmospora vasinfecta*, *Can. J. Bot. 57*: 491.
48. Budd, K. (1988). A high-affinity system for the transport of zinc in *Neocosmospora vasinfecta*, *Exp. Mycol. 12*: 195.
49. Starling, A. P., and Ross, I. S. (1990). Uptake of manganese by *Penicillium notatum*, *Microbios 63*: 93.
50. Starling, A. P., and Ross, I. S. (1991). Uptake of zinc by *Penicillium notatum*, *Mycol. Res. 95*: 712.
51. Pilz, F., Auhling, G., Stephan, D., Rau, U., and Wagner, F. (1991). A high-affinity Zn^{2+} uptake system controls growth and biosynthesis of an extracellular branched β-1,3-β-1,6-glucan in *Sclerotium rolfsii* ATCC 15205, *Exp. Mycol. 15*: 181.
52. Starling, A. P. (1990). The uptake of heavy metals by protoplasts and whole cells of filamentous fungi, Ph.D. thesis, University of Keele.
53. Failla, L. J., and Niehaus, W. G. (1986). Regulation of Zn^{2+} uptake and versicolorin A synthesis in a mutant strain of *Aspergillus parasiticus*, *Exp. Mycol. 10*: 35.
54. Boutry, M., Foury, F., and Goffeau, A. (1977). Energy-dependent uptake of calcium by the yeast *Saccharomyces pombe*, *Biochim. Biophys. Acta 464*, 602.
55. Eilam, Y., Lavi, H., and Grossowicz, N. (1984). Effects of plasma-membrane ATPase inhibitors on potassium and calcium fluxes, membrane potential and proton motive force in the yeast *Saccharomyces cerevisiae*, *Microbios 41*: 177.
56. Eilam, Y. (1984). Effects of phenothiazines on inhibition of plasma membrane ATPase and hyperpolarisation of cell membranes in the yeast *Saccharomyces cerevisiae*, *Biochim. Biophys. Acta 769*: 601.
57. Eilam, Y., and Chernichovsky, D. (1987). Uptake of Ca^{2+} driven by membrane potential in energy depleted yeast cells, *J. Gen. Microbiol. 133*: 1641.
58. Eilam, Y., and Othman, M. (1990). Activation of Ca^{2+} influx by metabolic substrates in *Saccharomyces cerevisiae*: role of membrane potential and cellular ATP levels, *J. Gen. Microbiol. 136*: 861.
59. Borbolla, M., and Pena, A. (1980). Some characteristics of Ca^{2+} uptake by yeast cells, *J. Membrane Biol. 54*: 149.

60. Ugalde, U. O., and Pitt, D. (1986). Calcium uptake kinetics in relation to submerged cultures of *Penicillium cyclopium*, *Trans. Br. Mycol. Soc. 87*: 119.
61. Pitt, D., Barnes, J. C., and Ugalde, U. O. (1988). Differential uptake of calcium by strains of *Penicillium notatum* and relationships to calcium-induced conidiation, *Trans Br. Mycol. Soc. 91*: 489.
62. LeJohn, H. B., Cameron, L. E., Stevenson, R. M., and Meuser, R. U. (1974). Influence of cytokinins and sulfhydryl group-reacting agents on calcium transport in fungi, *J. Biol. Chem. 249*: 4016.
63. Ponta, H., and Broda, E. (1970). Mechanismen der Aufnahme von Zink durch Bäckerhefe, *Planta 95*: 18.
64. Lawford, H. G., Pik, J. R., Lawford, G. R., Williams, T., and Kligerman, A. (1980). Hyperaccumulation of zinc by zinc-depleted *Candida utilis* grown in chemostat culture, *Can. J. Microbiol. 26*: 71.
65. Eilam, Y. (1982). Effect of monovalent cations on calcium efflux in yeast, *Biochim. Biophys. Acta 687*: 8.
66. Eilam, Y. (1982). Studies on calcium efflux in the yeast *Saccharomyces cerevisiae*, *Microbios 35*: 99.
67. Eilam, Y. (1982). The effect of potassium ionophores and potassium on cellular calcium in the yeast *Saccharomyces cerevisiae*, *J. Gen. Microbiol. 128*: 2611.
68. Stroobant, P., and Scarborough, G. A. (1979). Active transport of calcium in *Neurospora* plasma membrane vesicles, *Proc. Natl. Acad. Sci. USA 76*: 3102.
69. Ghislain, M., Goffeau, A., Halachmi, D., and Eilam, Y. (1990). Calcium homeostasis and transport are affected by disruption of cta3, a novel gene encoding Ca^{2+}-ATPase in *Schizosaccharomyces pombe*, *J. Biol. Chem. 265*: 18400.
70. Joho, M., Sukenobu, Y., Egashira, E., and Murayama, T. (1983). The correlation between Cd^{2+} sensitivity and Cd^{2+} uptake in the strains of *Saccharomyces cerevisiae*, *Plant Cell Physiol. 24*: 389.
71. Joho, M., Imai, M., and Murayama, T. (1985). Different distribution of Cd^{2+} between Cd-sensitive and Cd-resistant strains of *Saccharomyces cerevisiae*, *J. Gen. Microbiol. 131*: 53.
72. Ross, I. S., and Walsh, A. L. (1981). Resistance to copper in *Saccharomyces cerevisiae*, *Trans. Br. Mycol. Soc. 77*: 27.
73. Joho, M., Imada, Y., and Murayama, T. (1987). The isolation and characterisation of Ni^{2+} resistant mutants of *Saccharomyces cerevisiae*, *Microbios 51*: 183.
74. Joho, M., Tarumi, K., Inouhe, M., Tohoyama, H., and Murayama, T. (1991). Co^{2+} and Ni^{2+} resistance in *Saccharomyces cerevisiae* associated with a reduction in the accumulation of Mg^{2+}, *Microbios 67*: 177.
75. Venkateswerlu, G., and Sastry, K. S. (1979). Cobalt transport in a cobalt-resistant strain of *Neurospora crassa*, *J. Biosci. 1*: 433.
76. Stokes, P. M., and Lindsay, J. E. (1979). Copper tolerance and accumulation in *Penicillium ochro-chloron* isolated from copper-plating solution, *Mycologia 71*: 796.
77. White, C., and Gadd, G. M. (1986). Uptake and cellular distribution of copper, cobalt and cadmium in strains of *Saccharomyces cerevisiae* cultures on elevated concentrations of these metals, *FEMS Microbiol. Ecol. 38*: 277.

78. Belde, P. J. M., Kessels, B. G. F., Moelans, I. M., and Borst-Pauwels, G. W. F. H. (1988). Cd^{2+} uptake, Cd^{2+} binding and loss of cell K^+ by a Cd-sensitive and a Cd-resistant strain of *Saccharomyces cerevisiae*, *FEMS Microbiol. Lett. 49*: 493.
79. Bianchi, M. E., Carbone, M. L., and Lucchini, G. (1981). Mn^{2+} and Mg^{2+} uptake in Mn-sensitive and Mn-resistant yeast strains, *Plant Sci. Lett. 22*: 345.
80. Okorokov, L. A., Lichko, L. P., and Kulaev, I. S. (1980). Vacuoles: main compartments of potassium, magnesium and phosphate ions in *Saccharomyces carlsbergensis* cells, *J. Bacteriol. 144*: 661.
81. Lichko, L. P., Okorokov, L. A., and Kulaev, I. S. (1982). Participation of vacuoles in regulation of levels of K^+, Mg^{2+} and orthophosphate ions in cytoplasm of the yeast Saccharomyces carlsbergensis, *Arch. Microbiol. 132*: 289.
82. Okorokov, L. A., Kulakovskaya, T. V., Lichko, L. P., and Polorotova, E. V. (1985). H^+/ion antiport as the principal mechanism of transport systems in the vacuolar membrane of the yeast *Saccharomyces carlsbergensis*, *FEBS Lett. 192*: 303.
83. Ohsumi, Y., Kitamoto, K., and Anraku, Y. (1988). Changes induced in the permeability barrier of the yeast plasma membrane by cupric ions, *J. Bacteriol. 170*: 2676.
84. Kitamoto, K., Yoshizawa, K., Ohsumi, Y., and Anraku, Y. (1988). Mutants of *Saccharomyces cerevisiae* with defective vacuolar function, *J. Bacteriol. 170*: 2687.
85. Ohsumi, Y., and Anraku, Y. (1983). Calcium transport driven by a proton motive force in vacuolar membrane vesicles of *Saccharomyces cerevisiae*, *J. Biol. Chem. 258*: 5614.
86. Eilam, Y., Lavi, H., and Grossowicz, N. (1985). Cytoplasmic Ca^{2+} homeostasis maintained by a vacuolar Ca^{2+} transport system in the yeast *Saccharomyces cerevisiae*, *J. Gen. Microbiol. 131*: 623.
87. Anraku, Y., Ohya, Y., and Iida, H. (1991). Cell cycle control by calcium and calmodulin in *Saccharomyces cerevisiae*, *Biochim. Biophys. Acta 1093*: 169.
88. Ohya, Y., Umemoto, N., Tanida, I., Ohta, A., Iida, H., and Anraku, Y. (1991). Calcium-sensitive *cls* mutants of *Saccharomyces cerevisiae* showing a pet^- phenotype are ascribable to defects of vacuolar membrane H^+-ATPase activity, *J. Biol. Chem. 266*: 13971.
89. Cornelius, G., and Nakashima, H. (1987). Vacuoles play a decisive role in calcium homeostasis in *Neurospora crassa*, *J. Gen. Microbiol. 133*: 2341.
90. Miller, A. J., Vogg, G., and Sanders, D. (1990). Cytosolic calcium homeostasis in fungi: roles of plasma membrane transport and intracellular sequestration of calcium, *Proc. Natl. Acad. Sci. USA 87*: 9348.
91. Passow, H., and Rothstein, A. (1960). The binding of mercury by the yeast cell in relation to changes in permeability, *J. Gen. Physiol. 43*: 621.
92. Kuypers, G. A. J., and Roomans, G. M. (1979). Mercury-induced loss of K^+ from yeast cells investigated by electron probe x-ray microanalysis, *J. Gen. Microbiol. 115*: 13.
93. Ross, I. S. (1977). Effect of glucose on copper uptake and toxicity in Saccharomyces cerevisiae, *Trans. Br. Mycol. Soc. 69*: 77.

94. Gadd, G. M., and Mowll, J. L. (1983). The relationship between cadmium uptake, potassium release and viability in *Saccharomyces cerevisiae*, *FEMS Microbiol. Lett. 16*: 45.
95. Theuvenet, A. P. R., Kessels, B. G. F., Blankensteijn, W. M., and Borst-Pauwels, G. W. F. H. (1987). A comparative study of K^+ loss from a cadmium-sensitive and a cadmium-resistant strain of *Saccharomyces cerevisiae*, *FEMS Microbiol. Lett. 43*: 147.
96. Kessels, B. G. F., Theuvenet, A. P. R., Peters, P. H. J., Dobbelmann, and Borst-Pauwels (1987). Changes in ^{45}Ca and ^{109}Cd uptake, membrane potential and cell pH in *Saccharomyces cerevisiae* provoked by Cd^{2+}, *J. Gen. Microbiol. 133*: 843.
97. Parkin, M. J., and Ross, I. S. (1986). The regulation of Mn^{2+} and Cu^{2+} uptake in cells of the yeast *Candida utilis* grown in continuous culture, *FEMS Microbiol. Lett. 37*: 59.

5

Oligotrophic Growth of Fungi—Stress or Natural State?

Milton Wainwright

The University of Sheffield
Sheffield, England

I. INTRODUCTION

Oligotrophs are microorganisms that can grow when nutrients are present in very low concentrations, or even where they appear to be absent. However, the term "oligotrophy" is generally used to describe the strategy used by bacteria to grow on low concentrations of organic carbon (1). In fact, nearly all the literature on microbial oligotrophy relates to bacteria, and it has even been stated that this growth strategy is restricted to prokaryotes (2). However, as I hope this review will make clear, there is increasing evidence to show that fungi can grow oligotrophically, and that this ability provides an important survival strategy for these organisms.

The study of bacterial oligotrophy has been hindered by confusion about terminology and the lack of a clear definition of what level of nutrients can be regarded as constituting oligotrophic conditions. This means that any terms used in a discussion of microbial oligotrophy must first be clearly defined. In this review, an oligotrophic fungus will be defined as one that can grow at low levels of essential nutrients, or even in their apparent absence. Where organic carbon is the nutrient present in low concentration, then growth under these conditions is termed "oligocarbotrophy," and a fungus capable of growing in the presence of small amounts of nitrogen will be referred to as an "oligonitrotroph." This terminology can be modified for reference to fungal growth in low concentrations of any nutrient; for example, one could refer to oligophosphotrophs and oligoferrotrophs, i.e., fungi that can respectively grow in the presence of low concentrations of phosphorous and iron. On the basis of this approach to terminology, most authors, when discussing bacterial growth under low carbon conditions, have regarded the term oligotrophy as being synonymous with oligocarbotrophy (1,2). Here, however, I shall use oligotrophy in a generic sense to refer to fungal growth under generalized low nutrient conditions, rather than with specific reference to organic carbon. The concentration of organic carbon at which bacterial oligocarbotrophy occurs has been variously defined. However, a concentration of organic carbon of between 1 and 15 μg C ml^{-1} has been generally accepted; the Japanese Society for Microbiology has accepted the lower of these two values (2).

Oligotrophs can be further subdivided into obligate species, which can only grow at low concentrations of a given nutrient, but will not grow when transferred to media containing high concentrations of that nutrient, and facultative oligotrophs, which grow under both high and low nutrient conditions. Although claims for the existence of obligate oligocarbotrophic bacteria have been made (1,2), obligately oligotrophic fungi have not yet been reported, which may merely reflect the small number of reports on fungal oligotrophy that have appeared to date.

In at least one review (2) the growth of oligocarbotrophic bacteria has been discussed in relation to microbial growth in extreme environments, the obvious implication being that oligotrophic microorganisms are exposed to stress when growing under low nutrient conditions, such stress being overcome only when relatively large amounts of nutrients enter the system.

A wide variety of fungi have been shown to grow oligocarbotrophically on silica gel and where no nutrients have been added (Table 1), so oligotrophy is clearly widely distributed among the fungi and is not restricted to an individual species, genus, or group. However, some fungi, mainly wood-decomposing and mycorrhizal species, cannot grow oligotrophically on silica gel (Table 1).

My aim here is to review the literature on fungal oligotrophy and to examine the proposition that fungi are stressed when growing under low nutrient conditions. Stated simply, do fungi grow in low nutrient environments, and if so, are fungi stressed under these conditions?

II. EVIDENCE FOR FUNGAL OLIGOTROPHY

It can be readily shown that fungi can grow in liquid or on solid media containing otherwise normal concentrations of nutrients, but where carbon and nitrogen are provided at low concentrations (i.e., as oligocarbotrophs and

Table 1 Ability or Lack of Ability of Certain Fungi to Grow Oligocarbotrophically on Silica Gel

Some species that grow	Some species that fail to grow
Actinomucor	*Clitocybe fragans*
Aspergillus flavus	*Collybia butyracea*
A. niger	*Hypholoma fasciculare*
A. repens	*Lepista nuda*
Fusarium oxysporum	*Phanerochaete chrysosporium*
F. solani	*P. velutina*
Gliocladium virens	
Mucor rouxii	
Penicillium chrysogenum	
P. notatum	
Rhizopus stolonifer	
Sordaria fimicola	
Zygorhynchus	

Source: Refs. 3, 9, 15.

oligonitrotrophs). Fungi will also grow in the apparent absence of any nutrient whatsoever, e.g., in what is regarded as ultrapure water, to which no nutrients have been added (3–5).

Surprisingly few references to fungal oligotrophy can be found in the early mycological literature. However, Lockwood (6) noted that fungi often contaminate apparently nutrient-free laboratory reagents, while Castellani (7) and Stern et al. (8) demonstrated fungal growth in seemingly nutrient-free, bidistilled water.

The first modern observation of fungal growth under oligotrophic conditions was made by Mirocha and Devay (9). In the following year Tribe and Mabadeje (10) described the ability of fungi to grow on nutrient-free silica gel. More recently, studies aimed at further determining the nature of fungal oligotrophy have been reported (3,11–13).

Detailed studies on the ability of fungi to grow as oligonitrotrophs have yet to appear. There is, however, an extensive literature from the early part of this century that deals with the question of whether fungi fix atmospheric dinitrogen (14). When reassessed from the viewpoint of oligonitrotrophy, this literature points to the marked ability of fungi to scavenge traces of combined N from the atmosphere and growth medium.

III. ISOLATION AND GROWTH OF OLIGOTROPHIC FUNGI

Oligotrophic fungi can be isolated from soils or other natural ecosystems, such as the phylloplane, by using nutrient-free silica gel (3). Where oligocarbotrophs or oligonitrotrophs are sought, otherwise complete silica gel medium lacking a source of either C or N can be used (3,9). Fungi also readily grow from soil crumbs or other natural substrates onto nutrient-free or low nutrient gels (3,15). As colonies do not generally form under these conditions, fungal growth is not always obvious and needs to be confirmed using a microscope. Oligocarbotrophic fungi can also be readily isolated from the air by exposing otherwise complete medium to which no carbon source has been added.

Fungi can also be cultured under exacting oligotrophic conditions using media prepared so as to exclude trace contamination. Ultrapure chemicals are used under these conditions, and all glassware (which is used in preference to plastic containers, which may omit volatile organics) is acid-washed and heated in a muffle furnace to destroy all trace of potential organic nutrients (3,10).

IV. MORPHOLOGICAL OBSERVATIONS ON OLIGOTROPHICALLY GROWING FUNGI

Not surprisingly, compared to growth in high nutrient media, much smaller amounts of biomass are produced when a fungus is grown oligotrophically. Under such low nutrient conditions, fungi produces very fine hyphae, which form fine mycelial mats (or "gossamers"). These show numerous anastomoses and float just below the surface of the medium (11). Gossamers are conceivably formed because they provide a large surface area, thereby aiding nutrient scavenging from the atmosphere and the nutrient-poor medium. Most fungi, including soil isolates, produce spores normally, although not luxuriously, under oligotrophic conditions, and microcycle conidiation may also occur (16,17).

Fungi can grow oligotrophically without the apparent involvement of the lysis and utilization of preformed hyphae. While such lysis is occasionally seen when fungi grow oligotrophically on silica gel (10), Parkinson et al. (3) found that hyphae of *Fusarium oxysporum* growing under these conditions were not subject to lysis, but remained intact and full of cytoplasm from the point of inoculation to the actively growing hyphal tip. Since fungi can be grown oligocarbotrophically on silica gel and then transferred to fresh gel on as many as 20 occasions (3), it is clear that they can grow without relying on the lysis and utilization of preformed biomass, but instead utilize exogenous nutrients.

Cytoplasm-free hyphae have, however, been reported following fungal growth under low nutrient conditions. Dickinson and Bottomley (17), for example, showed that when spores of *Alternaria* and *Cladosporium* germinate in a nutrient-free solution, the hyphae that are initially formed become evacuated and remain empty. A recent model of fungal growth by Schnurer and Paustian (18) similarly predicts that, in soils, fungi grow by producing cytoplasm-free hyphae, along which cytoplasm can be translocated where necessary. This predicted growth pattern is conservative in terms of energy costs because the formation of cytoplasm is energetically less costly than is the production of wall. Schnurer and Paustian (18) calculated that the soils they studied lacked sufficient carbon to support the measured fungal biomass when both cytoplasm and walls were taken into account; however, sufficient carbon was present if only production of the latter was assumed. This model is interesting because it suggests that scavenged nutrients would be directed toward the cell wall.

V. PHYSIOLOGY OF FUNGI GROWING OLIGOTROPHICALLY

Although little is known about the physiology of fungi when growing under low nutrient conditions, it is likely that many of the physiological characteristics that allow bacteria to grow as oligotrophs will also operate in fungi.

Mirocha and Devay (9) suggested that fungi grow as autotrophs in the apparent absence of organic carbon. They suggested that under these conditions fungi fix CO_2 from the atmosphere, using energy obtained from the oxidation of atmospheric hydrogen (i.e., by the knall gas reaction). Although CO_2 fixation by fungi was demonstrated in these studies, no evidence was found for the presence of Calvin cycle enzymes, such as RUBP carboxylase. The lack of this enzyme in *F. oxysporum* grown in the apparent absence of nutrients has been confirmed by Parkinson et al. (12). By using stripping autoradiography, these workers also confirmed the findings of Mirocha and Devay (9) that CO_2 is in some way metabolized by fungi when growing under strict oligocarbotrophic conditions, the label being incorporated into hyphae. Killed mycelium, on the other hand, did not take up the label, showing the operation of an active process. As the labeled CO_2 could not be removed by acid washing, it was not present on the wall surface in the form of insoluble carbonates or adsorbed CO_2. By partitioning the cell wall components, it was shown that the label was associated with the wall or wall membranes. The active assimilation of CO_2 by oligocarbotrophically grown fungi could involve either autotrophic or heterotrophic pathways. Autotrophic CO_2 fixation, however, is an exceedingly energy-demanding process, which is not usually associated with fungi. Mirocha and Devay (9) pointed out that hydrogen oxidation might act as the source of energy for such autotrophic growth, citing supporting evidence to show that *Fusarium* spp. possess the necessary hydrogenases.

Parkinson et al. (13) studied the fate of the carbon assimilated by oligotrophically grown *F. oxysporum*. They found that approximately half of the fixed carbon was released into the medium, while 67% of that remaining in the mycelium was present in the protein fraction, i.e., a carbon distribution pattern that is similar to that found when fungi fix CO_2 anaplerotically. Some 0.68% of fungal cell carbon was derived from CO_2 under oligotrophic conditions. The specific activity of CO_2 assimilation was 259.5 μmol CO_2 g^{-1} dry wt mycelium. The fact that a biomass yield of 0.56 mg C^{-1} was achieved when *F. oxysporum* was grown on oligocarbotrophic medium confirmed that this fungus can scavenge airborne carbon sources.

It is now generally assumed that fungi grow under oligotrophic conditions by scavenging nutrients from the atmosphere and fixing CO_2 heterotrophically. As well as acting as a nutrient, CO_2 may have other functions in the fungal cell. Barinova (19), for example, found that CO_2 speeds up both the swelling and germination of fungal spores and also catalyzes the formation of protein and nucleoprotein polypeptides from amino acids.

It has been suggested that fungi may supplement their energy requirements by oxidizing inorganic ions such as ammonium or thiosulfate (5). Although such chemolithoheterotrophy is relatively common among heterotrophic bacteria, the ability of fungi to use this growth strategy has not yet been confirmed (5).

Wainwright and Grayston (11) provided evidence that fungi use energy gained from thiosulfate oxidation to help them grow oligotrophically, while Jones et al. (20) found that *F. oxysporum* could oxidize 13% of the thiosulfate provided to sulfate under oligotrophic conditions, this being sufficient to provide energy for 25% of the observed growth. Chemolithoheterotrophic growth by fungi under oligotrophic conditions, although not proven, would obviously be highly advantageous since it would allow a fungus to obtain energy from inorganic oxidations while directing toward biomass production the scarce supply of carbon available in oligotrophic environments.

Mirocha and Devay (9) have also suggested that fungi might gain energy by oxidizing ammonium to nitrite and/or nitrate when growing oligotrophically. However, since neither of these ions could be detected in growth medium, they concluded that ammonium oxidation did not provide energy for fungal growth under these conditions. It is worth noting, however, that Tolmstoff (21) found that conidia of *Verticillium albo-atrum* can oxidize ammonium to the equivalent of an amino group by a pathway that is sensitive to inhibitors such as cyanide. Ammonium salts were shown to stimulate respiration by this fungus in proportion to the amount added, and the reaction was found to be zero order for both ammonium and oxygen. Tolmstoff (21) concluded that two ammonium ions contribute 1 electron and 1 proton to the reduction of an atom of oxygen, leaving an excess of 1 proton per ion and an amino (NH_2) radical. If fungi can use this system to generate useful energy, then they may be capable of using atmospheric ammonia as both energy and nitrogen sources.

The crucial characteristic of oligotrophy, both in bacteria and in fungi, is that organisms utilizing this strategy must use whatever carbon sources or energy-generating systems are available, even if these, when considered individually, produce only small amounts of biomass or energy, the additive effect being sufficient to support growth. Mixed substrate growth, where the fungus utilizes a wide range of carbon and energy sources simultaneously, is similarly likely to be employed during oligotrophic growth.

Recent studies confirm that most filamentous fungi are incapable of fixing dinitrogen. Kostyaev et al. (22), for example, found that of 138 strains of fungi (representing 58 species from 25 genera) that they tested, none could fix N. However, recent claims for fungal N fixation, particularly by wood-rotting species like *Pleurotus*, continue to appear in the literature (23). It is likely then that, since the majority, if not all, of filamentous fungi cannot fix nitrogen, combined N, mainly as ammonia, will provide the main source of N for fungi when growing under oligonitrotrophic conditions.

Presumably a diffusion gradient is set up between the outside of the fungal hyphae and the surrounding air, thereby enabling gaseous nitrogen and carbon to continuously diffuse into the hyphae. In this way, the fungus could effectively scavenge trace amounts of airborne nutrients, a process that has been demonstrated in heterotrophic bacteria by Geller (24).

A considerable amount has been written about the physiological and biochemical attributes of oligocarbotrophic bacteria, much of which may be applicable to oligotrophic fungi. Poindexter (1) listed the physiological characteristics generally possessed by oligotrophic bacteria; these include: (1) a high surface:volume ratio; (b) the ability to use metabolic energy for nutrient uptake, with low endogenous metabolic rates; (c) the capacity for continual nutrient uptake; (d) a large proportion of inducible enzymes; (e) uptake systems that are preferentially used for reserve accumulation, so that net synthesis occurs only when rates of uptake and synthesis of metabolic pools become saturated, hence relatively low maximal growth rates; (f) uptake systems of high affinity with the ability to use several carbon and energy sources simultaneously.

Since aerobic respiration allows for the most efficient energy generation from the dissimilation of scarce organic substrates, model oligotrophs (both bacteria and fungi) would be expected to be strictly oxybiotic respiratory organisms. While many oligotrophic bacteria can grow as anaerobes, it is generally assumed that aerobic metabolism will be selected by microorganisms when growing in oligotrophic environments.

The mycelial growth form is particularly advantageous for organisms growing in low nutrient environments as it provides a high surface: volume ratio that favors an increase in nutrient uptake capacity. The thinner the hyphae, the larger the surface area, which could explain why the hyphae of oligotrophically growing fungi are so fine. In addition, it is unlikely that efficient nutrient transport by oligotrophs would involve substrate modification, e.g., substrate phosphorylation. Finally, low substrate specificity and selectivity would be expected to be advantageous for oligotrophs. When growing oligotrophically, fungi are also unlikely to use the scarce carbon sources available to produce secondary metabolites, unless, that is, the production of these compounds confirms some direct physiological advantage.

It is perhaps stating the obvious to point out that as a consequence of the above characteristics, fungi will not exhibit the same high growth rates produced under nutrient-rich culture conditions.

VI. UTILIZATION OF GASES AND VOLATILES BY FUNGI

The atmosphere appears to provide most of the nutrients that fungi use to support growth under apparently nutrient-free conditions. The indoor environment particularly contains a diverse range of gases and volatiles that are potential substrates for fungi, most being excreted by plants, animals, and humans or by the combustion of fossil fuels (25). Humans, for example, excrete C-rich compounds, mainly aliphatic and aromatic hydrocarbons and acetone. Microorganisms themselves also excrete carbon-containing gases, with the result that microbiology laboratories and growth rooms are likely to be enriched with a diversity of potentially useful airborne-microbial growth substrates.

A diverse range of carbon compounds is also found in the urban atmosphere, including aliphatic C1–C3 hydrocarbons (e.g., methane); monosaturated hydrocarbons (e.g., ethylene); polysaturated hydrocarbons (e.g., benzapyrene); aromatic hydrocarbons (e.g., benzene); halogenated hydrocarbons (halogenated methanes); aldehydes; ketones; and miscellaneous organic compounds such as phenols and polyols.

Volatiles also play a role in fungal nutrition in outdoor environments. Watson et al. (26), for example, reported that the biological growths occurring on, and in the vicinity of, bonded whiskey distillery warehouses in Scotland consist of a mixed fungal culture, which appears to use volatile distillery alcohols as a source of nutrients. Decomposing wood also emits volatiles that attract fungal mycelium (27).

Fungi can use an extremely diverse range of substrates, including carbon monoxide and hydrocarbons (5). The wide substrate specificity of fungi in relation to potential airborne carbon nutrients is further indicated by the ability of *Cunninghamella* spp. to use naphthalene, benzo(a)pyrene, benzo (a)anthracene, and methylcholanthrene (28). Recent studies by Oderna et al. (29) also show that *Scedosporium* sp. can use gaseous hydrocarbons as a sole source of both carbon and energy. As already mentioned, Mirocha and Devay (9) suggested that hydrogen may act as an energy source to support the growth of oligotrophically growing fungi. Certain bacteria can use hydrogen as an energy source while simultaneously using ethylene as a source of carbon (30). Both these gases occur in the soil atmosphere and laboratory air and may provide sufficient nutrients for fungal growth under oligotrophic conditions.

The importance of gases and volatiles in fungal nutrition was summarized by Fries (31) as follows: "To me it seems beyond doubt that the volatile organic substances must be included among the environmental factors which determine the distribution, rate of growth and mode of development of fungi in nature."

VII. EXAMPLES OF THE GROWTH OF FUNGI IN LOW NUTRIENT ENVIRONMENTS

In their natural unpolluted condition, the majority of the Earth's environments, the soil, oceans, and freshwater lakes and rivers, generally contain relatively low concentrations of available organic carbon and other nutrients. Certainly, the nutrient status of most natural, nonpolluted environments never approaches that found in laboratory media.

Fungi have been found growing on desert rocks and in water. The scarcity of moisture and nutrients, together with the prevailing high temperatures, make desert rocks one of the most inhospitable environments on earth. Despite this, microcolonies of fungi have been found on desert rocks, often growing at a distance from algal or lichen communities that might otherwise have provided nutrients via photosynthesis (32). Fungal microcolonies have been found in desert areas, on various rock surfaces, throughout the world (32–34). Most are black or brown and are less than 100 μm in diameter; they are generally made up of spheroid subunits of approximately 5 μm in diameter. Staley et al. (32) found that rock inhabiting fungi are metabolically active and do not merely survive on rocks in a state of dormancy. They appear to scavenge nutrients either from the atmosphere or from windblown soil and dust. Fungi have also been shown to colonize desert varnishes (i.e., clay-rich ferromanganese oxide coatings found on desert rocks), where they appear to coexist with bacteria, from which they may obtain nutrients (32).

Fungi have also been found to be a major component of the microflora of sandstone, and to be associated with the weathering of historic buildings (35), a process that seems to depend on cation chelation, rather than the production of organic acids.

Fungi can grow on inert surfaces including glass (e.g., binocular lenses); such growth is particularly common in the tropics, where fungi can lead to severe pitting of optical lenses (36,37). Fungi such as *Fusarium semitectum*, *Paecilomyces varioti*, and *Rhizopus nigricans* have also been found contaminating highly polished silicon wafers used to produce microchips (38). Fungi appear to grow under these conditions by scavenging nutrients from the atmosphere or from grease transferred from the user's fingers. Smith and Nadim (39) also showed that filamentous fungi colonize electrical equipment by using nutrients in exhaled breath and cigarette smoke, as well as in traces of volatile vegetable and mineral oils (e.g., gas turbine oil).

Fungi can be readily isolated from drinking waters, water supply pipes (40,41), and the surface microlayer of freshwater ponds (42), although it is not clear whether hyphal growth occurs under these conditions, or whether fungi merely survive as spores.

Mention has already been made of the fact that fungi can grow in distilled water, even when stringent precautions have been taken to avoid nutrient contamination. Nutrient limitation should therefore not limit growth of fungi in marine or freshwaters, which, although not usually rich in nutrients, clearly contain more nutrients than does purified water.

The ability of fungi to grow in apparently nutrient-free water could have major industrial and medical consequences. Montagnac et al. (43), for example, reported that potentially pathogenic fungi can contaminate water delivery systems used to supply hospital dialysis centers. Further biotechnological consequences of the oligotrophic growth of microorganisms have recently been detailed by Wainwright et al. (44).

VIII. OLIGOTROPHIC GROWTH OF FUNGI IN SOILS

The ability of fungi, like most other microorganisms, to grow in soils is generally thought to be limited by the presence of low levels of available carbon (45), with continuous growth occurring only where sufficient available carbon is present, e.g., in the rhizosphere or adjacent to plant residues. The amount of organic carbon available to microorganisms when growing in soils has been variously estimated to be 0.17–0.67 mM, 0.14 mM, and below 8.3 mM, i.e., below 100 mg C l^{-1} (20). Smith et al. (45) gave a value of 60–102 mg of available carbon per kilogram of soil.

Microorganisms are generally thought to lie dormant in soils as spores, or else live off endogenous metabolites, while they await the input of fresh nutrients, which then stimulate fresh growth and nutrient depletion. Morita (46) has recently argued that microbial cells exist in the environment in a state of suspended animation, which he terms "metabolic arrest." This view of microbial growth in soils has a long history, and soil microorganisms have traditionally been divided into (a) autocthonous types, which are considered to be true soil inhabitants and thus capable of slow, but continuous, growth in soils, and (b) zymogenous types, which are adapted to rapid growth and substrate utilization, followed by a period of dormancy to survive low nutrient conditions.

Parkinson et al. (3) showed that fungi will readily grow from soil particles and plant roots onto nutrient-free agar, an observation that suggests that soils should contain sufficient nutrients to support fungal growth (and microbial growth in general) in situ. Barakah (15) has also shown that fungi can be isolated from soils using capillary pedoscopes containing nutrient-free silica gel, a finding that also suggests that active mycelial growth of fungi in soils is not limited by a lack of available organic carbon.

The view that soils contain insufficient available carbon to support continuous fungal/microbial growth is based on calculations of carbon input to soils, the amount of fungal biomass present, and estimates of fungal growth rates. Although it seems likely that such models rely on the input of unreliable data (45), they certainly appear to ignore the fact that fungi can grow oligotrophically by scavenging carbon from the soil solution or soil atmosphere. While soils clearly do not contain large amounts of available carbon, even small amounts of scavenged substrates can support fungal growth.

Fungi will obviously grow in soils at a far slower rate than in nutrient-rich media and will tend to produce fine hyphae and fewer reproductive structures and spores. Probably the most approximation to date of how fungi grow in soils was provided by Sheenan and Gochenaur (16), who showed that when *Penicillium* spores germinate in soils, very fine hyphae are formed, and microcycle conidiation is common.

Fungi have been classified in the past on the basis of the substrates they can utilize; thus, for example, we refer to sugar fungi, cellulose, or lignin decomposers. However, successful soil fungi are likely to be able to use a diverse range of different substrates, on occasion at the same time, i.e., mixed substrate growth. For example, species of *Mucor* can grow oligotrophically despite being generally regarded as sugar fungi and thought to need substantial amounts of simple sugars for growth. It seems, therefore, that to group fungi in relation to the utilization of a single substrate is artificial and does not truly reflect the ability of fungi to utilize substrates in nature.

The ability of fungi to efficiently scavenge nutrients is obviously of paramount importance to this view of fungal growth in soils. Evidence for the importance of such an ability in soils was provided by Menzies and Gilbert

(47), who showed that volatiles from alfalfa residues overcame nutrient-dependent fungistasis, allowing spores to germinate, with hyphal growth being attracted toward the source of the volatiles. Fungal growth in soils can therefore be envisaged as involving the production of fine hyphae, numerous chlamydospores (48), and relying on microcycle conidiation (16). By growing in air spaces, soil fungi can efficiently scavenge volatiles or gases or obtain nutrients from the soil solution. Under these conditions, fungi will doubtless use a wide variety of nutrients simultaneously and also employ a variety of metabolic strategies to gain energy and anabolic carbon (5).

It is essential that, when growing oligotrophically, a fungus should be in no way limited either (a) by substrate specialization, or (b) by being obligately oligotrophic. Since in order to survive soil fungi must rapidly exploit any new substrate that enters the system, it comes as no surprise to find that most, if not all, fungi are facultative oligotrophs and can switch from oligotrophy to growth under nutrient-rich conditions without suffering an extended lag phase (3).

IX. FUNGAL OLIGOTROPHY IN RELATION TO METHODS CURRENTLY USED TO STUDY FUNGAL ECOLOGY

In the past, the study of fungi in the environment, particularly soils, has been biased toward a phytocentric approach, centered on the role of fungi as pathogens and as agents of litter degradation. Fungal ecologists have generally uncritically used the physiological principles developed using nutrient-rich media. For the many natural environments that are oligotrophic, such physiological principles are likely to be of dubious value in interpreting the growth of fungi in these environments. In addition, the use of such concepts may be misleading since it inevitably leads the investigator to assume that a fungus can achieve in nature what it is capable of achieving in nutrient-rich culture. Fungi can participate in all manner of transformations in nutrient-rich culture. For example, they produce antibiotics and organic acids; they also nitrify, oxidize sulfur, and also bring about the dissolution of insoluble phosphates (5). Which, if any, of these products are formed or transformations mediated when fungi grow in the natural environment? An example of the confusion that can result from applying physiological principles derived from nutrient-rich cultures to oligotrophic environments is provided by recent studies on the microbial decay of stone. It is generally assumed that stone decay (as well as phosphate solubilization) results from the production of organic acids (e.g., citrate) that are produced in quantity when fungi are grown in nutrient-rich culture. However, as indicated above, Palmer et al. (35) found that organic acids were not responsible for the in situ weathering of stone, a process that instead involves the production of cation chelating agents.

X. OLIGOTROPHY IN RELATION TO FUNGISTASIS

Fungal spores that are susceptible to the inhibitory effects of fungistasis will obviously be prevented from germinating in soils except when available carbon is present to overcome this limitation (49). As a result, a fungus whose spores are subjected to fungistasis will be at a disadvantage compared to actively growing mycelium in the race to utilize such nutrient inputs. This is obvious, since such a spore will need to germinate and grow into a mycelium before it can begin to exploit the newly available nutrients. An oligotrophically growing fungus (provided it is facultative in this respect), in contrast, already exists in the mycelial state and is therefore in an ideal position to immediately use any nutrients that enter the system. It is worth noting that the concept of fungistasis has been developed using spores that have been grown in high nutrient media. It may well be that this has produced artifacts and that fungal spores grown under oligotrophic conditions, similar to those found in natural environments, will not be subjected to fungistasis.

XI. DOES OLIGOTROPHY REPRESENT A CONDITION OF STRESS?

In the introduction to this chapter it was pointed out that microbial oligotrophy, i.e., with reference to oligocarbotrophy in bacteria, has often been discussed in monographs devoted to the microbiology of extreme environments, the obvious implication being that microbial growth in low nutrient environments represents a condition of stress. Before discussing the merits of this view, it is necessary to provide a definition of stress. An organism can obviously be regarded as being under stress if it is unable, due mainly to limiting external forces, to achieve the same level of growth, activity, and reproduction that can be achieved in the absence of such limitation. The limitation can be a chemical toxicant (e.g., a heavy metal) or a physical effect (e.g., unfavorable temperature or pH). Many organisms, including fungi, can adapt to such stresses and become tolerant. They then become adapted to an otherwise extreme environment, which result from the existence of similar stress-inducing conditions and, freed from the limitations imposed by the stress, reach a higher level of growth and/or reproductive potential than was originally achieved when the organism was susceptible to the stress.

At first sight, a low nutrient environment might appear to be highly stressful. Under such conditions, a fungus clearly cannot produce the large amounts of biomass typically associated with growth in nutrient-rich media and as a result might be regarded as being stressed. On the other hand, Foster (50) has eloquently argued that a fungus shows signs of stress when grown under nutrient-rich conditions; secondary metabolite production, for example, can be seen as a means by which the fungus can cope with excess substrate that the organism never meets in such quantity in nature. The fact that only small amounts of biomass are produced when fungi grow oligotrophically can be regarded erroneously as indicating stress. Such a view, however, is clearly influenced by our experience of growing fungi under nutrient-rich conditions in the laboratory. In fact, the small amount of biomass produced by a fungus growing oligotrophically represents the maximum biomass that can be produced under the prevailing low nutrient conditions. As a result, the organism can be seen to be in equilibrium with the prevailing nutrient regime in that environment. Should even trace amounts of an essential nutrient (carbon and nitrogen) cease to be available, then the fungus would need to fall back on the utilization of reserves. Once these become used up, then the fungus will obviously suffer stress and eventually die.

Equally, stress is imposed, however, if a fungus is grown on a high nutrient medium and then access to nutrients is denied. This situation can be described as starvation, rather than oligotrophy. These two terms are frequently used as if they were synonymous. However, the term "starvation" is best reserved for the condition when nutrient shortage is imposed following a period of nutrient excess. Under these conditions a fungus is clearly stressed, since it can no longer support the metabolism and growth rates that it previously maintained under high nutrient conditions. Similarly, a fungus would be stressed if, having grown on a nutrient-rich resource such as wood, this resource is used up; quite clearly, the biomass produced would then no longer be sustainable. In these conditions the older parts of the mycelium must lose viability and die, the younger parts utilizing the products of autolysis, to a point where the biomass is sustainable by use of the available nutrients. Whether a fungus exposed to such starvation conditions can return to an oligotrophic growth pattern is not clear.

It seems that, in the absence of added nutrients, most soil fungi can grow as oligotrophs. Oligotrophy would seem to be the rule, then, rather than the exception, among common soil molds, and we have yet to find a common soil fungus that is incapable of growing oligotrophically. On the other hand, this is not so for yeasts, ectotrophic mycorrhizal fungi, and wood-decomposing fungi, which seem incapable of growing on nutrient-free silica gel. The inability of such fungi to grow as oligotrophs, however, may prove not to be a hard and fast rule, especially since the oligotrophic abilities of only a relatively few fungi have been evaluated. However, if this rule does hold, then it is tempting to suggest that organisms that are adapted to growth in the presence of continuous supplies of relatively large amounts of substrates (provided by rotting wood or the direct supply of photosynthate from higher plants) do not grow as oligotrophs simply because they have never had the need to do so. Common soil fungi, in contrast, grow in an environment where carbon is at a premium for long periods, and as a result, they must have become adapted to growth under these conditions by including oligotrophy in their physiology.

If it is true that common soil fungi, but not ectotrophic mycorrhizal fungi and wood-decomposing species, can grow oligotrophically, then we are provided with an ideal opportunity to compare the biochemistry of these two groups to determine what factors make an oligotroph.

Obligate oligotrophy would appear to be distinctly disadvantageous to a soil-inhabiting fungus or other microorganism, since in this environment nutrients are in a state of continual flux. Under these conditions, an obligate oligotroph would be limited to low rates of biomass production attuned to low nutrient supply and so would be unable to respond by growing rapidly when new substrates appear. Obligately oligotrophic fungi, should they exist, will presumably be stressed when exposed to high nutrient conditions; as a result, they would not appear on nutrient-rich isolation media.

It is worth reemphasizing that many so-called sugar fungi can grow oligotrophically in the absence of added substrate. Rather than being an ecological advantage, restriction to a single substrate limits a fungus, which explains why cellulose-decomposing fungi (e.g., *Fusarium* sp.) can use other substrates and also grow oligotrophically. Mixed-substrate growth and metabolic diversity tend to be the norm, rather than the exception, in microbiology.

To return to the main question posed in this chapter—does fungal growth under oligotrophic conditions represent a stressed or nonstressed condition? The answer is somewhat rhetorical. Where a fungus can grow as oligotroph, then oligotrophy is not a stressful condition; however, if a fungus does not possess the mechanisms that enable it to scavenge nutrients under low nutrient conditions, then exposure to oligotrophic conditions results in stress, culminating in death. Prior to this terminal stage, the fungus can sporulate or enter some other dormant state, perhaps existing in endogenous metabolism on storage products. Fungi capable of oligotrophic growth, in contrast, do not need to sporulate, except as a means of dispersing their progeny in time and space and as a means of inducing genetic variation. As a result, such fungi need not be exposed to the limitation of fungistasis, since they can maintain themselves indefinitely in a hyphal form. Nor does the oligotrophic fungus need to build up large supplies of endogenous reserves, which are typically found when fungi are grown under nutrient-rich conditions.

In conclusion, then, oligotrophy among free-living soil fungi appears to be the norm rather than the exception. When fungi grow under these conditions, biomass production and growth rates are attuned to the low level of available nutrients, and as a result, the organism is not stressed. On the other hand, nonoligotrophic fungi will be unable to grow in low nutrient environments, and should such conditions be imposed, they will suffer stress —a condition which, if not relieved, will eventually lead to their death.

REFERENCES

1. Poindexter, J. S. (1981). Oligotrophy fast and famine existence, *Adv. Microbial Ecol. 5*: 63.
2. Fry, J. C. (1990). Oligotrophs, in *Microbiology of Extreme Environments* (C. Edwards, ed.), Open University Press, Milton Keynes, p. 93.
3. Parkinson, S. M., Wainwright, M., and Killham, K. (1989). Observations on oligotrophic growth of fungi on silica gel, *Mycol. Res. 93*: 529.
4. Wainwright, M. (1987). Can fungi grow on fresh air? *Mycologist 21*: 182.
5. Wainwright, M. (1988). Metabolic diversity of fungi in relation to growth and mineral cycling in soil—a review, *Trans. Br. Mycol. Soc. 90*: 159.
6. Lockwood, L. B. (1936). Fungi from laboratory reagents, *Mycologia 28*: 10.
7. Castellani, A. (1939). Viability of some pathogenic fungi in distilled water, *J. Trop. Med. Hyg. 42*: 225.
8. Stern, M., Da Silva Lacaz, C., and Halbe, H. W. (1956). Growth of contaminating fungi in bidistilled water, *Hosp. Rio de Janeiro 49*: 501.
9. Mirocha, C. J., and Devay, J. E. (1972). Growth of fungi on an inorganic medium, *Can. J. Microbiol. 17*: 1373.
10. Tribe, H. T., and Mabadeje, S. A. (1972). Growth of moulds on media prepared without organic nutrients, *Trans. Br. Mycol. Soc. 58*: 127.
11. Wainwright, M., and Grayston, S. J. (1988). Fungal growth and stimulation by thiosulphate under oligotrophic conditions, *Trans. Br. Mycol. Soc. 91*: 149.
12. Parkinson, S. M., Killham, K., and Wainwright, M. (1990). Assimilation of $^{14}CO_2$ by *Fusarium oxysporum* grown under oligotrophic conditions, *Mycol. Res. 94*: 959.
13. Parkinson, S. M., Jones, R., Meharg, A. A., Wainwright, M., and Killham, K. (1991). The quantity and fate of carbon assimilated from $^{14}CO_2$ by *Fusarium oxysporum* grown under oligotrophic and near oligotrophic conditions, *Mycol. Res. 95*: 1345.
14. Senn, G. (1928). The assimilation of the molecular nitrogen of the air by lower plants—especially fungi, *Biol. Rev. 3*: 77.
15. Barakah, F. N. I. (1992). Observations on the oligotrophic growth of fungi, PhD thesis, University of Sheffield, England.
16. Sheenan, P. L., and Gochenaur, S. E. (1984). Spore germination and microcycle conidiation of two penicillia in soil, *Mycologia 76*: 523.
17. Dickinson, C. H., and Bottomley, D. (1980). Germination and growth of *Alternaria* and *Cladosporium* in relation to their activity in the phylloplane, *Trans. Br. Mycol. Soc. 74*: 309.
18. Schnurer, J., and Paustian, K. (1986). Modelling fungal growth in relation to nutrient limitation in soil, in *Perspectives in Microbial Ecology* (F. Meguson and M. Ganter, eds.), Slovene Soc. Microbiol, Ljubljana, p. 123.
19. Barinova, S. A. (1962). Carbon dioxide utilization and its role in metabolism of mould fungi, *Mikrobiliologiya 31*: 10.
20. Jones, R., Parkinson, S., Wainwright, M., and Killham, K. (1991). Oxidation of thiosulphate by *Fusarium oxysporum* grown under oligotrophic conditions, *Mycol. Res. 95*: 1169.

21. Tolmstoff, W. J. (1969). Ammonium linked respiration in conidia of *Verticillium albo-atrum, Phytopathology 59*: 1053.
22. Kostyaev, V. Y., Milko, A. A., and Izvekova, G. I. (1983). Capability of fungi to fix molecular nitrogen, *Mikrobiologiya 52*: 667.
23. Ginterova, A., and Gallon, J. (1979). *Pleurotus ostreatus*: a nitrogen fixing fungus? *Biochem. Soc. Trans. 7*: 1293.
24. Geller, A. (1983). Growth of bacteria in inorganic medium at different levels of airborne organic substrates, *Appl. Environ. Microbiol. 46*: 1258.
25. Johansson, I. (1978). Determination of organic compounds in indoor air, *Atmospher. Environ. 12*: 1371.
26. Watson, R. D., Minter, D. W., and McKelvie, A. D. (1984). Dense growth of Deuteromycetes on and around bonded distillery warehouses in Scotland, *Bull. Br. Mycol. Soc. 18*: 57.
27. Mowe, G., King, B., and Senn, S. J., (1983). Tropic responses of fungi to wood volatiles, *J. Gen. Microbiol. 129*: 779.
28. Gibson, D. T., and Subramanian, V. (1984). Microbial degradation of aromatic hydrocarbons, in *Microbial Degradation of Aromatic Hydrocarbons* (D. T. Gibson, ed.), Marcel Dekker, New York, p. 181.
29. Oderna, M., Endo, Y., and Ogasawara, N. (1989). Oxidation of gaseous hydrocarbons by a gaseous hydrocarbon assimilating mold, *Scedosporium* sp, A-4, *Agric. Biol. Chem. 53*: 1947.
30. Slabova, O. I., and Nikitin, V. I. (1986). Characteristics of combined utilization of hydrogen and ethylene by oligotrophic bacteria, *Mikrobiologiya 57*: 373.
31. Fries, N. (1973). Effects of volatile organic compounds on the growth and development of fungi, *Trans. Br. Mycol. Soc. 60*: 1.
32. Staley, J. T., Palmer, F., and Adams, J. B. (1982). Microcolonial fungi; common inhabitants of desert rocks, *Science 215*: 1093.
33. Weirich, G., and Schweisfurth, R. (1985). Extraction and culture of micro-organisms from rock, *Geomicrobiol. J. 4*: 1.
34. Taylor-George, S., Palmer, F., Staley, J. T., Borns, D. J., Curtiss, B., and Adams, J. B. (1983). Fungi and bacteria involved in desert varnish formations, *Microbial Ecol. 9*: 227.
35. Palmer, R. J., Siebert, J., and Hirsch, P. (1991). Biomass and organic acids in sandstone of a weathering building: production by bacterial and fungal isolates, *Microbial Ecol. 21*: 253.
36. Hutchinson, W. G. (1947). The fouling of optical glass. *J. Bacteriol. 54*: 45.
37. Richards, O. W. (1949). Some fungus contaminants of optical instruments, *J. Bacteriol. 58*: 453.
38. Inoue, M. (1988). The study of fungal contamination in the field of electronics, *Biodeterioration 7*: 580.
39. Smith, R. N., and Nadim, K. (1983). Fungal growth on inert surfaces, *Biodetermination 5*: 538.
40. Burman, N. P. (1965). Taste and odour due to stagnation and local warming in long lengths of pipes, *Proc. Soc. Water Treat. Exam. 14*: 125.
41. Nagy, L. A., and Olsen, B. H. (1982). The occurrence of filamentous fungi in drinking water distribution systems, *Can. J. Microbiol. 28*: 667.

42. Maki, J. S., Danon, S. C., and Remsen, C. C. (1984). Qualitative changes in fungal colony forming units in the surface monolayers of two freshwater ponds, *Can. J. Microbiol. 30*: 578.
43. Montagnac, R., Schillinger, F., Roquebert, M. F., Croix, J. C., and Eloy, C. (1991). Contamination par des champignons d'un de traitment d'eau en auto-dialyse, *Nephrologie* 27.
44. Wainwright, M., Barakah, F., Adam Ali, T., and Alturk, I. (1992). Oligotrophic microorganisms in medicine, industry and the environment, *Sci. Progr.* 75: 313–322.
45. Smith, J. L., McNeal, B. L., Cheng, H. H., and Campbell, G. G. (1986). Calculation of microbial maintenance rates and net nitrogen mineralization, *Soil Sci. Soc. Am. J. 50*: 332.
46. Morita, R. Y. (1988). Bioavailability of energy and its relationship to growth and starvation survival in nature, *Can. J. Microbiol. 34*: 436.
47. Menzies, J. D., and Gilbert, R. G. (1967). Responses of the soil microflora to volatile compounds in plant residues, *Proc. Soil Sci. Soc. Am. 31*: 495.
48. Park, D. (1954). Chlamydospores and survival in soil fungi, Nature (*Lond.*) *173*: 454.
49. Mitchell, C. P., and Dix, N. J. (1975). Growth and germination of *Trichoderma* spp. under the influence of soil fungistasis, *Trans Br. Mycol. Soc. 64*: 235.
50. Foster, J. W. (1949). *Chemical Activities of Fungi*, Academic Press, New York, p. 162.

6

Tolerance to Freezing and Thawing

David Smith

International Mycological Institute
Egham, Surrey, England

I. INTRODUCTION

A. Fungi in Nature at Low Temperature

Fungi are ubiquitous, managing to colonize most natural and man-made environments, including those where high and low temperatures are encountered. Many organisms that are considered psychrophilic not only tolerate, but actually grow, at low temperature. Schmidt-Nielson (1) proposed the term "psychrophilic," and it is now considered that such organisms have optimum growth temperatures up to 16 °C and a maximum growth temperature of ca. 20 °C (2). However, many fungi are capable of growing and tolerating much lower temperatures; lichens from the Arctic and Antarctic exhibit optimum metabolism between 0 and 5 °C. Species of *Aureobasidium* that have been isolated from rocks in Antarctica are able to grow and reproduce in this harsh environment and in addition may have to tolerate temperatures perhaps as low as −70 to −80°C. Fungi have been found as spoilage organisms not only of foodstuffs stored at domestic refrigeration temperatures (5–10 °C), where *Penicillium* sp. abound, but also in deep freezers at temperatures ca. −20 °C. Spores of *Cladosporium* sp. and *Sporotrichum* sp. will germinate and grow on meat in cold storage at temperatures below freezing (3).

Phacidium infestans causes snow blight of conifers, proliferating at temperatures of 15 °C, grows down to 3 °C, and endures temperatures of −20 °C. Other plant pathogens also do well at low temperatures; among these are *Fusarium* species, *Sclerotinia borealis*, and *Typhula* species. Although the latter organisms are considered psychrophilic, others, such as *Monographella nivalis*, which will cause infection at 3 °C and has an optimum growth temperature of 20 °C, are considered to be cold-tolerant mesophiles.

Lichens, the symbiotic combination of fungus and algal cells, are capable of withstanding low temperatures, being found in, among many other extreme environments, the cold deserts of polar and alpine regions. It is thought that they are capable of withstanding an ice age. Kappen (4) reviewed their ecophysiological relationships in different climatic regions. Many species survive temperatures of −30 to −50 °C in the winter. Although several require thawing to commence photosynthesis, others are able to photosynthesize at low subzero temperatures, *Sterocaulum alpinum* at −24 °C and *Cetratia nivalis* at −14 °C. While photosynthesis and respiration have been recorded at such temperatures, *Cladonia rangiferina, Cetraria nivalis,* and *Umbilicaria vellea* have survived temperatures of −196 °C when cooled dry and −75 °C when cooled wet.

B. Overwintering

Large numbers of soil-borne fungi survive severe winters. Several of these are plant pathogens, which appear again and infect their host in the following or subsequent years. They probably remain viable as a resting stage or reproductive propagule, which normally has less water than the vegetative state. Fungi are capable of producing many such structures, but perhaps most information is available on the spore. Studies carried out by Mazur (5) showed that spores of *Aspergillus flavus* were resistant to freezing in water to −73 °C. Such survival was attributed to the lower water content (6–25%) where little ice, if any, would form (6). The ability of fungi to withstand extremes of environments makes them successful pioneers in exploiting new environments and products created by humans.

C. Cryopreservation

The use of ultralow temperature for long-term preservation of viability and stability of fungi has been considered the best method available for some time. Fungi have proven to be resistant to adverse environments and not surprisingly showed the ability to survive cryopreservation protocols. Early methodology was based on the success of Polge et al. (7) with avian spermatozoa. This method was adopted by Hwang (8) for fungi with great success. The aim of all methods is to minimize the two damaging factors of dehydration and intracellular freezing. Early work established that slow cooling in the presence of a cryoprotectant such as glycerol or dimethyl sulfoxide (DMSO) and storage at temperatures below −139°C, quite often in liquid nitrogen or the vapor above it, gave best results when applied routinely (8). Recovery is normally higher after rapid warming, usually achieved by immersion in a water bath at 35–40°C.

There are obvious problems when trying to apply one particular preservation procedure to a large variety of organisms. There are many reasons why individual fungi will respond differently to low-temperature stress, including their physiological condition, structure, constituent chemicals, means of osmoregulation, and degree of hydration. The immense taxonomic variety of fungi and the variability of response between strains of the same species make successful cryopreservation more difficult. It is essential that we understand the response of the fungus cells to the stresses of freezing and thawing in order that appropriate preservation protocols can be applied to give optimum survival and stability.

II. PRINCIPLES OF RESPONSE OF LIVING CELLS TO FREEZING AND THAWING

A. Low Temperature and Freezing

1. Low Temperature

This is a relative term. In biology it is usually identified with subzero temperatures with a lower limit of −70 °C below which no life processes persist (9). Cooling to low temperatures reduces the rate at which chemical reactions occur, increases the viscosity of water, denatures protein, increases the relative permittivity of water, which enables it to reduce the attraction between ions of opposite charge, and acidic and basic residues of proteins are markedly affected. The manifestation of such changes contribute to chill injury, to which only a few fungi are apparently susceptible, as most survive cooling and storage at low temperature.

2. Freezing

In the terminology of thermodynamics, freezing is referred to as a first-order phase change. This phase change in water requires low temperatures but will not occur without a nucleation event (9). A critical group of water molecules adopts a configuration that can be recognized by other molecules that condense on it forming an ice crystal; such events become more likely as the temperature decreases below 0 °C. The configuration of water molecules may arise due to density and energy fluctuations at low temperatures, or nucleation may be aided by a solid particle around which the water crystallizes.

The initial ice crystals that form are normally unstable and recrystallize, which in turn can be injurious to a living organism. Without a nucleation event, water supercools (undercools).

It is at a cellular level that an understanding of the freezing process must begin. Steponkus (10) has presented a clear picture of events occurring during the freezing of plant cells and protoplasts. Initially supercooling of both the cell and suspending medium occurs during cooling. Subsequently, ice nucleation occurs in the suspending medium at a temperature depending on the freezing point of the solution and the presence of effective ice nucleating agents. Freezing occurs until the eutectic point is reached where the chemical potential of water in the unfrozen portion is in equilibrium with ice. In turn, the chemical potential of the intracellular solution must also come into equilibrium with the extracellular ice. Equilibrium is achieved by either intracellular ice nucleation or cell dehydration, the plasma membrane acting as a barrier to extracellular ice and as a semipermeable membrane allowing dehydration.

The rate of cooling affects the cell's response. At rates where the efflux of water allows the cell contents to concentrate and to come into equilibrium with the extracellular solution, the cell shrinks. At rates where the efflux is not sufficiently rapid, extensive supercooling occurs, and if the cell cools to −39 °C or below, homogeneous nucleation of ice occurs. If a suitable particle is present to act as a nucleus, heterogenous nucleation will occur before this temperature is attained. Seeding by extracellular ice requires perturbation or penetration of the plasma membrane.

After extracellular ice nucleation, both the intracellular and extracellular solutes are concentrated and in some instances their solubility limits may be exceeded. Precipitation of buffers may result in pH changes. Gas bubbles are formed during freezing as gases are excluded from ice. These bubbles are often seen as a blackening or granulation of the specimen, and it is this event that is usually observed on intracellular ice nucleation in plant cell protoplasts. A rapid blackening of the cell interior occurs; this is often termed "flashing." Electrical potentials are also created during freezing by charge separation, thereby giving rise to potential differences between the liquid and the solid phases.

B. Mechanisms of Cryoinjury

The combined effect of low temperature and freezing gives rise to changes in positions of chemical and biochemical equilibria, reduces the rates of motion of molecules, and increases the concentration of components in the mixture resulting from phase separation and the possibility of irreversible aggregation or dissociation events involving macromolecules. These events will inevitably affect the structure and metabolism of cells.

Mazur (11) proposed a two-factor hypothesis of injury: at relatively fast cooling rates injury is caused by intracellular ice formation, whereas at slow cooling rates injury is caused by prolonged exposure to "solution effects" resulting from the concentration of the extracellular solution or cell dehydration. However, Steponkus (10) states that there is no evidence that intracellular ice causes injury. An alternative view is that ice forms as a consequence of injury to the plasma membrane. Many hypotheses have been put forward for freezing injury caused by cellular dehydration and categorized under "solution effects" (11–14). These include volumetric and area contraction, concentration of inter- and extracellular solutes, possible pH changes because of different solubilities of buffering compounds, crystallization and possible removal of water of hydration from macromolecules (15–17).

Cells shrink during cooling and are said to have a minimum critical volume. However, Steponkus (10) observed that this hypothesis is not applicable to isolated protoplasts. The demise of many hypotheses based on a single stress occurs when the proposed mechanism does not apply under all conceivable conditions. Given that freezing results in a multitude of stresses, it is more reasonable to suppose that injury is the result of the interaction of several stresses. Cryomicroscopic studies have been used to discover what constitutes cellular injury in plant cell protoplasts during freeze-thaw cycles.

Steponkus (10) carried out extensive studies of rye leaf cell protoplasts during cooling and observed many manifestations of freezing injury.

1. Expansion-induced lysis during warming occurs as dilution of the suspending medium allows the expansion of the protoplasts. The duration of time in the plasmolized state influencess the incidence of injury. Large contractions in cell surface area were found to be irreversible and survival was directly related to the degree of injury incurred. When plasma membranes have to reduce in area, there is a loss of material into the cell interior and numerous vesicles become apparent in the cytoplasm. The reduction in plasma membrane area becomes irreversible when the material necessary for reincorporation cannot match the increasing rate of the process of area increase. However, others state that membrane lesions are the result of other lethal stresses and that this is a pathological symptom (18,19).
2. Loss of osmotic responsiveness during slow cooling and warming is a common occurrence. Electrolyte concentration, changes in pH, removal of water from the membrane due to crystallization, electrical perturbations, and thermotropic phase transitions in membrane lipids destabilize the plasma membrane, causing a loss of osmotic responsiveness. A membrane is a heterogeneous system of lipids and proteins that exist in dynamic equilibrium with each other. Lowering of temperature reduces the fluidity within the membrane; subsequently the phase transition temperature is

reached where the phospholipids change from the liquid to the gel phase. This inevitably affects movement of water and solutes across the membranes (20).

3. Alterations in osmotic behavior during warming can be observed, which is suggestive of a prior transient loss of intracellular solutes or leakage through the plasma membrane. The failure of the cell to attain its original volume on return to ambient temperature occurs at low frequency in both cold-acclimated and nonacclimated rye cell protoplasts and is thought to be a sublethal injury.
4. Intracellular ice formation occurs during rapid cooling ($> -3\,°C\ min^{-1}$). There are two prerequisites for intracellular ice formation: supercooling and seeding of the intracellular solution. It is commonly assumed that intracellular nucleators are not responsible for intracellular ice formation but it is a consequence of seeding by extracellular ice when the plasma membrane is damaged.

Cryoinjury is dependent on the rate of cooling, the cell type (whether it is sensitive or resistant to cold), and the composition of the suspending medium. The types of injury described above for plant cell protoplasts have also been observed in whole cells (17).

Steponkus (10) has reported how plant protoplasts from acclimated and nonacclimated plants differ in their properties. Cold acclimation may be an acquisition of progressive cold hardiness, not by becoming more resistant to one stress but by overcoming each stress barrier in turn. In nonacclimated protoplasts, there are deletions of membrane material to a much greater extent and the tension necessary to effect area increase is considerably smaller than in acclimated protoplasts. In nonacclimated rye cell protoplasts with reductions in membrane area, endocytotic vesiculation of membranes is observed consequent upon deletions of plasma membrane material. This predisposes the protoplast to other events such as intracellular ice nucleation by the extracellular ice. However, in cold-acclimated protoplasts, which are more resistant to cooling, exocytotic extrusions form without membrane breakage.

A second difference is that the freezing point of the cytoplasm within the protoplast of an acclimated plant is considerably lower than that of a nonacclimated plant. In rye the intracellular ice nucleation temperature is $-15\,°C$ for nonacclimated and $-45\,°C$ for acclimated. This was invariably due to a more concentrated cell solution. However, this second difference is less important, as intracellular ice nucleation usually occurs after a breakdown of the membrane. Therefore, it is more likely that cold acclimation is affected by an increase in membrane stability. The plasma membrane stability is increased by the inclusion of unsaturated fatty acids and sterols. In bacteria there is an increase in the proportion of unsaturated phospholipids in the walls of cold-tolerant cells (19). However, only small changes in content have been seen in cold-acclimated protoplasts. Very little information has been available on these mechanisms in fungi but they are considered to be similar to those of plant cells.

III. CRYOPRESERVATION

The knowledge that living cells can withstand freezing has been with us for many years. Morris (21) recalls the work of Henry Power in 1663 when he successfully froze and thawed nematodes. However, it was not until the middle of the 20th century, when liquefied gases became more readily available, that storage at ultralow temperature started to become a more common means of long-term preservation.

Following the work of Hwang (8), little work was done to discover how fungi responded to these techniques. Preservation was initially considered very successful, and it was not until more recent times that there became a need to understand the mechanisms of injury in this group of organisms. It was always considered that they responded like all other cells to the stresses of freezing and thawing. The preparation of the fungus, the cryoprotectant used, the rate of cooling, the storage temperature, and the rate of warming all influence final recovery.

A. Methodology

1. Sample Preparation and Cryoprotectant(s)

The physiological condition of the culture influences its survival rate. More healthy samples can withstand the stresses of cryopreservation better than those grown under limiting conditions. However, growth at lower temperatures, for the organisms that are not cold sensitive, can develop a degree of cold hardiness that can acclimatize them to lower temperature resulting in greater recovery. Providing the fungus with all its nutritional needs and selecting cells from actively growing portions of the colony usually result in highest survival. The vegetative hyphae are more sensitive to treatments than the structures produced for propagation or for withstanding adverse conditions such as overwintering, lack of nutrients, or drought. However, from a culture collection point of view, it is necessary to preserve the strain in a manner that retains all its original properties. Replication errors made in the genome during asexual or sexual reproduction could result in change and the loss of critical characteristics of a production or reference strain. It is therefore essential that techniques that are used allow preservation of the vegetative hyphae.

Despite other mechanisms of natural resistance to freezing, it is the compounds produced within the cell that impart protection to low temperature and freezing that have received most attention and use in the laboratory. The lowering of the freezing point by the production of "antifreeze" compounds such as glycoprotein, which is also seen in Arctic and Antarctic fish, and the formation of water-binding molecules in many microorganisms (22) enable the resistance to dehydration caused not only by the cold, but also by heat and salinity.

Methods of osmoregulation in fungi are important in their resistance to cryoinjury, and the mechanisms have been discussed by several workers (23–29). The ability to withstand exposure to hypertonic solution encountered as external ice nucleates will increase survival. The presence of compounds within the cell can endow tolerance; these substances are sometimes termed osmoregulators or stress chemicals. Several of these are listed in Table 1; this list is not exhaustive but shows that such chemicals are found in a wide range of

Table 1 Compounds Produced by Fungi Under Stress that May Act as Osmoregulators and Provide Some Protection or Resistance to Freezing

Compounds	Species	Reference
Arabitol	*Debaryomyces hansenii*	18
	Dendryphiella salina	18
	Neocosmospora vasinfecta	27
Erythritol	*N. vasinfecta*	27
Glycerol	*Chrysosporium fastidium*	23–25
	Penicillium chrysogenum	25
	Saccharomyces cervisiae	30
	Zygosaccharomyces rouxii	30
	D. hansenii	28
	Pichia miso	28
	N. vasinfecta	27
	Mucor (23 species)	31
Proline	*Phytophthora cinnamomi*	25
	Pythium debaryanum	25
	Saprolegnia sp.	25
	Mucor hiemalis	25
Trehalose	*Dictyostelium discoideum*	32
	Phycomyces blakesleeanus	32
	S. cerevisiae	32, 33
	Torulaspora delbrueckii	34
	Z. rouxii	34

Source: After Jennings (26).

fungi. These compounds are quite often synthesized when the organism is placed under stress. The most common of these are glycerol and trehalose, which are quite often used as chemical additives, cryoprotectants, in cryopreservation. Trehalose has been shown to impart cryotolerance on the baker's yeast *Saccharomyces cerevisiae*, enabling it to withstand temperatures of −20 to −70°C (33).

Meryman and Williams (35) discuss how colligative cryoprotectants reduce the amount of ice formed at any temperature, and this in turn reduces the water loss from cells. They state that properties of a suitable cryoprotectant are that it is nontoxic, it will penetrate the cell membrane, and it must be able to permeate the cell rapidly.

During kinetic freezing, excessive shrinkage caused by dehydration causes cell injury due to membrane stress. In *Phytophthora nicotianae* the diameter of hyphae was reduced up to 60% of that of unfrozen controls at rates of cooling between −0.5 and −100°C min^{-1} (Fig. 1). Shrinkage to this extent killed all hyphae (36; Fig. 2). A similar loss in viability was also seen in *S. cerevisiae* when it was exposed to hypertonic solution without cooling (37). The loss in viability was directly related to the reduction in cell volume (37). When the cell freezes internally, injury could be due to mechanical distortion of intracellular structures or physical rupture of the membrane due to the 10% increase in volume of the water when ice is formed.

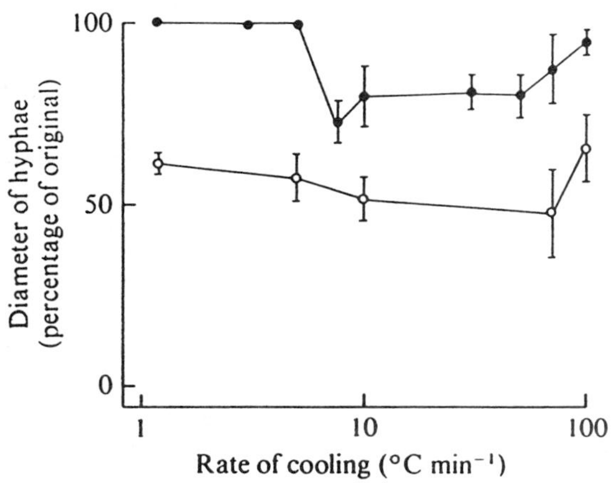

Figure 1 Diameter of hyphae (percentage unfrozen) of *Phytophthora nicotianae* after different rates of cooling to −30°C. Hyphae in either growth medium (o) or glycerol (10% v/v) (●). The bars represent one SD.

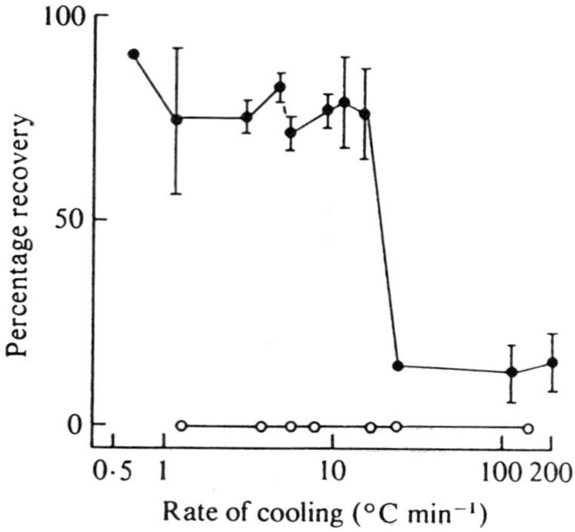

Figure 2 Recovery of *Phytophthora nicotianae* after different rates of cooling to −196 °C. Colonies were frozen in either growth medium (o) or glycerol (10% v/v) (●). The bars represent one SD.

Cryoprotection is achieved by:

1. Noncritical volume loss by the reduction of ice formation
2. An increase in viscosity, which slows down ice crystal growth and formation and solute effects
3. Reduction of the rate of diffusion of water caused by the increase of solute concentrations

Different cryoprotectants give different cooling rate optima. In the case of blood platelets the optimum is $-30\,°C\ min^{-1}$ in glycerol and $-1\,°C\ min^{-1}$ in DMSO. This is also apparent in fungi, where DMSO often allows optimum recovery at slower cooling rates than glycerol. This may be linked to the chemical's ability to penetrate the cell membranes more rapidly.

Extracellular cryoprotectants, for example, polyvinylpyrrolidone (PVP), dextran, and hydroxethyl starch, protect by the rapid increase in viscosity at low temperatures reducing the cooling rate for optimum recovery. PVP can protect the cell by affecting the structure of water at low temperatures. Ice either forms homogeneously where water molecules become orientated, immobilized, and bonded at low subzero temperatures down to −39 °C, or it may form heterogeneously where a solid particle acts as a nucleus (see Section II.A.2). The latter normally occurs at higher subzero temperatures than homogeneous nucleation. PVP may prevent the heterogeneous formation of extracellular ice by preventing particles coming into contact with water.

This will stop dehydration that would normally occur due to the increase in concentration of the extracellular solution when ice forms. The viscous nature of PVP enhances the possibility of homogeneous nucleation of ice which is noncrystalline, amorphous, and is less damaging to the outside of the cell. PVP is not normally found inside the cell and therefore does not affect the structure of ice in the cytoplasm. Consequently, ice crystals may form that put the cell membranes and organelles under osmotic stress and may cause mechanical damage. However, despite homogeneous ice formation outside the cell, on thawing crystalization of extracellular ice occurs as the rapid rate of thawing necessary cannot be achieved in this viscous material. PVP is also thought to increase surface tension sufficiently to prevent protein loss through damaged membranes, though when restored to normal media, if the membrane is not repaired, lysis will occur. Despite these beneficial properties during cooling, PVP does not protect all cells; for example, red blood cells (35) and some fungi (38) fail to recover.

The use of cryoprotective additives in suspending media was introduced by Polge et al. (7), who used glycerol to protect spermatozoa. Glycerol lowers the concentration of salts in equilibrium with ice at any temperature below freezing whether in the medium or when it has penetrated the cell (39). If enough glycerol is present, the salt concentration does not rise to damaging levels until the temperature is so low that the rate of damage is tolerable. This theory of the mechanism of cryoprotection and others have been discussed by Calcott (40). It is evident that different substances protect by different mechanisms. The large-molecular-weight compounds such as PVP are nonpenetrating but are said to change the cell's permeability to ions and allow influx and eflux of sodium and potassium ions during freezing and thawing as well as the alteration to the viscosity of the extracellular solution as discussed above. Thus they avoid excessive osmotic stress. Another theory is that they alter the structure of water both in and around cells so that on freezing an amorphous noncrystalline structure is formed and no crystallization occurs. There are many compounds that impart protection during freezing and thawing, and the mechanisms by which they all protect is not fully understood (40,41). However, they are generally considered to reduce ice crystal size, shift the rate of cooling where intracellular ice forms, or reduce shrinkage effects by penetrating the cell and replacing water. On the other hand, addition of a so-called cryoprotective agent may have adverse effects. Thus Holden and Smith (42) showed that spores of *Puccinia abrupta* var. *partheniicola* gave better recovery in terms of viability and infectivity when frozen and thawed without additives (i.e., dry) than when frozen in the presence of the cryoprotective agents, DMSO, glycerol, proline, trehalose, and PVP. Again, it is apparent that there is a great need for information on the mechanism of low-temperature stress and response of the fungus cell in order that the most suitable protective agent can be applied.

2. Cooling Rate

The early studies of the response of fungi to cooling (8) showed that slow cooling rates gave better results with the few test strains examined. Slow cooling allows the cell to adapt to its changing environment by exosmosis of water to the concentrated environmental solution. This loss of water is dependent on the water permeability of the membrane. The cell will consequently shrink, putting the membrane under stress. Where the rate of cooling does not allow the efflux of water of a sufficient magnitude to prevent supercooling, and when the temperature is low enough or a suitable particle is present to act as a nucleus, intracellular ice nucleation will occur at any temperature below 0 °C down to −30 °C where homogenous nucleation of ice occurs. For over 20 years, it was considered that fungi responded similarly to plant and animal cells. The loss of water was assumed to be beneficial because it reduces the amount of ice within the cell and therefore results in less mechanical damage. Cooling at −1 °C min^{-1} has been the recommended rate for fungi for many years (43). Although many fungi survive, many others were proving difficult to preserve with high viability and in good condition, particularly those of the Oomycota *Phytophthora* and *Pythium*, and the Basidiomycota, for example, *Volvariella volvacea*. Attempts were made to improve results but these empirical experiments rarely involved an investigation of cooling rate effects.

3. Storage Temperature

There is a wide array of equipment providing a variety of temperatures for storage. From the icebox in the refrigerator at ca. −10 °C, through mechanical freezers, −20 to −150 °C, to the temperature of liquefied gases or the vapor above them. To have complete stability, the temperature must be reduced sufficiently not only to minimize metabolism, but also to prevent ice crystal growth, which can in turn cause damage in storage. The limit of ice crystal growth is −139 °C (21), and such a temperature is approaching the limits of mechanical freezers but is surpassed in liquid nitrogen or its vapor.

It has been reported that, if a simple chemical reaction could occur at −196°C, it would be some eight million times slower than that same reaction at 0 °C (21). It is also reported that background irradiation could cause mutations, and at the temperature of −196 °C they would not be repaired and would be expressed on thawing. However, it has been calculated that if background irradiation remains at current levels, it would take 10,000 years before the median lethal dose would be reached (44).

A collaborative project partially sponsored by the Commission of the European Community (CEC) under the Biotechnological Action Programme (BAP) concerned the preservation of fungi and their response to freezing and thawing (45,46). The study aimed to improve methods of cryopreservation and freeze drying and extend the knowledge to a wider taxonomic range of fungi. Seventy-five strains were preserved by a variety of protocols and stored at several different subzero temperatures. Many fungi, particularly those of the Oomycota, showed rapid loss of viability during storage at temperatures between −20 and −40 °C (see Table 2). Many failed to recover after 2 years of storage. Over these 2 years, no significant difference was seen in viability or stability of properties of the same strains stored at −80, −135, and −196 °C. It would be expected that at −80 and −135 °C changes may occur, albeit slowly, and in time these would be manifested in diminished recovery and loss of stability after storage.

B. Survival

A wide range of fungi and fungal structures have been preserved successfully by liquid nitrogen storage (47). Mycelial fragments and hyphal tips of *Neurospora crassa* and *Sordaria fimicola* (48) and many plant pathogens have been frozen and stored successfully in liquid nitrogen. These include *Phytophthora phaseoli* (49), *Sclerospora* species (50), and some rust and smut spores (51,52).

It is the stability conferred by ultralow-temperature storage that makes cryopreservation in liquid nitrogen such a valuable means of long-term preservation. The requirement of a good preservation technique is not only that it results in viability being retained in the long term, but also there must be a retention of high stability of all cells. Fungi produce special structures for reproduction and survival. Replication can result in new strains with different properties; therefore, it is essential that as many cells of all kinds survive in order to retain the full genetic potential of the organism. Many drying and freezing techniques are available, but only carefully optimized and controlled cooling to temperatures below −139 °C can give high recovery of all cell types including the vegetative hyphae of the fungi.

The BAP study showed the value of cryopreservation techniques. Freeze drying and storage at subzero temperatures were investigated and compared (45,46). Of 75 strains of wide taxonomic and economic importance studied, cryopreservation techniques were developed for 73 strains. Although freezing protocols were found for the two remaining strains, *Phytophthora humicola*, −3 °C min^{-1} to give 69% recovery of propagules, and *Volvariella volvacea*, −1 °C min^{-1} to give 38% recovery, they failed to recover following 1 year of storage. Storage temperature was shown to be quite critical. The study compared storage at −20, −40, −80, −135, −196°C in liquid nitrogen and in the vapor above it in the range −90 to −190 °C. Although in the short term little difference was seen in viability or stability at storage temperatures below −80 °C, a fall in viability and morphological changes were noted within

Table 2 Fungi Failing to Survive or Showing Reduced Survival During Storage at −20 and −40°C Suspended in 10% (v/v) Glycerol

Fungi	IMI strain number	Storage at −20°C: % viability immediately after freezing	Storage at −20°C: % viability after storage (years)	Storage at −40°C: % viability immediately after freezing	Storage at −40°C: % viability after storage (years)
Oomycota					
Achlya ambisexualis	143524	89	0 (1)	81	0 (1)
Phytophthora cactorum	021168	0	0	80	0 (1)
P. cambivora	077374	0	0	0	0
P. citricola	021173	0	0	100	0 (1)
P. erythroseptica	034684	0	0	100	0 (1)
P. humicola	302303	0	0	0	0
P. infestans	181530	100	0 (1)	100	0 (1)
P. nicotianae	158733	100	0 (1)	100	0 (1)
Pythium aphanidermatum	104962	0	0	100	0 (1)
Saprolegnia parasitica	169621	14	0	20	0 (1)
deuteromycetes					
Alternaria alternata	089343	100	1 (2)	100	30 (2)
Aureobasidium sp.	291019	95	1 (2)	90	3 (2)
Epicoccum nigrum	068797	79	0 (1)	77	18 (2)
Fusarium avenaceum	309555	75	0 (1)	11	1.2 (2)
F. equiseti	316832	100	0 (1)	100	87 (2)
F. moniliforme	316821	100	0.9 (2)	100	30 (2)
F. oxysporum	313776	100	0 (2)	100	10 (2)
Wallemia sebi	037388	100	20 (2)	100	45 (2)

1 year at -20 and 40 °C, and several strains failed to survive 2 years (Table 2). Storage in the vapor phase of liquid nitrogen can place a tremendous stress on cells. Fluctuation of storage temperature can influence water structure and could induce ice crystal growth, which will cause mechanical damage that could be lethal. This may explain the death of *P. humicola* and *V. volvacea* in storage. At the International Mycological Institute (IMI) vapor-phase storage continues to be used, but to ensure stable low temperature the cultures are stored in metal drawer racks that remain in contact with liquid nitrogen. Optimum cooling protocols and low stable storage temperatures are essential to retain high viability and stability of fungi.

IV. CRYOGENIC LIGHT MICROSCOPY

Observation of cells by transmission electron microscopy and scanning electron microscopy can reveal the ultrastructural changes caused by freeze/thaw cycles by comparing frozen and thawed cells with untreated cells. Such techniques enable close examination of the cell and will reveal more information at the ultrastructural level. However, they will reveal the effect at one point in time, and several treatments will have to be followed to get the complete picture of the events during freezing and thawing. The observations are usually made on dead fixed cells. Cryogenic light microscopy enables the observation of living cells during freezing and thawing. The events can be recorded and quantitative optical techniques can then be applied to analyze the response of the cells. The ability to accurately control the temperature of the sample is an important benefit to cryogenic light microscopy.

A. Methodology

Temperature-controlled cryogenic light microscopy enables the observation of living cells during freezing and thawing. It provides a tool to examine the response of the fungus cells to the stresses imposed by lowering temperatures to below the freezing point of the cell cytoplasm and rewarming to ambient. Such methodology has been available for many years and is improving with time. McGrath (53) describes the significant advances made in low-temperature microscopy.

At IMI studies have been carried out on species of fungi from a wide taxonomic range and from many different environments. During freezing and thawing hyphae have been observed on the CM-3 cryomicroscope system (Planer Biomed). An Amstrad PC1640SD is used for temperature control of the stage heater. Hyphae are cooled from 20 to 5 °C at a rate of -10 °C min^{-1}, held at 5 °C for 0.5 min, and then cooled at different linear rates to temperatures down to -60 °C. Samples are placed on a quartz glass slide with a heater embedded. The slide is in intimate contact with a hollow stage through which cold nitrogen gas generated from a self-pressurized liquid nitrogen vessel is circulated. The stage heater is controlled to allow cooling to temperatures close to the coolant gas temperature.

The conduction stage is mounted on a Zeiss Axioskop H-DIC microscope fitted with Plan-Neofluar 40/0.75 (PH2) objective. Data are recorded on video (Panasonic AG6200 recorder, JVC TK870E camera) using a video character generator (CM3200-00). The recorded data can be played back for analysis on a video monitor (Sony Trinitron KX14CP1 14″).

During slow freezing, extracellular ice can be clearly seen. The formation of ice concentrates the remaining unfrozen solution in which the cell is trapped. A reduction in cell size can be seen and measured. During fast cooling rates, cells do not lose water and intracellular freezing can be visualized by a darkening or granulation of the cell contents. The degree of shrinkage, ice nucleation temperatures, the rate of cooling at which these events occur, and the effect of cryoprotectant additives can be recorded.

B. Observations

Until 1986, few quantitative studies on the biochemistry and biophysics of freezing injury to vegetative fungal cells had been carried out. The response of filamentous fungi to freezing and thawing was first investigated using a cryogenic light microscope in 1986 (54). It was assumed that a fungus would respond in a similar manner to other cell types examined, cellular injury at slow rates of cooling being due to prolonged exposure to hypertonic solutions while fast rates of cooling induce intracellular ice, which results in death (11). Study of *Penicillium expansum* demonstrated shrinkage of hyphae at slow rates of cooling, a response to exposure to concentrated solutions, and intracellular ice formation at fast rates of cooling (54). It also showed that the hyphal septa were not barriers to intracellular ice nucleation. A further study (36) revealed that *Pen. expansum* was extremely resistant to these stresses of freezing and thawing, whereas another fungus *Phytophthora nicotianae*, was very sensitive and failed to recover without cryoprotection. Intracellular ice was not observed in *Phy. nicotianae*, but extensive and rapid shrinkage occurred at all rates of cooling up to $-120\,°C\ min^{-1}$. Indeed, fungi can respond in several different ways to cooling depending on the species, age of culture, and growth conditions (55). In general, fungi can be divided into two groups based on their morphological response to freezing. The first group includes the ascomycete *Sordaria fimicola*, hyphomycetes, for example *Pen. expansum*, and oomycete species including *Pythium aphanidermatum* and *Saprolegnia parasitica*, responding as Mazur (11) predicted, with much shrinkage at low rates of cooling and less shrinkage and the formation of intracellular ice as cooling rate increased. Slow rates of cooling and fast rates vary from fungus to fungus and have no absolute values.

The second group of fungi shrink at all rates of cooling (up to $-100\,°C\ min^{-1}$) and no intracellular ice is observed. Such a response is seen in *Pythium anandrum* (Fig. 3). The responses of fungi cannot be linked taxonomically; some species of oomycetes, for example *Phy. nicotianae*, the hyphomycetes, for example *Aschersonia alleyrodis*, and several basidiomycetes, for example *Lentinus edodes* and *Volvariella volvacea*, respond without intracellular ice nucleation (Table 3).

When the hypha shrinks, it tends to reduce in diameter and not in length. This can be a rapid collapse and involve both the cell wall, the plasmalemma, and cyloplasmic contents, or it may result in plasmolysis. In many cases the cell recovers its original dimensions on thawing. When intracellular ice nucleation is observed, it usually occurs rapidly throughout the entire hyphal length regardless of septa being present. However, in a small number of cases shrinkage can be seen up to a septum with no visible ice, whereas in the remainder of the structure ice forms. This has been seen in *Sap. parasitica* (55). Ice is generally seen as a granulation rather than a blackening of the cell. However, the large hyphae of *Phycomyces blakesleeanus* (Fig. 4) do go black under the light microscope; again, this response has not been observed with many other fungi and is probably a result of the size of the cell rather than its composition. In many cell types freezing is detected by an optical darkening of the cells (53). The formation of large intracellular gas bubbles can accompany thawing following freezing. Following ice formation within the cytoplasm at rates of cooling $> -40\,°C\ min^{-1}$, the saturation point of many solutes and gases is reached and they come out of solution (54). This adds to the optical darkening of the cells. The ice matrix excludes these solutes and gases and will concentrate them in gaps between ice crystals. On thawing the solutes and gases will redissolve. However, small gas bubbles can coalesce to form large bubbles that do not dissolve immediately (Fig. 5). These bubbles have been seen to adhere to the ice interface and move rapidly through the cell, collapsing septa as they do so. Others have been seen to emerge through the hypha wall.

The events of freezing and thawing can be linked to the recovery of the fungus cell. Generally the use of cooling rates that avoid the more dramatic responses results in a higher recovery of cells following freezing and thawing. This will inevitably save time and expense in carrying out extensive viability studies in order to discover the protocol for optimum recovery. A viability study will not always reveal the best technique to follow as fungi can retain viability equally well at many different cooling rates, surviving shrinkage and ice nucleation (55). Ice nucleation can damage the cell and recovery

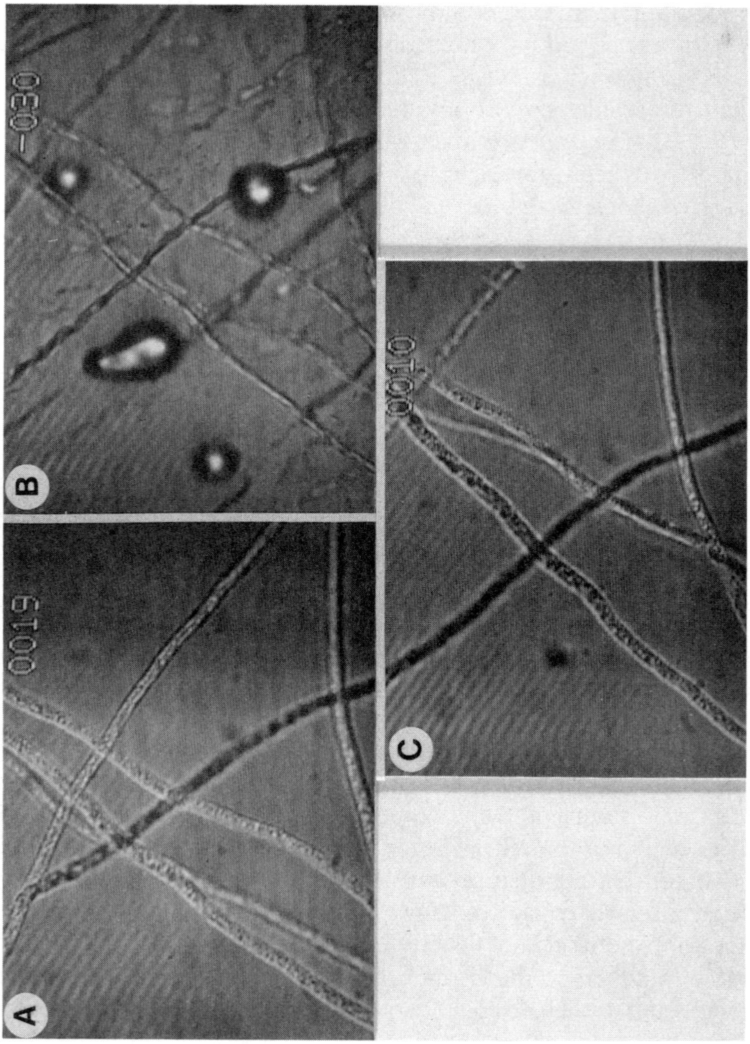

Figure 3 Cryogenic light microscopy of *Pythium anandrum* during cooling at a rate of $-10\,°C\ min^{-1}$. (A) Unfrozen control, (B) at -30, showing shrunken hyphae, (C) intact hyphae after thawing.

Table 3 Quantitative Cryogenic Light Microscopy of Fungal Hyphae

Species	Critical cooling rate (°C min^{-1})[a]	Nucleation temperature (°C)
Oomycota		
Achlya ambisexualis	6	−6 to −18
Phytophthora humicola	53	−8 to −17
P. nicotianae	>120	No ice seen
Pythium aphanidermatum	16	−6 to −14
Saprolegnia parasitica	4	−2 to −7
Zygomycota		
Mortierella elongata[b]	18.5	−9
Mucor racemosus	10	−11 to −14
Ascomycota		
Sordaria fimicola	4	−8.5
Basidiomycota		
Lentinus edodes	>100	No ice seen
Schizophyllum commune	9	−25 to −30
Serpula lacrymans		
(i) Hyphae from 7–21-day cultures	15.5	−10
(ii) Hyphae from 28-day cultures		
(a) <4.45 m in diameter	>100	No ice seen
(b) >4.45 m in diameter	1	−15.5
Sporobolomyces roseus	>100	No ice seen
Volvariella volvacea	>100	No ice seen
deuteromycetes		
Alternaria alternata	12.5	−14.5
Aschersonia alleyrodis	>100	No ice seen
Aureobasidium sp.	>100	No ice seen
Penicillium expansum	18	−14
Trichoderma viride	5	−5 to −11
Trichophyton rubrum	18	−18
Wallemia sebi	9	−18

[a]The rate of cooling at which intracellular ice formed in 50% of the hyphae.
[b]The results presented here are those after growth in liquid medium.
Source: After Morris et al. (55).

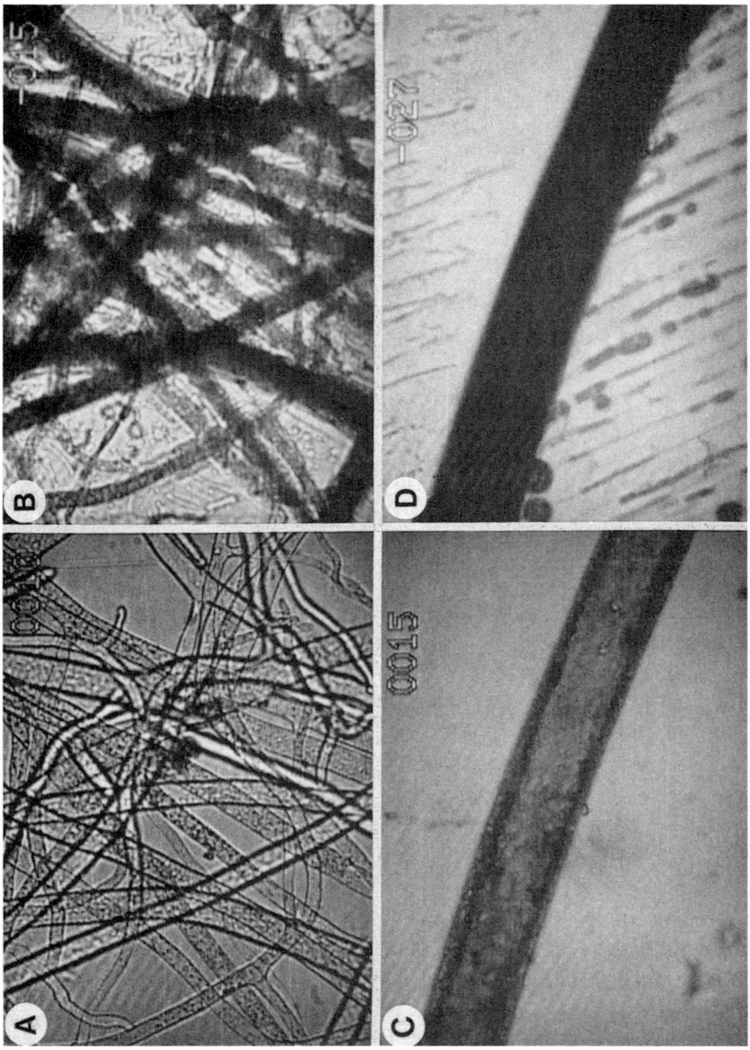

Figure 4 Cryogenic light microscopy of *Aspergillus repens*: (A) unfrozen control, (B) at −30 °C; and *Phycomyces blakesleeanus*: (C) unfrozen control, (D) at −30 °C.

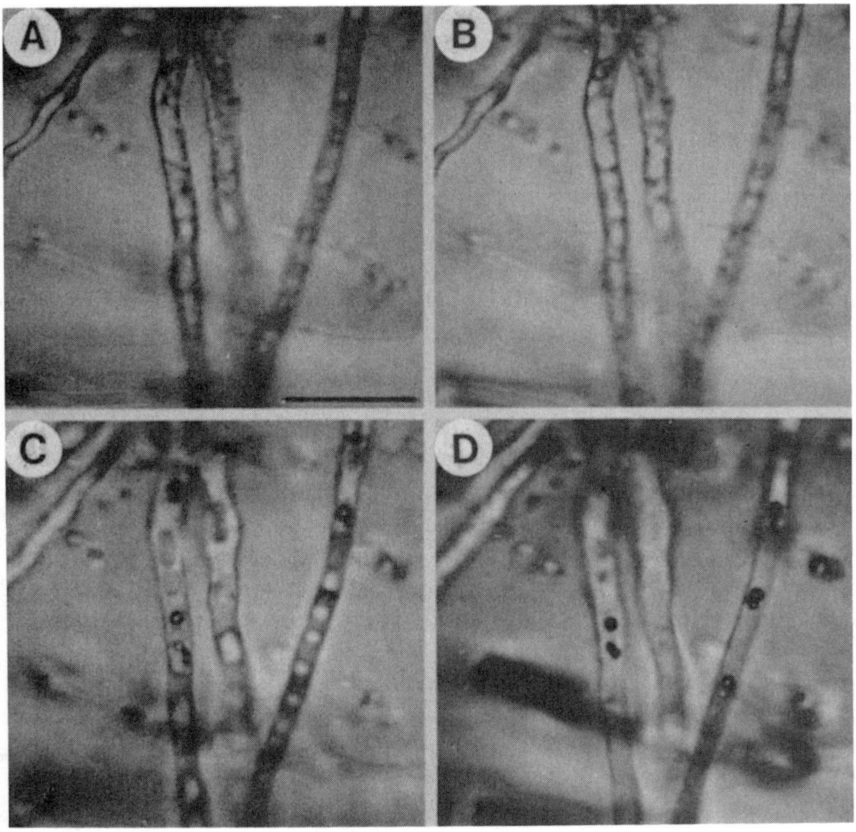

Figure 5 Cryogenic light microscopy of *Penicillium expansum* after cooling at a rate of $-100\,°C\ min^{-1}$ to $-25\,°C$ and during rewarming at a rate of $50\,°C\ min^{-1}$. Showing the formation of large intracellular gas bubbles at (A) −25, (B) −20, (C) −12, (D) −10 °C.

of full metabolism may be delayed (46). This delay in recovery of a particular property may be costly if a production strain fails to exhibit its main attribute. Although cryogenic light microscopy cannot prevent this, it will indicate which cooling rates should be used to minimize the stress of freezing and thawing.

V. TOLERANCE MECHANISMS IN FUNGI

Fungi are capable of withstanding freezing and thawing in nature and when it is applied for preservation purposes. Although data have been gathered on the occurrence and survival of fungi, in particular lichens, the tolerance of fungi is not well understood. Very few direct studies have been carried out with fungi and knowledge has been based on the response of other organisms such as blood or plant cells.

Studies with fungi have revealed similar responses to plant protoplasts although the benefits of cold acclimation are not quite as clear cut. Shrinkage of the hyphae clearly requires the deletion of membrane material, which has been seen in many fungi, for example, *Phy. nicotianae* and *Pen. expansum* (36). Morris et al. (55) reported that *Lentinus edodes* shrank at all rates of cooling investigated. Roquebert (56) examined hyphae of this organism following freezing on a transmission electron microscope. Extracellular membrane-bound vesicles, deletions in the plasma and nuclear membranes, and strands of membrane and cytoplasm between the cell wall and membrane were observed. On thawing the hyphae reexpanded. In most cases the hyphae of fungi regain their original size and the organisms remain viable (36), unlike the response of *L. edodes*, where the cell does not attain its original size (56). Despite this loss of membrane and cellular material, cells recover and eventually grow normally. This ability allows *Lentinus* to survive slow cooling and rapid thawing (55).

The membrane is a critical structure in tolerance to freezing and thawing. It is not elastic and does not fold to any great extent; it simply loses material. Shrinkage of hyphal cells of *Phy. nicotianae* causes its death (36). Intracellular ice is not observed at any rate of cooling up to $-120\,^{\circ}\mathrm{C\ min^{-1}}$ but extensive shrinkage occurs (Fig. 1). This suggests that there is a high permeability to water, which enables a fast response to osmotic stress, but such high permeability is extremely damaging. When colonies of *Phy. nicotianae* are placed on a diffusion stage of a microscope and exposed to hypertonic solution, the shrinken hyphae suffer the loss of plasma membrane function (36). When placed in isotonic solution, the hyphae reexpand. However, a second exposure to hypertonic solution produces no effect on the cell. The hyphae appear intact, there being no leakage of material, but they are nonviable. The addition of cryoprotectant to suspending media reduces the extent of shrinkage in this organism and cells remain viable. The ability of the membrane on expansion to reabsorb material or for the hyphae to function in a reduced volume on thawing conveys resistance to freezing at slow rates.

At fast rates of cooling, the hyphae does not have time to lose water and it freezes internally, normally a lethal event. However, many fungi can survive intracellular ice formation; the hyphomycetes, zygomycetes, the ascomycete *Sordaria fimicola*, and the basidiomycete *Schizophyllum commune* are particularly resistant, recovering in the absence of cryoprotective agents (55). The nature of the ice formed may contribute to resistance. The granular

Table 4 The Optimum Cooling Rate and Recovery of 20 Species of Fungi Suspended in Either Growth Medium or Glycerol (10% v/v)

	Growth medium		Glycerol	
Fungus	Optimum cooling rate (°C min^{-1})	Recovery (%)	Optimum cooling rate (°C min^{-1})	Recovery (%)
Oomycota				
Achlya ambisexualis	0	0	9	88
Phytophthora humicola	0	0	3	69
P. nicotianae	0	0	0.5–11	72–91
Pythium aphanidermatum	0	0	8–29	52–53
Saprolegnia parasitica	0	0	10	32.5
Zygomycota				
Mortierella elongata[a]	10–200	95	1–200	100
Mucor racemosus	25	36	5	31
Ascomycota				
Sordaria fimicola	1–200	ca. 100	1–200	ca. 100
Basidiomycota				
Lentinus edodes	1	27	1–3.5	100
Schizophyllum commune	1	92	23	92
Serpula lacrymans				
(i) Hyphae from 7–21-day cultures	0	0	0.5	99
(ii) Hyphae from 28-day cultures				
(a) <4.45 m in diameter	0	0	13	20
(b) >4.45 m in diameter	0	0	13	20
Sporobolomyces roseus	4.5	ca. 70	1	100
Volvariella volvacea	0	0	1	38
deuteromycetes				
Alternaria alternata	21.5	88	20	97
Aschersonia alleyrodis	10.5	86	1	100
Aureobasidium sp.	10.2	83	200	91
Penicillium expansum	0.5–200	85–100	1–12	75–90
Trichoderma viride	3.8	ca. 100	3.8 to 10	ca. 100
Trichophyton rubrum	4	70.5	9	86
Wallemia sebi	83	ca. 100	77	ca. 100

[a]The results presented here are those after growth in liquid medium.
Source: After Morris et al. (55).

appearance of the fungal cell following ice formation suggests that ice crystals are small, causing less damage intracellularly. Many fungi freeze internally (Table 3) and survive despite this (Table 4).

Fungi withstand freezing and thawing, withstanding cell shrinkage at slow rates of cooling and ice formation at fast rates.

REFERENCES

1. Schmidt-Nielson, S. (1902). Ubereinige psychrophile Mikroorganismen und ihr Vorkommen, *Zentral. Bakteriol. Parasitenk. D. Infektionskr. Hyg. Abt. II 19*: 145.
2. Morris, G. J. and Clarke, A. (1987). Cells at low temperature, in *The Effects of Low Temperatures on Biological Systems* (B. W. W. Grout and G. J. Morris, eds.), Edward Arnold, London, p. 72.
3. Cochrane, V. W. (1958). *Physiology of Fungi*, Wiley, New York.
4. Kappen, L. (1988). Ecophysiological relationships in different climatic regions, in *Handbook of Lichenology*, Vol II (M. Galun, ed.), CRC Press, Boca Raton, FL, p. 37.
5. Mazur, P. (1956). Studies of the effects of subzero temperatures on the viability of spores of *Aspergillus flavus*. I. The effect of rate of warming, *J. Gen. Physiol. 39*: 869.
6. Mazur, P. (1966). Physical and chemical basis of injury in single celled microorganisms subjected to freezing and thawing, in *Cryobiology* (H. T. Meryman, ed.), Academic Press, London and New York, p. 213.
7. Polge, C., Smith, A. U., and Parkes, S. (1949). Revival of spermatozoa after dehydration at low temperatures, *Nature (Lond.) 164*: 666.
8. Hwang, S-W. (1960). Effects of ultralow temperature on the viability of selected fungus strains, *Mycologia 52*: 527.
9. Franks, F. (1981). Biophysics and biochemistry of low temperatures and freezing, in *Effects of Low Temperature of Biological Membranes* (G. J. Morris and A. Clarke, eds.), Academic Press, London, p. 3.
10. Steponkus, P. L. (1984). Role of the plasma membrane in freezing injury and cold acclimation, *Annu. Rev. Plant Physiol. 35*: 543.
11. Mazur, P. (1970. Cryobiology: the freezing of biological systems, *Science 168*: 939.
12. Mazur, P. (1969). Freezing injury in plants, *Annu. Rev. Plant Physiol. 20*:419.
13. Mazur, P. (1977). Slow-freezing injury in mammalian cells, in *The Freezing of Mammalian Embryos* (K. Elliot, and J. Whelan, eds.), Ciba Foundation Symposium 52, p. 19.
14. Meryman, H. T., Williams, R. J., and Douglas, M. St. J. (1977). Freezing injury from "solution effects" and its prevention by natural or artificial cryoprotection, *Cryobiology 14*: 287.
15. Meryman, H. T. (1974). Freezing injury and its prevention in living cells, *Annu. Rev. Biophys. Bioeng. 3*: 341.

16. Steponkus, P. L., Dowgert, R. Y., and Gordon-Damm, W. J. (1983). Destabilization of the plasma membrane of isolated plant protoplasts during a freeze-thaw cycles: the influence of cold acclimation, *Cryobiology 20*: 448.
17. Steponkus, P. L., Evans, R. Y., and Singh, J. (1982). Cryomicroscopy of isolated rye mesophyll cells, *Cryo-Lett. 3*: 101.
18. Grout, B. W. W., and Fuller, B. J. (1982). Hepatocyte plasma membrane response to hypertonic stress and related cell viability, *Cryo-Lett. 3*: 51.
19. McLellan, M. R., Morris, G. J., and Kalinina, L. V. (1982). Membrane evagination from *Amoeba proteus* in response to stress under isotonic conditions, *Cryo-Lett. 3*: 35.
20. Pringle, M. J., and Chapman, D. (1981). Biomembrane structure and effects of temperature, in *Effects of Low Temperature on Biological Membranes* (G. J. Morris and A. Clarke, eds.), Academic Press, London, p. 21.
21. Morris, G. J. (1981). *Cryopreservation: An Introduction to Cryopreservation in Culture Collections*, Culture Centre of Algae and Protozoa, Cumbria, UK.
22. Gould, G. W., and Measures, J. C. (1977). Water relations in single cells, *Philos. Trans. R. Soc. B278*: 151.
23. Luard, E. J. (1982a). Accumulation of intracellular solutes by two filamentous fungi in response to growth at low steady state osmotic potential, *J. Gen. Microbiol. 128*: 2563.
24. Luard, E. J. (1982b). Effect of osmotic shock on some intracellular solutes in two filamentous fungi, *J. Gen. Microbiol. 128*: 2575.
25. Luard, E. J. (1982c). Growth and accumulation of solutes by *Phytophthora cinnamomi* and lower fungi in response to changes in external osmotic potential, *J. Gen. Microbiol. 128*: 2583.
26. Jennings, D. H. (1990). Osmophiles, in *Microbiology of Extreme Environments* (C. Edwards, ed.), Open University Press, Milton Keynes, p. 117.
27. Kelley, D. J. A., and Budd, K. (1991). Polyol metabolism and osmotic adjustment in the mycelial ascomycete *Neocosmospora vasinfecta* (E. F. Smith), *Exp. Mycol. 15*: 55.
28. Jennings, D. H. (1983). Some aspects of the physiology and biochemistry of marine fungi, *Biol. Rev. 58*: 423.
29. Paton, F. M., and Jennings, D. H. (1988). Effect of sodium and potassium chloride and polyols on malate and glucose 6-phosphate dehydrogenases from the marine fungus *Dendryphiella salina*. *Trans. Br. Mycol. Soc. 91*: 205.
30. Brown, A. D. (1978). Compatible solutes and extreme water stress in eukaryotic microorganisms. *Adv. Microb. Physiol. 17*: 181.
31. Takahashi, T., and Asai, T. (1933). Fermentation products of *Mucor*. III. Production of glycerol and the effect of the addition of $NaHSO_3$ and Na_2CO_3, *J. Agric. Chem. Soc. Jpn. 9*: 443.
32. van Laere, A. (1989). Trehalose, reserve and/or stress metabolite?, *FEMS Microbiol. Rev. 63*: 201.
33. Gelinas, P., Fiset, G., Leduy, A., and Goulet, J. (1989). Effect of growth conditions and trehalose content on cryotolerance of bakers' yeast in frozen doughs, *Appl. Environ. Microbiol. 55*: 2453.

34. D'Amore, T., Crumplen, R., and Stewart, G. G. (1991). The involvement of trehalose in yeast stress tolerance, *J. Ind. Microbiol. 7*: 191.
35. Meryman, H. T., and Williams, R. J. (1985). Basic principles of freezing injury to plant cells; natural tolerance and approaches to cryopreservation, in *Cryopreservation of Plant Cells and Organs*, (K. K. Cartha, ed.), CRC Press, Boca Raton, FL, p. 13.
36. Smith, D., Coulson, G. E., and Morris, G. J. (1986). A comparative study of the morphology and viability of hyphae of *Penicillium expansum* and *Phytophthora nicotianae* during freezing and thawing, *J. Gen. Microbiol. 132*: 2013.
37. Morris, G. J., Winters, L., Coulson, G. E., and Clarke, K. J. (1983). Effect of osmotic stress on the ultrastructure and viability of the yeast *Saccharomyces cerevisiae, J. Gen. Microbiol. 129*: 2023.
38. Smith, D. (1983). Cryoprotectants and the cryopreservation of fungi, *Trans. Br. Mycol. Soc. 80*: 360.
39. Nash, T. (1966). Chemical constitution and physical properties of compounds able to protect living cells against damage due to freezing and thawing, in *Cryobiology* (H. T. Meryman, ed.), Academic Press, London, p. 179.
40. Calcott, P. H. (1978). *Freezing and Thawing Microbes*, Meadowfield Press, UK.
41. Grout, B. W. W., and Morris, G. J. (1987). *The Effects of Low Temperatures on Biological Systems*, Edward Arnold, London.
42. Holden, A., and Smith, D. (1992). Effect of cryopreservation methods in liquid nitrogen on viability of *Puccinia abrupta* Diet & Holw var. *partheniicola* urediniospores, *Mycol. Res. 96*: 473.
43. Smith, D., and Onions, A. H. S. (1983). *The Preservation and Maintenance of Living Fungi*, CAB Mycological Institute, Kew, UK.
44. Ashwood-Smith, M. J., and Grant, E. (1977). Genetic stability in cellular systems stored in the frozen state, in *The Freezing of Mammalian Embryos*, Ciba Foundation Symposium 52 (New Series) (K. Elliot and J. Whelan, eds.), Elsevier, Amsterdam, p. 25.
45. Smith, D., Tintigner, N., Hennebert, G. L., de Bievre, C., Roquebert, M. F., and Stalpers, J. A. (1990a). Improvement of preservation techniques for fungi of biotechnological importance, in *Biochemistry R & D in the EC. I. Catalogue of BAP Achievements* (A. Vassarotti and E. Magnien, eds.), Elsevier, Paris, p. 115.
46. Smith, D., Tinignere, N., Hennebert, G. L., de Bievre, C., Roquebert, M. F., and Stalpers, J. A. (1990b). Improvement of preservation techniques for fungi of biotechnological importance, in *Biotechnology in the EC. II. Detailed Final Report of BAP Contractors* (A. Vassarotti and E. Magnien, eds.), Elsevier, Paris, p. 89.
47. Smith, D. (1982). Liquid nitrogen storage of fungi, *Trans. Br. Mycol. Soc. 79*: 415.
48. Wellman, A. M., and Walden, D. B. (1964). Qualitative and quantitative estimates of viability for some fungi after periods of storage in liquid nitrogen, *Can. J. Microbiol. 10*: 585.

49. San-Antonio, J. P., and Blount, V. (1973). Use of liquid nitrogen to preserve downy mildew (*Phytophthora phaseoli*) inoculum, *Plant Dis. Rep. 57*: 724.
50. Gale, A. W., Schmitt, C. G., and Bromfield, K. R. (1975). Cryogenic storage of conidia of *Sclerospora sorghi, Phytopathology 65*: 828.
51. Kilpatrick, R. A., Harmon, D. L., Loegering, W. Q., and Clark, W. A. (1971). Viability of urespores of *Puccinia graminis* f. sp. *tritici* stored in liquid nitrogen, *Plant Dis. Rep. 55*: 871.
52. Prescott, J. M., and Kernkamp, M. F. (1971). Genetic stability of *Puccinia graminis* f. sp. *tritici* in cryogenic storage, *Plant Dis. Rep. 55*: 695.
53. McGrath, J. J. (1987). Temperature controlled cryogenic light microscopy—an introduction to cryomicroscopy, in *The Effects of Low Temperatures in Biological Systems* (B. W. W. Grout and G. J. Morris, eds.), Edward Arnold, London, p. 234.
54. Coulson, G. E., Morris, G. J., and Smith, D. (1986). A cryomicroscopic study of *Penicillium expansum* hyphae during freezing and thawing, *J. Gen. Microbiol. 132*: 183.
55. Morris, G. J., Smith, D., and Coulson, G. E. (1988). A comparative study of the morphology of hyphae during freezing with the viability upon thawing of 20 species of fungi, *J. Gen. Microbiol. 134*: 2897.
56. Roquebert, M. F. (1993). Freezing of *Lentinus edodes, World J. Microbiol. Biotech.* in press.

7

Tolerance of Fungi to Sulfur Dioxide

Naresh Magan

Cranfield Institute of Technology
Cranfield, Bedford, England

I. INTRODUCTION

Sulfur dioxide (SO_2) is one of the oldest food additives known to mankind and was employed by the Romans, ancient Greeks, and Egyptians to preserve wine. Historically, it was also used for centuries as a disinfectant, predominantly by the burning of elemental sulfur and use of the resultant flames. With time, and the development of inorganic chemistry, SO_2 and its salts became commonly used as preservatives, particularly of food and beverages (1).

SO_2 is a colorless gas, with a characteristic pungent odor and smell, which is readily soluble in water. Exposure to SO_2, even at low concentrations (<20 ppm), can cause coughing and both throat and eye irritation. Indeed, in some instances, exposure to SO_2 has resulted in the onset of asthmatic attacks, which have sometimes proved to be fatal. However, although it is a very reactive food additive, because of its versatility, it has been used extensively to prevent food spoilage microorganisms, enzymic and nonenzymic browning, been used as a bleaching agent for flour, silk, and straw, for the stabilization of vitamin C, and to inhibit fungal growth in preserving fruits, fruit juices, and grain (2). However, some yeasts and molds found in food and beverages appear to be very tolerant of SO_2.

The relationship between exposure to SO_2 and activity of fungi has received increased attention in recent years because of the release of SO_2 from anthropogenic sources, particularly the burning of fossil fuels containing elemental sulfur into the atmosphere, and subsequent dry and wet deposition. Although the concentrations involved are much smaller (<100 ppb) than that used in food preservation (100–500 ppm), the relationship between such deposition and microbial activity in terrestrial and aquatic environments has focused much more attention on the effects of SO_2 on fungal populations of both cultivated crops and plants generally (3). This has added to the knowledge of the sensitivity and tolerance of fungi to SO_2.

This chapter will consider the chemistry of SO_2, which underpins any understanding of its effect on fungi (Section II), experimental methods employed (Section III), effects on spore germination (Section IV), yeast and mycelial growth (Section V), and mechanisms of transport and tolerance (Section VI).

II. CHEMICAL CONSIDERATIONS

The structure of SO_2 is $O \leftarrow S = O$ (4) and it dissolves readily in water to produce sulfurous acid, H_2SO_3, and various other ionic forms dependent on the pH of the solution (1,5). The equilibria of SO_2 with water can be expressed as follows:

$$\underset{\text{molecular sulfur dioxide}}{H_2O + SO_2} \rightleftharpoons \underset{\text{undissociated sulfurous acid}}{H_2SO_3} \rightleftharpoons \underset{\text{bisulfite ion}}{HSO_3^-} \rightleftharpoons \underset{\text{sulfite ion}}{SO_3^{2-}} + 2H$$

$$1 \longleftarrow \text{pH} \longrightarrow 7$$

King et al. (5) found that the two dissociation constants for SO_2 were at pH 1.77 and 7.2, when the solution consisted of 50:50 H_2SO_3/HSO_3 and HSO_3/SO_3, respectively. However, Wedzicha (6) more accurately determined the equilibria as 1.86 and 7.18 at 25°C. The relationship, therefore, between the concentration of SO_2 and effects on fungi is influenced by the pH. It is also affected by the ability of some of the SO_2 to become bound to substrates, e.g., glucose (5). Based on these pK_a values, calculations have been made of the proportions of each species present in solutions as a function of pH (Table 1). At the present time, a range of compounds are used to generate SO_2 and related ions when gas is dissolved in water. These include sodium metabisulfite (Na_2SO_5), potassium and sodium salts of sulfite and bisulfite, and potassium metabisulfite. Thus, the type of SO_2 used in experiments, the pH, and also the time of contact all critically determine the effect on growth of fungi.

Table 1 The Percentage Distribution of Molecular Species of Sulfurous Acid as a Function of pH

pH	% H_2SO_3	% HSO_3^-	% SO_3^{2-}
1.5	64.35	35.65	0
1.77	50.00	50.00	0
2.0	37.03	62.97	0
2.5	15.72	84.28	0.002
3.0	5.56	94.43	0.006
3.5	1.83	98.15	0.019
4.0	0.59	99.35	0.063
4.5	0.19	99.61	0.20
5.0	0.058	99.31	0.63
5.5	0.018	98.05	1.93
6.0	0.006	94.15	5.84
6.5	0.0015	83.55	16.45
7.0	0.0002	61.30	38.70

Source: Adapted from Ref. 5.

III. EXPERIMENTAL METHODS

The interpretation of data from the literature has to be assessed in the context of the methods and the form of the SO_2 used. Exposure of fungi to SO_2 has involved (a) gaseous treatments of cultures grown on agar or in liquid media; (b) direct addition of SO_2 in various forms and at different pH values to determine the exposure time required for germination or growth of fungi. Studies have involved either chronic exposure at low concentrations for long periods of time (days/weeks) or acute exposure at very high concentrations of atmospheric SO_2 (minutes or hours, see 7).

Sulfurous or sulfuric acid has sometimes been incorporated into liquid broth cultures to determine effects on yeast growth by measurements of optical density of solutions (5,8). However, studies in liquid cultures are not relevant for simulating potential effects on growth of fungi in solid substrates. Few studies have addressed the problem of binding of the SO_2 in its various forms to agar, which may influence and modify the results obtained (9). Also, few studies have attempted to differentiate between pH and SO_2 effects (5,10) or to examine the effect of interaction with other environmental factors like temperature and water availability, as must be the case when considering spore germination and mycelial growth of fungi.

Table 2 Summary of the Major Studies Carried Out on the Effect of Different Forms of Sulfur Dioxide (SO_2), and Its Concentration, on Spore Germination of Fungi

Fungal species	Concentration (ppm) (form)	Effect	Reference
Alternaria alternata	0.2 (gaseous SO_2)	Germination unaffected, germ tube growth significantly inhibited, 10–20°C	13
Alternaria sp. (unknown)	500 (gaseous SO_2)	Interaction of relative humidity and SO_2; wet spores more sensitive than dry (80% vs. 90% r.h.)	16
Arthrinium sphaerospermum	100 (SO_2 in solution)	Germination inhibited	8
Aspergillus niger	200 (SO_2 in solution)	Germination inhibited	18
Botrytis cinerea	>100 (gaseous SO_2)	Germination inhibited, increased r.h., increased toxicity, toxicity increased 1.5 times for each 10°C rise between 0–30°C	14
	100 (SO_2 in solution)	Germination inhibited	8, 20
Byssochlamys nivea	300 (sodium metabisulfite, pH 3)	Ascospore germination inhibited	19
Cladosporium herbarum	0.2 (gaseous SO_2)	Germination unaffected, germ tube growth significantly decreased	13
	200 (SO_2 in solution)	Germination inhibited	8
Didymella macrospora	150 (SO_2 in solution)	Inhibition after 48 hr exposure	17
Diplocarpon rosae	35 (sulfite solution)	Germination inhibited, malformation of germ tubes	17
	100 (SO_2 in solution)	Germination inhibited	
Epicoccum nigrum	0.2 (gaseous SO_2)	Germ tube extension decreased	13
	200 (SO_2 in solution)	Germination inhibited	8
Fusarium culmorum	0.2 (gaseous SO_2)	Germ tube extension decreased	13
Neurospora crassa	60 (SO_2 in solution)	Inhibition after 10 min exposure	44
Phytophthora infestans	60 (SO_2 in solution)	Zoospores inhibited after 48 hr exposure	17

Table 2 Continued

Fungal species	Concentration (ppm) (form)	Effect	Reference
Aspergillus terreus *Botrytis allii* *Chaetomium sp.* *Colletotrichum lagenerium* *Fusarium oxysporum* *Penicillium egyptiacum* *Rhizopus stolonifer* *Trichoderma viride*	10–100 (gaseous SO_2)	1–6 hr exposure, no effect on germination	15

r.h., relative humidity.

IV. EFFECTS OF SO_2 ON SPORE GERMINATION

Spores of most filamentous fungi appear to be very resistant to direct exposure to gaseous SO_2. The range of studies carried out to determine the efficacy of SO_2 against spores of fungi is summarized in Table 2. It is noticeable that either gaseous SO_2 or SO_2 in solution has been predominantly used in the majority of studies. With some exceptions, many studies failed to accurrately measure or report the pH of the medium used. This is important because of what has already been described in Section II, and because lowered pH alone can significantly decrease germination of a range of deuteromycetes (11). The effect of pH must be taken into account when investigating the dissociation of SO_2 and its effect on germination. For example, Saunders (10) showed that at pH 3.0 and in the presence of SO_2 no germination of *Diplocarpon rosea* occurred, though at pH 2 in the absence of SO_2, some germination still occurred. Indeed, it has to be admitted that solutions of SO_2 prepared in unbuffered water can lose a significant amount by volatilization, which is dependent on temperature, pH, and its actual concentration (12).

Despite these problems, one can highlight some findings obtained when spores of fungi have been exposed to gaseous SO_2. Generally, yeasts such as *Sporobolomyces roseus* and *Cryptococcus* spp. have been found to be very sensitive to low concentrations (0.1–0.2 ppm) of SO_2 (13) whereas spores of filamentous fungi were very tolerant (13–15). Interestingly, Couey and Uota (14) found that the reduction in spore germination of *Botrytis cinerea* was quantitatively proportional to the SO_2 concentration (up to 340 ppm). Germination also decreased linearly with increasing time of exposure (up to 60 min). Interactions between moisture, temperature, and gaseous SO_2 were, however, found to influence the ability of spores to germinate. Moist spores were found to be more sensitive to SO_2 than dry spores. For example, germination of wet *Alternaria* spores of an unknown species decreased by 60% after exposure to 50 ppm SO_2 for 24 min, whereas 100 ppm SO_2 was required for dry spores, even at 98% relative humidity. Between 100 and 400 ppm SO_2 was necessary to impair germination of *Alternaria* spores. With both *Alternaria* and *Bot. cinerea* spores, the SO_2 dose required for a given amount of inhibition was inversely related to relative humidity (14,16). Couey and Uota also found that toxicity of SO_2 increased by about 1.5 times with every 10°C rise in temperature. Experimental exposures to gaseous SO_2 have also indicated that *Dip. rosae* was adversely affected by increasing concentration in an almost linear way, and that the presence of rainfall decreased this toxicity (17). This suggests quite important relationships between form of SO_2, water, and other environmental factors.

Early studies of SO_2 in aqueous solution were mainly carried out in relation to inhibiting yeasts in fruit and vegetable extracts and have been reviewed by Saunders (18). He (17) also showed that spores of a filemantous fungus, *Dip. rosae*, were affected by the form of SO_2 used in experiments. He found that aqueous sulfite solutions of 10–30 ppm for 24 hr stimulated germination, whereas those of 35 ppm caused abnormal or complete inhibition. By contrast, more than 100 ppm SO_2 in solution was required to achieve a similar effect. Nevertheless, there was an increase in abnormal germination with time of exposure in the range 1–9 hr. Ascospores of the thermotolerant spoilage fungus *Byssochlamys nivea* have been found to survive up to 300 ppm SO_2 in fruit juice when treated with sodium metabisulfite at pH 3 (19).

Recently, Kirkwood (8) found that both temperature and concentration affected germination of spores of filamentous fungi. Spore germination at 15 and 25°C and 100 ppm SO_2 in solution were, respectively, 45 and 70% for *Epicoccum nigrum*, 43 and 90% for *Cladosporium herbarum*, and 85 and 3% for *Bot. cinerea.* These studies were carried out at pH 3.5. Previously, 50–100 ppm SO_2 at pH 5.0 was shown to have little effect on these fungi as well as on *Trichoderma viride* (20).

V. EFFECTS ON YEAST AND FILAMENTOUS FUNGAL GROWTH

Generally, in terms of growth, fungi have been found to vary considerably in their tolerance to SO_2. Table 3 summarizes the studies carried out and the effects observed on different fungal species in response to various forms of SO_2. As with studies on spore germination, many studies, particularly those involving exposure of fungi to atmospheric SO_2, have neglected to control pH adequately. Thus, the form of SO_2 and its efficacy cannot easily be distinguished.

Table 3 Summary of the Major Studies in the Laboratory in Which the Effect of Sulfur Dioxide (SO_2) on Yeast and Mycelial Growth Has Been Determined

Fungal species	Concentration (ppm) (form)	Effect on growth	Reference
Hansenula anomala	10–500 (SO_2 in solution)	Inhibited	5
Saccharomyces cerevisiae	1.5 (HSO_3 form)	Interaction of pH and SO_2	5
Aureobasidium pullulans *Sporobolomyces roseus* *Torulopsis* sp.	150 (SO_2 in solu-	Reduced significantly	8
Cryptococcus sp. *Sporobolomyces roseus*	0.2 (gaseous SO_2)	Very sensitive, decreased	13
Botrytis cinerea	1–1000 (gaseous SO_2)	Unaffected	14
Alternaria brassicicola *Aspergillus niger* *Didymella macrosporum*	100–150 (SO_2 in solution, pH 3–5)	Unaffected	17
Alternaria alternata *Aspergillus* spp. *Cladosporium cladosporioides* *Epicoccum nigrum* *Penicillium* spp.	2.66 (gaseous SO_2)	Exposure for 60 min, stimulated	23
Byssochlamys nivea	50–75 (sodium metabisulfite, pH 3.0)	Inhibited	25
Neosartoya fischeri	100 (sodium metabisulfite, pH 3.0)	Inhibited	26
Schirhia acicola	1 (gaseous SO_2)	4 hr exposure, unaffected	43
Rhizoctonia solani	4 (gaseous SO_2)	960 min exposure, inhibited	44

Some important studies have been carried out on yeasts, such as *Saccharomyces cerevisiae* and *Hansenula anomala*, responsible for spoilage of fruit juices and filamentous fungi involved in spoilage of perishables. Detailed experiments with *S. cerevisiae* and *H. anomala* showed that there was a linear relationship between free SO_2 concentration (10–500 ppm) and growth. The effect is, however, pH dependent and profoundly influenced by the form of the SO_2. Undissociated H_2SO_3 concentrations of only about 1.5 ppm are necessary to achieve a similar effect to that with the free SO_2 (5). These studies also demonstrate that duration of contact between yeasts and SO_2, pH, and any binding of SO_2 with the substrate all influence the tolerance of yeasts. This confirmed and supported work carried out previously by Vas and Ingram (21) and Babich and Stotzky (22) that fungi are more sensitive to HSO_3^- than SO_3^{2-}, particularly at pH values less than 4.5.

Recently, the growth of a range of filamentous fungi from cultivated cereals (wheat) including *Alternaria alternata, Cladosporium cladosporioides, Epi. nigrum,* and *Aspergillus* and *Penicillium* spp. were found to be stimulated by a 10-min exposure to SO_2 (only 2.66 ppm) on PDA plates, while after 60 min they were found to be significantly inhibited (23). Although the inhibitory effect of atmospheric SO_2 was correlated with exposure times, mycelial growth was only measured for 24 hr. Furthermore, no attempt was made to buffer the pH of the medium, making it very difficult to draw any useful conclusions from this type of study. Both accurate pH control and longer periods of exposure and measurement are necessary to obtain useful data. Studies with *Aspergillus* spp. isolated from grain on malt extract agar (MEA) supported the fact that a number are very tolerant of SO_2, growing in media containing up to 100 ppm SO_2 (24). Kirkwood (8), in detailed studies of phyllosphere fungi of conifer needles, similarly found that a number of filamentous fungi grew with up to 100 ppm SO_2 incorporated into 1% MEA at 25°C. However, this changed with both temperature and pH.

A number of excellent studies have been carried out on thermotolerant filamentous fungi found to contaminate fruit juices and other food products (25,26). In these studies, pH was effectively controlled at pH 2.5–3.5 so that the form of the SO_2 is known. For example, *Neosartorya fischeri* was found to grow with 100 ppm SO_2 and also produce toxic secondary metabolites (mycotoxins) such as fumitremorgin A and C and verruculogen (26). *Byssochlamys nivea* grew in the presence of 50–75 ppm SO_2 in fruit juices, the extent depending on temperature (25,27). This tolerance was accompanied by the production of patulin with up to 50 ppm SO_2, particularly at 21 and 30°C.

Few studies have considered the influence that interactions between SO_2 concentration and other environmental factors might have on growth. Kirkwood (8) examined the relationship between SO_2 concentration and water availability (water activity) and found that for some species, particularly *Aureobasidium pullulans* and *Epi. nigrum*, at 50 ppm SO_2 and a reduced water activity, stimulation of growth occurred. Interestingly, laboratory studies at pH 3.5 demonstrated that interactions between a range of conifer phyllosphere fungi were markedly influenced by 25–50 ppm SO_2 incorporated into MEA. Fungi such as *Aur. pullulans* were more dominant than other species in the presence of SO_2. This is supported by open-air fumigation studies where cereals were exposed to different levels of atmospheric

SO_2 (27). This study showed that a significant decrease in yeast populations, particularly *Sporobolomyces roseus* and a *Cryptococcus* sp., on the flag leaves was accompanied by a significant increase in *Aur. pullulans*. Changes in such populations on plant surfaces may also influence fungal pathogen activity in the presence of atmospheric SO_2. For example, Rai and Pathak (28) suggested an inverse relationship between populations of phyllosphere microorganisms and the severity of the pathogen *Alternaria solani* on potato leaves. More recently, Bos (29) studied the relationship between phyllosphere populations of rapeseed and *Alternaria brassicae* and found that less infection occurred in the presence of yeasts when plants were exposed to low concentrations of SO_2 (<0.05 ppm), while at elevated pollutant levels (>0.1 ppm) infection was markedly increased. These studies suggest that complex interactions can occur between SO_2 concentration, phyllosphere fungal populations, and fungal pathogen infection of plants.

The sensitivity of phyllosphere fungi, particularly yeasts, to SO_2 has also been used as an indicator of atmospheric pollution. Surveys by Dowding and Carvill (30) and Dowding and Richardson (31) found that the leaf yeast, *Spo. roseus*, was very sensitive to a mixture of atmospheric pollutants, including SO_2, using the spore fall method. This sensitivity of *Spo. roseus* to SO_2 was confirmed in laboratory exposure of *Spo. roseus* and a *Cryptococcus* sp. to atmospheric SO_2, and in controlled open-air fumigation studies of cereals (13,28). However, the activity of these yeasts was predominantly measured by their ability to discharge ballistospores onto the agar medium. Thus, this method is primarily by measuring the effect of SO_2 on sporulation ability, and not growth per se.

Indeed, an area of research that has been seriously neglected is the effect of atmospheric SO_2 and SO_2 in solution on sporulation processes of fungi, which may be particularly important in relation to epidemiology of fungal pathogens.

VI. MECHANISMS OF TRANSPORT OF AND TOLERANCE TO SO_2

Studies of the mechanisms by which fungi are able to tolerate high concentrations of SO_2 have to a large extent been carried out only on yeasts, often food and soft drink spoilage organisms (see 32). Because the relationship between tolerance and SO_2 concentration is so intimately linked with pH, it is critical to be able to differentiate between effects of these factors. However, some studies of SO_2 transport by fungi have failed to specify the pH of experimental conditions, making it difficult to draw conclusions from the data.

Detailed studies by Macris (33) on the spoilage yeast *S. cerevisiae* found that of the three aqueous forms of SO_2, only the molecular SO_2 species was actively transported. Macris and Markakis (34), using radioactive SO_2, suggested that transport by *S. cerevisiae* var. *ellipsoideus* was a carrier-mediated process. This was based on the specificity and kinetics of the uptake and the pH dependence of the process. They suggested that the carrier may be located in the plasma membrane, but no supporting evidence is available for this supposition. Interestingly, they found that uptake was significantly affected by temperature, increasing from 20–40°C to a maximum at 50°C. However, more recent work by Stratford and Rose (35) has shown that in *S. cerevisiae* SO_2 transport is by free diffusion, raising the possibility that sulfite resistance in this yeast may be related to lipid composition of the plasma membrane. The lack of involvement of proteins in active transport was supported by these findings; (a) the near vertical Woolf-Hofstee plots for transport of SO_2 at pH 3.0 and 4.0, (b) the inability of inhibitors such as iodoacetamide and *p*-chlormercuric benzoate to affect the initial velocities of transport, and (c) excluding glucose from the reaction mixture had no effect on the initial velocity of accumulation. In experiments carried out with *Hansenula anomala* and *S. cerevisiae* (5), it was found that SO_2-binding substances, mostly acetaldehyde, enable the yeasts to grow with much higher concentrations of SO_2. However, contact time and SO_2 concentration both played a role in determining the ability to tolerate SO_2. Some resistance is attributable to production of acetaldehyde, as well as other sulfite-binding compounds such as 2-oxoglutarate and pyruvate (36–38). The role of acetaldehyde is supported by studies on both *S. cerevisiae* and *Saccharromycodes ludwigii* in liquid culture, which demonstrated that as sulfite declined there was a concomitant increase in the acetaldehyde excreted into the medium (39).

It is possible that the mechanism of transport for filamentous fungi may be different from that of yeasts. Tweedie and Segel (40), working with wild types and sulfate permease mutants of *Penicillium chrysogenum, Penicillium notatum,* and *Aspergillus nidulans*, suggested that for these fungi there is a common active transport system for inorganic SO_4, S_2O_3 (and SeO_4 and MnO_4). In addition, the mycelia appeared to possess distinct permeases for sulfite and tetrathionate. Benitez et al. (41) suggested that uptake in *Candida utilis* may be under endogenous control, with sulfate uptake being competitively inhibited if either sulfite, thiosulfate, selenate, chromate, or molybdate was present. Alonso et al. (42) showed that sulfite was transported into *C. utilis* on the same system as sulfate and thiosulfate, but carried out their experiments at pH 6.1, where most of the SO_2 would be in a form known to be very ineffective against microorganisms (5). Their data also suggested that for *C. utilis*, sulfate transport may be an active, energy-dependent process and that other sulfate analogs could competitively inhibit sulfate incorporation. Evidence for a sulfite-specific permease has been questioned by Rose (32), because the uptake of sulfite by sulfate permease–deficient mycelia could have been equally due to simple diffusion. He also rightly suggests that where HSO_3 exists in solution, a proportion will be present as molecular SO_2 depending on the pH and therefore that molecular SO_2 could just as easily diffuse into filamentous fungi, as occurs in *S. cerevisiae*.

Sensitivity or tolerance of fungi to SO_2 is therefore governed by the SO_2 species, their rate of penetration, and the interactions that can occur after they cross the cell membrane. SO_2, depending on its form, may affect cell functioning and metabolism directly. It may act as an oxidizing agent or be reduced to elemental sulfur; SO_2 itself can be oxidized to produce sulfate; addition compounds can become established with organic substances, particularly aldehydes and ketones; and it can become bound, affecting enzyme and coenzyme activity and forming bisulfite compounds. Reaction with aldehydes and ketones can occur in the following way:

$$R1(R2)CO + HSO_3^- \rightleftharpoons R1(R2)C(OH)SO_3^-$$

The equilibrium position is in favor of the sulfonate, especially in the case of aldehydes. These reactions can affect biochemical functioning by (a) trapping metabolic intermediates that have aldehyde or ketone groups and (b) forming α-hydroxysulfonates, which are inhibitors of enzyme-catalyzed reactions. Inhibition of enzyme activity by sulfite can occur via the following reaction:

$$R\text{-}S\text{-}S\text{-}R + SO_3^{2-} \rightleftharpoons R\text{-}S\text{-}S\text{-}O_3^- + RS^-$$

The sulfite reacts with the disulfide bonds affecting its catalytic ability. Coenzymes and prosthetic groups, particularly those involved in oxidation-reduction reactions, can also be inactivated by sulfite.

In sensitive yeasts such as *Spo. roseus* and *Cryptococcus* sp. (13) it is most probable that the disulfide bonds that stabilize the tertiary structure of enzymes may be broken by the SO_2 and thus inactivate enzyme function. Rose (32) also suggested that reaction of SO_2 with the pyridine residues of nucleic acids could occur by addition reactions, thus interfering with the ability to polymerize new DNA as well as affecting transcription of DNA into RNA. Furthermore, interaction of uracil residues in mRNA molecules would markedly affect translation into polypeptides.

In tolerant yeasts and filamentous fungi, binding or oxidation of SO_2 is probably the primary method of inactivation of the antimicrobial properties of the preservative. SO_2 can act as an oxidizing agent, being then reduced to elemental sulfur. The formation of addition compounds with, for example, acetaldehyde or D-glucose may be important (1). The high affinity for acetaldehyde is an important reason for the ability of yeasts such as *S. cerevisiae* to tolerate high concentrations of SO_2. However, both these addition reactions are influenced by pH. For example, in the D-glucose/sulfite interaction the proportion of the combined SO_2 is at its maximum between pH 2 and 5. Stratford et al. (39) have suggested that in tolerant yeasts, molecular SO_2 enters by diffusion into an environment that is at a higher pH than outside the cell. A proportion of the SO_2 is converted to HSO_3^-, which forms an effective sulfite trap, depending on intracellular pH (pH 6–7). These authors also suggest that intracellular systems for ATP generation of these yeasts may also be particularly tolerant of exogenous molecular SO_2. SO_2 can also become bound, forming bisulfite compounds with D-xylose, L-xylose, D-galacturonic acid, and pyruvic acid. No comparative studies have been carried out on the relationship between pH, SO_2 species, and their rates of penetration. More detailed studies on the metabolism of SO_2 by filamentous fungi are also needed, particularly in relation to the mechanisms that enable growth and production of toxic secondary metabolites.

Because of environmental concerns, there is pressure to reduce the actual concentrations of SO_2 and other preservatives used by the food, beverage, and agricultural sectors. However, the use of suboptimal concentrations of such chemicals can sometimes result in stimulation, instead of inhibition, of fungal growth and possibly also toxin production. It is therefore critical that research on the mechanisms of sensitivity and tolerance of fungi to SO_2 is continued, so that if concentrations are reduced they are still effective. Because of concerns about atmospheric pollutants and their interaction with plants, particularly cultivated crops, one can also expect knowledge on the relationship between SO_2 and plant pathogen activity, including germination, growth, and epidemiology, to continue to expand.

REFERENCES

1. Green, L. F. (1978). Sulphur dioxide and food preservation, *Food Chem. 1*: 103.
2. Hanssen, M., and Marsden, J. (1987). *The New E for Additives*, Thorsens, Wellingborough, Northhants, UK.
3. Magan, N., and McLeod, A. R. (1991). Effects of atmospheric pollutants on phyllosphere microbial communities, in *Microbial Ecology of Leaves* (J. H. Andrews and S. S. Hirano, eds.), Springer-Verlag, New York, pp. 379–400.
4. Sidgwick, N. V. (1952). *The Chemical Elements and Their Compounds*, Oxford University Press, Oxford.
5. King, A. D., Jr., Ponting, J. D., Sanshuck, D. W., Jackson, R., and Mihara, K. (1981). Factors affecting death of yeast by sulphur dioxide, *J. Food Protect. 44*: 92–97.
6. Wedzicha, B. L. C. (1984). *Chemistry of Sulphur Dioxide in Foods*, Elsevier, London and New York.

7. Babich, H., and Stotzky, G. (1982). Gaseous and heavy metal pollutants, in *Experimental Microbial Ecology* (R. G. Burns and J. H. Slater, eds.), Blackwell, Oxford, p. 631.
8. Kirkwood, I. A. (1991). Phyllosphere microflora of conifer needles and the effects of atmospheric sulphur dioxide and ozone, Ph.D. thesis, Cranfield Institute of Technology, Cranfield, Bedford, UK.
9. Ko, W. H., Kleijunas, J. T., and Shimooka, J. T. (1976). Effect of agar on inhibition of spore germination by chemicals, *Phytopathology 66*: 363.
10. Saunders, P. J. W. (1971). Modification of the leaf surface and its environment by pollution, in *Ecology of Leaf Surface Microorganisms* (T. F. Preece and C. H. Dickenson, eds.), Academic Press, London, p. 81.
11. Magan, N., and Lacey, J. (1984). The effect of temperature and pH on the water relations of field and storage fungi, *Trans. Br. Mycol. Soc. 82*: 71.
12. Gilbert, D. L. (1968). Bryophytes as indicators of air pollution in the Tyne Valley, *New Phytol. 67*: 15.
13. Magan, N., and McLeod, A. R. (1986). In vitro growth and germination of phylloplane fungi in atmospheric sulphur dioxide, *Trans. Br. Mycol. Soc. 90*: 571.
14. Cuoey, H. M., and Uota, M. (1961). Effect of concentration, exposure time, temperature and relative humidity on the toxicity of sulphur dioxide to spores of *Botrytis cinerea, Phytophatology 51*: 815.
15. Hibben, C. R. (1966). Sensitivity of fungal spores to sulphur dioxide and ozone, *Phytopathology 56*: 880.
16. Uota, M. (1965). Inhibition of germination of *Alternaria* spores by sulphur dioxide under various moisture conditions, *Phytopathology* 525–527.
17. Saunders, P. J. W. (1966). The toxicity of sulphur dioxide to *Diplocarpon rosea Wolf* causing blackspot of roses, *Ann. Appl. Biol. 58*: 103–114.
18. Saunders, P. J. W. (1970). Air pollution in relation to lichens and fungi, *Lichenologist 4*: 337–349.
19. Beuchat, L. R. (1976). Effectiveness of various food preservatives in controlling the outgrowth of *Byssochlamys nivea* ascospores, *Mycopathologia 59*: 175.
20. Kirkwood, I. A., Magan, N., and McLeod, A. R. (1989). Effect of atmospheric sulphur dioxide on the phylloplane mycoflora of conifer needles, in *Air Pollution and Forest Decline*, Vol. II (J. D. Bucher and I. Bucher-Wallin, eds.), Swiss Federal Institute for Forestry Research, Birmensdorf, Switzerland, p. 457.
21. Vas, K., and Ingram, M. (1949). Preservation of fruit juice with less sulphur dioxide, *Am. J. Enol. Vinicult. 20*: 32.
22. Babich, H., and Stotzky, G. (1978). Influence of pH on inhibition of bacteria, fungi and coliphages by bisulphite and sulphite, *Environ. Res. 15*: 405.
23. Singh, A. K., and Rai, B. (1990). Effect of sulphur dioxide and ammonia on growth behaviour of some phylloplane fungi from wheat in vitro, *Water Soil Air Pollut. 49*: 343.
24. Lacey, J., and Magan, N. (1991). Fungi in cereal grains: their occurrence and water and temperature relations, in *Cereal Grain: Mycotoxins, Fungi and Quality in Drying and Storage* (J. Chelkowski, ed.), Elsevier, New York, p. 77.

25. Roland, J. O., and Beuchat, L. R. (1984). Biomass and patulin production by *Byssochlamys nivea* in apple juice as affected by sorbate, benzoate, sulphur dioxide and temperature, *J. Food. Sci. 49*: 402–406.
26. Nielsen, P. V., Beuchat, L. R., and Frisvad, J. C. (1989). Growth and fumitremorgin production by *Neosartorya fischeri* as affected by food preservatives and organic acids, *J. Appl. Bact. 66*: 197.
27. Magan, N., and McLeod, A. R. (1991). Effect of open-air fumigation with sulphur dioxide on the occurrence of phylloplane fungi on winter barley, *Agric. Ecosystems Environ. 33*: 245.
28. Rai, B., and Pathak, K. K. (1981). Studies on the phylloplane microflora of potato in relation to air pollutants, *Environ. Pollut. 26*: 153–166.
29. Bos, M. (1986). Effect of sulphur dioxide on the antagonism between saprotropic phyllosphere organisms and the pathogen *Alternaria brassicae* on agar plates and summer rape, M.Sc. thesis, Institute for Plant Pathology Research, Wageningen, Holland.
30. Dowding, P., and Carvill, P. H. (1980). A reduction of counts of *Sporobolomyces roseus* Kluyve on ash leaves in Dublin City, *Irish J. Environ. Sci. 1*: 65.
31. Dowding, P., and Richardson, M. (1990). Leaf yeasts as indicators of air quality in Europe, *Environ. Pollut. 66*: 223.
32. Rose, A. H. (1989). Transport and metabolism of sulphur dioxide in yeasts and filamentous fungi, in *Nitrogen, Phosphorous and Sulphur Utilization by Fungi* (L. Boddy, R. Marchant, and D. J. Read, eds.), Cambridge University Press, Cambridge, Chapter 3, p. 59.
33. Macris, B. J. (1972). Transport and toxicity of sulphur dioxide in the yeast *Saccharomyces cerevisiae* var ellipsoideus, Ph.D. thesis, Michigan State University.
34. Macris, B. J., and Markakis, P. (1974). Transport and toxicity of sulphur dioxide in *Saccharomyces cerevisiae* var *ellipsoideus, J. Sci. Food. Agric. 25*: 21.
35. Stratford, M., and Rose, A. H. (1986). Transport of sulphur dioxide by *Saccharomyces cerevisiae, J. Gen. Microbiol. 132*: 1.
36. Rankine, B. C. (1968). Formation of α-ketoglutaric acid by wine yeasts and its oenocological significance, *J. Sci. Food. Agric. 19*: 624.
37. Rankine, B. C., and Pockock, K. F. (1969). Influence of yeast strain on binding of sulphur dioxide in wines and on its formation during fermentation, *J. Sci. Food. Agric. 20*: 104.
38. Weeks, C. (1969). Production of sulphur dioxide binding compounds and of sulphur dioxide by two *Sacchoromyces* yeasts, *Am. J. Enol. Vinicult. 20*: 32.
39. Stratford, M., Morgan, P., and Rose, A. H. (1987). Sulphur dioxide resistance in *Saccharomyces cerevisiae* and *Saccharomycodes ludwigii, J. Gen. Microbiol. 133*: 2173.
40. Tweedie, J. W., and Segel, I. W. (1970). Specificity of transport processes for sulphur, selenium and molybdenum anions by filamentous fungi, *Biochim. Biophys. Acta 196*: 95.
41. Benitez, J., Alonso, A., Delgado, J., and Kotyk, A. (1983). Sulphate transport in *Candida utilis, Folia Microbiol. 28*: 6.
42. Alonso, A., Benitez, J., and Diaz, M. A. (1984). A sulphate, sulphite and thiosulphate incorporating system in *Candida utilis, Folia Microbiol. 29*: 8.

43. Ham, D. L. (1971). The biological interactions of sulphur dioxide and *Scirrhia acicola* on loblolly pine, Ph.D. thesis, Duke University, Durham, NC.
44. McCallan, S. E. A., and Weedon, F. R. (1940). Toxicity of ammonia, chlorine, hydrogen cyanide, hydrogen sulphide and sulphur dioxide gases. II. Fungi and bacteria, *Contr. Boyce Thomp. Inst. Plant Res. 11*: 331.

8

Tolerance of Fungi to Ethanol

Prashant Mishra

Indian Institute of Technology
New Delhi, India

I. INTRODUCTION

Ethanol is produced as a result of microbial fermentation of carbohydrates and, once accumulated in growth media, acts as a chemical stress. Thus, a number of ethanologenic microorganisms, including bacteria and fungi, have to face this lethal environmental challenge. Among fungi, yeasts have traditionally been employed for ethanol production. Hence most of the studies on ethanol stress are concentrated on yeasts, especially *Saccharomyces* species. The deleterious effects of ethanol on fungal cells and strategies developed by cells to deal with this stress will be the main focus of this chapter. Ethanol, as a major substitute for petroleum, has recently attracted considerable interest of many biotechnologists, and hence implications of ethanol tolerance in increasing productivity of ethanol are emerging. Since lipids are the main targets of ethanol action, their role in mediating tolerance will be included in this discussion. In addition, ethanol acts as a chemical stress affecting fungal growth, survival, and morphological transitions. These changes will be discussed in detail.

II. ETHANOL AS A CHEMICAL STRESS

Ethanol at higher concentrations acts as a chemical stress to microbial cells analogous to extremes of pH or temperature (1). Generally, microorganisms growing under extreme environmental conditions modify their membrane composition and structure as well as enzymatic profile. Fungal cells, which are ubiquitous in their occurrence, also have the ability to bring about these adaptive changes when challenged with extremes of chemical environment. Selection under stressed conditions has brought about adaptive changes in the enzymatic profile and membrane of the organism. Adaptation to ethanol involves mainly changes in properties of the plasma membrane (2–6). Most of the cells that can produce ethanol are subject to high concentrations of ethanol. Nevertheless, the ability of organisms to withstand ethanol stress is affected both by their physiological state and by their ability to induce protective responses. It is therefore not surprising to note that various microorganisms differ in their ability to tolerate ethanol present in their surrounding media (1,7). Among fungi, yeasts include those species that are most tolerant to ethanol. However among yeasts, the ability to tolerate ethanol is species specific; for example, sake yeasts, which produce wine, are the most tolerant (20% v/v) and brewers' yeasts, employed in beer production, are the least tolerant (4–5% v/v) to ethanol (8). Other fungi, such as *Candida albicans* and *Candida tropicalis*, have also been studied with regard to their tolerance to ethanol (9–11). Although these fungi are less tolerant than *Saccharomyces* species to ethanol, they are interesting because they exhibit changes in their morphology once exposed to ethanol. *C. albicans* and *Neurospora crassa* also synthesize specific proteins similar to heat shock proteins under this stress (9,12) (discussed in Section IV.B).

Ethanol causes inhibition of fermentation at higher concentrations even when the substrate concentration is not limiting (7). Such inhibition of fermentation by ethanol has been widely debated and is thought to be primarily due to cytotoxic effects of ethanol on the plasma membrane (1,7,8). The other possible effect of ethanol is inhibition of metabolic enzymes, which is unlikely as most of the glycolytic enzymes are relatively resistant to ethanol produced during the process of fermentation (8). Ethanol, being a small molecule and soluble in lipids, diffuses very rapidly across the biological membrane in response to a concentration gradient equilibrating the intra- and extracellular ethanol (13), thereby eliminating the possibility of accumulation of ethanol intracellularly. Since ethanol is amphipathic, the lipids of plasma membrane are the possible site of its interaction. Further, the inhibitory potency of alcohols can be directly correlated with their lipid solubility, implying a hydrophobic site of membrane as the primary target of ethanol interaction. Table 1 summarizes some of the effects of ethanol on fungal cell physiology. The following section will enumerate inhibitory effects of ethanol on fungal cells.

Table 1 Effect of Ethanol on Cellular Physiology of Fungi

Fungi	Effect on cellular physiology	Ref.
1. *Saccharomyces cerevisiae*	Inhibits glucose, maltose, ammonium, and amino acid uptake	30–34
	Enhances passive influx of protons and decreases glucose induced proton efflux	40–42
	Decreases intracellular pH	41,50
	Causes transient hyperpolarization of plasma membrane and transient efflux of potassium ions	92
	Inhibits plasma membrane ATPase activity	43
	Increases membrane fluidity	35,36
	Alters fatty acid and sterol composition	5,13
	Causes leakage of amino acids, nucleosides, and nucleotides	93
	Leads to accumulation to cytochrome P-450	74–76
	Induces stress proteins	58
2. *S. carlsbergensis*	Increases anion and proton permeability of plasma membrane	47
3. *Schizosaccharomyces pombe*	Alters phospholipid and fatty acid composition	54
4. *Candida albicans*	Induces stress proteins and morphological transitions	9,23
5. *C. tropicalis*	Induces morphological transitions	10,11
6. *Metschnikowia reukaufii*	Activates electron transfer from cytosolic NADH to oxygen	94
7. *Neurospora crassa*	Induces stress proteins	12

III. EFFECT OF ETHANOL ON FUNGI

A. Growth and Survival

Ethanol is a well-known antimicrobial agent and has been employed as a disinfectant for many years. However, various fungi differ in their ability to grow in the presence of ethanol. In general, fungi known to produce ethanol, namely yeasts, include species that are comparatively resistant to ethanol and can grow at concentrations up to 12% (v/v) and survive at concentrations up to 15% (v/v), depending on the species employed and culture conditions. Although the kinetics of growth of yeasts in the presence of extracellular ethanol has been studied by many workers, it still remains controversial (14–17). A linear relationship between growth rate and concentration of extracellular ethanol has been suggested by Ghosh and Tyagi (15), while Holzberg et al. (16) have suggested a threshold concentration of ethanol (about 0.6 M) below which there is no growth inhibition. Ethanol-induced inhibition above this concentration follows a linear pattern. However, Toda et al. (17) suggested that there is a critical concentration of ethanol above which yeast cells are unable to grow. This growth inhibitory concentration varies among yeasts, ranging from 45 g/L for *Kluyveromyces fragilis*, to 95 g/L for *Saccharomyces carlsbergensis*, to 125 g/L for *Saccharomyces cerevisiae*. Some yeasts, e.g., *C. albicans*, are sensitive to ethanol and can grow only up to 3% (v/v) concentration (9). Thus the growth of yeasts in the presence of ethanol has been taken as a criterion for determining ethanol tolerance in yeasts (18,19). The concentration at which growth of a particular yeast is just prevented is taken as a measure of ethanol tolerance in yeasts (20,21).

The viability of fungal cells in the presence of ethanol is another characteristic for determining ethanol tolerance. This method is mainly employed to evaluate nutritional and environmental effects on ethanol tolerance (19). Based on the study of cells of *S. cerevisiae* enriched with various fatty acids or sterols, Rose and his co-workers (22) have suggested that cell survival in the presence of ethanol depends on the nature of fatty acids or sterols of the plasma membrane. *S. cerevisiae* enriched with ergosterol or stigmasterol and linoleyl residues were more resistant to ethanol than those enriched with campesterol or cholesterol and linoleyl residues.

B. Morphogenesis

Some fungi, such as *C. albicans* and *C. tropicalis*, exhibit remarkable variations in their morphological forms when grown in the presence of ethanol (9–11,23). Ethanol promotes germ tube formation in various strains of *C. albicans* at concentrations ranging from 0.015 to 2% that can be generated metabolically by the fungus (9). Maximum germination was reported at ethanol concentrations ranging from 0.25 to 1%. However, no germination occurred at concentrations above 4% or below 0.015%. Surprisingly, in fermentation experiments germ tube formation was not detected, and this discrepancy has been attributed to difference in the effect of ethanol produced in growing cultures and ethanol added to germination media (9). *C. tropicalis* also grows in a filamentous form in a defined medium containing 1.5% ethanol or more (10). Interestingly, ethanol-induced morphological changes in *C. tropicalis* were prevented by addition of myo-inositol (10,11). On the basis of such studies, a role for membrane phosphoinositides in fungal morphogenesis has been suggested (11,24,25). Morphological transition of various fungi has also been reported in the presence of other alcohols such as phenethyl alcohol (7). Phenethyl alcohol has two important effects on microorganisms: inhibition of DNA synthesis (26) and alteration of membrane structure and function (27). Both these effects lead to mitochondrial mutations (28). In contrast to ethanol, which causes germ tube/mycelia formation in *C. albicans* and *C. tropicalis*, phenethyl alcohol blocks dimorphic transition reversibly in various fungi, viz. *C. albicans* (29), *Mucor rouxii*, *Rhizopus arrhizus*, and *Zygorhynchus moelleri* (7).

C. Nutrient Transport

Ethanol primarily inhibits the uptake of various nutrients and therefore inhibits growth and survival. Although both prokaryotic and eukaryotic microorganisms show reduction in their transport activities in the presence of ethanol, this aspect has been most frequently studied in *S. cerevisiae* (7). Studies have shown that ethanol noncompetitively inhibits uptake of glucose (30,31), maltose (32), amino acids (30,33), and ammonium ions (34). Further studies have also confirmed that uptake of amino acids is noncompetitively inhibited by ethanol and such an inhibition is dependent on the plasma membrane lipid composition (35,36). With the exception of glucose transport in *S. cerevisiae* in particular, which is via facilitated diffusion (37), all other transport systems studied in the presence of ethanol are active, depending on the proton motive force (37–39). Thus it is possible that ethanol may affect the formation and maintenance of proton motive force across the plasma membrane. The membrane potential and proton gradients may be influenced either by inhibition of proton extrusion or by enhancement of passive proton influx or by a combination of both. This notion has been strengthened by the observation that ethanol and other alkanols accelerate passive reentry of protons into *S. cerevisiae* (40), thereby uncoupling the secondary transport (41,42).

Ethanol at higher concentrations reverses the glucose-induced acidifying ability of yeast cells by retarding both the initial net rate of proton efflux and the final extent of acidification (3,35,36,40,41). Indeed, transmembrane flow of protons has been found to be very sensitive to ethanol and has been suggested as the primary target of ethanol action (42). However, the other possibility is that ethanol may inhibit secondary transport by inhibiting plasma membrane ATPase activity (43), which is known to act as an electrogenic pump coupling ATP hydrolysis to translocation of protons across the membrane and thereby establishing a transmembrane proton electrochemical gradient (44,45). This transmembrane electrochemical gradient is known to energize active transport of solutes in yeasts and fungi (46). The plasma membrane ATPase activity is indeed inhibited by ethanol (43). However, the fact that inhibition of ATPase is the primary cause of inhibition of nutrient uptake could not be confirmed. Using plasma membrane vesicles of *S. carlsbergensis*, Petrov and Okorokov (47) observed that even at low concentration of ethanol, when ATPase activity is uninhibited, ΔpH and membrane potential decline significantly. In addition, ethanol also increased plasma membrane permeability to protons as well as other ions. Thus, it was suggested that ethanol increases permeability of protons and other anions across the plasma membrane and thereby blocks the secondary transport (47).

The possibility of ethanol acting as a nonspecific inhibitor of protein synthesis in yeasts leading to inhibition of permease activity has been investigated (48,49). It has been suggested that ethanol binds weakly to ribosomes and inhibits protein synthesis in a yeast cell-free system by changing their electric properties reversibly. Thus ethanol acts by reducing protein synthesis-dependent supply of permeases to the plasma membrane. Thus Ferreras et al. (48) have shown that general amino acid permease activity was reversibly deactivated by ethanol in growing cultures of *S. cerevisiae* var. *ellipsoideus*. More recently, observations by the same group have suggested that ethanol inhibits the transport of methylamine and causes a reduction in the number of active carrier molecules located in the plasma membrane (49). Although a direct interaction of ethanol at the level of protein synthesis is an interesting proposition, it needs further careful evaluation before any generalized action of ethanol on nutrient transport can be proposed.

D. Intracellular pH

Ethanol inhibits glucose-induced proton efflux in energized organisms and enhances influx of protons under deenergized conditions, as has been discussed in Section III.C. Thus increased concentrations of ethanol result in a decrease in intracellular pH, which limits the fermentation rate of the organism (41,42). On the other hand, increased production of acetic acid has also been implicated as the primary cause of cytoplasmic acidification in yeast (50,51). Using a respiratory-deficient mutant of *S. cerevisiae*, a correlation between inhibition of fermentation and decrease in internal pH has been observed (50). Ethanol potentiated the effect of acetic acid with respect to inhibition of fermentation as well as internal acidification. Modification of lipid composition also influences proton flux. Both phosphatidylserine (35) and linolenyl residues (36) are known to provide protection against ethanol-induced reversal of proton flux. Since ethanol-induced reversal of proton efflux is a rapid and sensitive process, it has great potential as a means of evaluating organisms with regard to their ethanol tolerance (19).

E. Membrane Structure

Ethanol interacts with plasma membrane and perturbs the organization of membrane lipids by lowering the transition temperature and increasing the membrane fluidity (52). Based on the unsaturation index, Rose and his co-workers have correlated increased fluidity of plasma membrane to ethanol tolerance (13). Later, using a membrane integrated lipophilic fluorescent probe 1,6-diphenylhexatriene (DPH), Mishra and Prasad (35,36) observed that increasing concentrations of ethanol result in decreased fluorescence anisotropy, i.e., increased membrane fluidity. The fluidizing effect was dependent on the type of phospholipid head group or fatty acyl group present in the plasma membrane. The fluidizing effect of ethanol was less when cells were enriched with phosphatidylserine (35) or linolenyl residues (36). Using a similar probe, Walker-Caprioglio et al. (53) observed a thermally induced discontinuity in the anisotropy profile of yeast plasma membrane in the presence of ethanol. However, no such discontinuity was detected in the presence or absence of ethanol once cells were grown in the presence of ethanol. This suggested that cells exposed to ethanol modify their plasma membrane in a way that they become refractory to ethanol-induced changes in physical properties of the membrane (5). A detailed account of how cells adapt to ethanol is dealt with in the following section.

IV. ADAPTATION TO ETHANOL

Once fungi are exposed to ethanol, like other stresses, there are mechanisms that can become functional to combat the deleterious effects of ethanol by various adaptive mechanisms, serving either to metabolize ethanol at a faster rate or allowing survival in the presence of high concentrations of ethanol by altering the physical properties of the plasma membrane. The stressed condition leads to changes in membrane lipid composition and enzymatic activity, and most often there is synthesis of specific proteins called "stress proteins." There may also be morphological changes as part of the adaptation, as discussed in Section III.B. In this section the mechanisms utilized by fungi to overcome ethanol stress and physiological changes thereupon will be discussed.

A. Modification of Lipid Composition

The presence of ethanol is known to alter the plasma membrane lipid composition of various microorganisms (1,7), and the findings for yeasts are in keeping with this (5,13). Changes in the membrane fluidity are achieved by changes in lipid composition of the plasma membrane. Rose and his co-workers have amply demonstrated that increasing concentrations of ethanol ranging from 0.5 to 1.5 M leads to a progressive decrease in proportions of saturated fatty acids (mainly 16:0) and a corresponding increase in unsaturated fatty acids (mainly 18:1). This alteration in fatty acid composition also leads to an increase in membrane fluidity, as indicated by the unsaturation index (13). The plasma membrane of *Schizosaccharomyces pombe* also exhibits ethanol-induced changes in fatty acyl composition (54). However, the effect of ethanol was also dependent on the culture conditions. A decrease in oleic acid content was observed in ethanol-grown aerobic cultures of *Sch. pombe*; on the other hand, ethanol-grown anaerobic cells exhibited an increase in content. Further, phospholipid:protein ratio increased under both anaerobic and aerobic conditions in the presence of ethanol, probably via increased synthesis of phosphatidylinositol. These observations suggested that ethanol tolerance in *Sch. pombe* is associated with a high content of oleic acid and ability to maintain a high rate of phospholipid biosynthesis (54).

Sterols, which are known rigidifiers of plasma membrane, are also susceptible to changes in their cell concentration in the presence of ethanol. Cells of *S. cerevisiae* exposed to ethanol exhibit a significant reduction in their total ergosterol content (5). However, ergosterol as a percentage of total sterols increased in the presence of ethanol owing to decreased levels of other minor sterols such as zymosterols, fecosterols, and lanosterols (5).

B. Induction of Stress Proteins

It is becoming fairly clear that under environmental stress the synthesis of a set of proteins occurs in all organisms from bacteria to humans. These proteins are called stress proteins. Although it might seem obvious that the function of stress proteins is to protect cells from potentially lethal challenges, their exact function is still enigmatic. Recent studies have indicated that all stresses may act through a common trigger, possibly by the intracellular accumulation of aberrant or misfolded proteins (55). A link between stress proteins and 'chaperoned' protein folding has also been suggested (56,57).

Early studies of Plesset and his co-workers suggested that ethanol can induce the same set of proteins as that induced by heat shock in *S. cerevisiae* (58). They also observed a correlation between induction of heat shock proteins (HSPs) and the acquisition of thermotolerance even if the proteins were induced by ethanol. In a subsequent study, it was observed that presence of HSPs is associated with increased viability of *S. cerevisiae* in the presence of ethanol (59). Although functional mitochondria were found to be necessary for greater tolerance to ethanol in *S. cerevisiae* (60), this heat shock acquisition of ethanol and thermotolerance was not inhibited by inhibitors of mitochondrial or cytoplasmic protein syntheses (61). Respiratory-deficient mutants of *S. cerevisiae* also exhibited this response. Thus it was suggested that mitochondrial and cytoplasmic protein syntheses are not required for heat shock acquisition of ethanol and thermotolerance in yeasts (61).

The accumulation of HSPs is enhanced in *N. crassa* in the presence of ethanol and under carbon source starvation (12). About 8–11 HSPs are generally synthesized in *N. crassa*, of which HSP 80 is most abundant and the first to be synthesized in a medium supplemented with ethanol. Stimulation of the synthesis of HSP 80 occurs at normal growth temperature of 28 °C. A new, high-molecular-weight polypeptide tentatively designated alc "80" was synthesized in mycelium stressed concomitantly with ethanol and acetate, lactate, or low sucrose concentration as well as hyperthermia. It is interesting to note that alc "80" was not observed under heat shock alone or heat shock coupled with metal ion or arsenite treatment. The exact functional identity of alc "80" has not been established, but it does not appear to represent alcohol dehydrogenase activity (12).

Induction of stress proteins by ethanol has also been reported in *C. albicans* (9). Short-term exposure of *C. albicans* to 7% ethanol led to synthesis of six ethanol stress proteins (ESPs) of 34, 38, 46, 64, 68, and 84 kDa. Four of these were similar in size to HSPs (46, 64, 68, and 82 kDa) synthesized by cells shifted from 23 to 37 °C. Two ESPs (68, 82 kDa) were highly conserved HSPs seen in many cells (62–65), and two others (34 and 38 kDa) were similar to a minor and a major HSP of *N. crassa* (66). Since, as we have seen, ethanol induces germ tube formation in *C. albicans* (9,23) and *C. tropicalis* (10,11), an interesting query arises as to whether stress proteins induced in response to ethanol lead to morphological changes in these fungi. Differentiation-specific proteins are synthesized in uredospore germlings of the wheat stem rust *Puccinia graminis tritici* under heat stress (67), and in *Achlya klebsiana* nutritional stress proteins are involved in development of asexual sporangia and sporulation (68). However, synthesis of ESPs and their direct involvement in morphogenetic events are not established. One of the previous studies has indicated similarities in cellular inactivation by hyperthermia and ethanol (69).

The question whether synthesis of various stress proteins occurs via a common mechanism was addressed by Iida (70). He observed that heat shock resistance in mutants of *S. cerevisiae*, isolated by their ability to withstand lethal heat treatment, was due to insertion of retrotransposons (Ty) into either 5′ coding or noncoding regions close to the putative insertion codon of adenylate cyclase gene *CYR1* or *CDC35*. These heat shock–resistant mutants had about threefold lower adenylate kinase activity than their wild-type strain. These mutants were also resistant to other stresses such as ultraviolet (UV) light and ethanol (70). Insertion of Ty element into CYR region generates resistance to various stresses, including exposure to 15% ethanol. Further genetic analysis showed that heat shock resistance, UV resistance, and ethanol resistance cosegregated with the Cyr1 T1 mutation. Thus it is possible that under various stresses, cells may respond via similar mechanisms or develop multistress resistance. In this regard it is interesting to note that the plasma membrane ATPase activity exerts a major influence over the ability of a cell to tolerate several types of stress (71). PMA 1.1 mutants of *S. cerevisiae* and *Sch. pombe*, which have reduced ATPase activity, have been shown to exhibit greater tolerance to ethanol than their wild types. These mutants exhibited tolerance to heat and osmotic shock as well (71).

C. Induction of Cytochrome P-450

One of the adaptive mechanisms by which *S. cerevisiae* can adapt to grow under higher concentrations of ethanol is to metabolize ethanol at a faster rate. This mechanism is supported by the observation that during growth under strict fermentation conditions, i.e., at glucose concentrations in the range of 5–20% (w/v), cells of *S. cerevisiae* accumulate large amounts of cytochrome P-450, larger than that apparently required for its endogeneous role in 14 α-demethylation of lanosterols (72,73). A number of other workers have also reported induction of cytochrome P-450 in *S. cerevisiae* in response to ethanol (74–76). Further studies have indicated that in nongrowing yeasts the existing contents of cytochrome P-450 were increased by 50% by glucose over a limited range of concentrations (8–12%) in which ethanol is produced (77). Alkanols, other than ethanol, when added to the culture caused rapid degradation of cytochrome P-450 in nongrowing yeasts, and the rate of loss was directly proportional to the lipid solubility of alkanols. Thus it was proposed that ethanol favors accumulation of cytochrome P-450 and this isoenzyme may play an important role in oxidative removal of ethanol from endoplasmic reticulum of the cell (77). However, this mechanism of adaptation to higher concentrations of ethanol needs further study.

V. EFFECTS OF LIPID SUPPLEMENTATION ON ETHANOL TOLERANCE

As the prime target of ethanol action, lipids have attracted considerable interest (7). Hence modulatory effects of lipids on ethanol tolerance have been studied in some detail (Table 2). Early studies of Hayashida and his co-workers have suggested that a factor present in *Aspergillus oryzae* is responsible for resistance of cells of *S. sake* to ethanol (78,79). Later this factor was designated as proteolipids (80) and further lipid analysis revealed that its phosphatidylcholine fraction is mainly involved in conferring ethanol tolerance to cells of *S. sake* (81,82). Other workers have also suggested the modulatory role of lipids in ethanol tolerance. Jin et al. observed an increased ethanol

Table 2 Lipids as Modulators of Ethanol Tolerance

Lipid	Yeast strain	Ethanol tolerance	Ref.
Proteolipids (*Aspergillus oryzae*)	*Saccharomyces sake*	Increased growth, fermentative activity, and endurability	78,80 81
Crude egg yolk PC	*S. sake*	Increased growth and endurability	82
Purified PC	*S. sake*	Increased growth and fermentative activity	82
Protein-phospholipid complex	*S. cerevisiae*	Increased ethanol productivity	83
PC-albumin	*S. sake*	Increased ethanol production	78
PC, palmitic acid, and cholesterol	*S. cerevisiae*	Increased growth	95
PS	*S. cerevisiae*	Increased alanine uptake, H^+ efflux, and fermentative activity	35
Ergosterol	Grape must	Increased growth and viability but restricted fermentative activity	85
Ergosteryloleate or ergosterol + oleic acid	*S. sake*	Increased growth and endurability but no enhancement in fermentative activity	82
Ergosteryloleate + egg yolk PC	*S. sake*	Increased growth, endurability, and fermentative activity	82
Ergosterol + oleic and linoleic acid	*Kluyveromyces fragilis*	Increased growth rate, biomass production	96
Ergosterol or oleic acid	*S. sake*	No effect on alcohol endurability or fermentative activity	82
Yeast hull (mixture of sterols + UFAs)	Wine yeast	Enhanced growth rate and fermentative activity	97
Tween 80, ergosterol + albumin	*S. sake*	Increased fermentative activity and reduced fermentation time	98
Ergosterol or campesterol + linoleic acid	*S. cerevisiae*	Increased viability and nutrient uptake	22
Linoleic acid	*S. cerevisiae*	Increased viability and nutrient uptake	30

Table 2 Continued

Lipid	Yeast strain	Ethanol tolerance	Ref.
Linoleic acid or Tween 80	*S. uvarum*	Increased ethanol production	99
Oleic acid or linoleic acid or linolenic acid	*S. cerevisiae*	Sequential increase in alanine uptake, H^+ efflux, and fermentative activity	36

PC, phosphatidylcholine; PS, phosphatidylserine; UFAs, unsaturated fatty acids.
Source: Adapted from Ref. 19.

productivity after addition of protein-phospholipid complex to fermentor cultures (83). Anaerobically grown cells of *S. cerevisiae* supplemented with different sterols/fatty acids differ in their resistance to ethanol. Cells enriched with polyunsaturated fatty acids or sterols containing unsaturated alkyl chain acquired greater resistance to ethanol than those enriched with monounsaturated fatty acids or sterols with saturated alkyl chain (22,30). Cells grown under aerobic conditions and enriched with different fatty acyl residues or phospholipid head groups also exhibited increased resistance to ethanol when the plasma membrane contained linolenyl residues or phosphatidylserine (35,36).

In addition to phospholipids and fatty acyl residues, sterols also play a major role in the resistance of the cell to ethanol (19). Ergosterol generally acts as a rigidifier of plasma membrane, as evidenced by the increased osmotic stability of spheroplasts of cells of *S. cerevisiae* enriched with different sterols (84). This prompted Thomas et al. (22) to study the role of sterols in modulating ethanol tolerance. Enrichment of cells with either cholesterol, ergosterol, or campesterol has specific effects on ethanol tolerance. Cells enriched with a Δ^{22} unsaturated alkyl chain (i.e., ergosterol and stigmasterol) rather than a saturated chain at C-17 (i.e., cholesterol and campesterol) exhibited increased tolerance to ethanol with respect to their viability. This improved ethanol tolerance of cells enriched with sterols containing unsaturated alkyl chain is thought to underly the increased "barrier forming" ability against the entry of ethanol into cells (22). The role of ergosterol in modifying ethanol tolerance of cells has been emphasized by Hayashida and Ohta (82). They observed that ergosterol oleate slightly modified the growth in the presence of ethanol but had no effect on fermentative activity of cells. However, their longevity was markedly enhanced. Thus it was suggested that sterols play a significant role in promoting longevity of cells by providing rigidity to the membrane. This observation was supported by similar findings where a decrease in ergosterol content of membrane led to a decrease in viability of cells in the presence of ethanol (85,86).

VI. GENETIC BASIS OF ETHANOL TOLERANCE

Although knowledge of ethanol production has evolved over many centuries, the genetic basis of ethanol tolerance is still obscure. For a long time, there had been a general consensus that the maximum level of ethanol produced or the ability of cells to tolerate ethanol is an intrinsic ability, but lately it has been observed that a number of environmental factors also influence ethanol tolerance (for review see 8). Early studies of Ismail and Ali (87) suggested that ethanol tolerance is regulated by a polygenic system. They observed that segregants derived from diploid *S. cerevisiae* differ in their ability to tolerate ethanol. While none of the haploids exceeded the parental level of tolerance, some of the crosses between haploids exceeded parental level of tolerance. One problem is that it is exceedingly cumbersome to isolate ethanol-tolerant mutants by conventional procedures; isolation requires a long-term selection procedure in continuous culture (88). The genetic analysis of ethanol-tolerant wine yeast strains has led to the conclusion that a number of genes are involved in limiting proliferation of cells in the presence of ethanol (87,89,90). Since these genes are different in nonisogenic strains, nonisogenic hybridization (genetic complementation) might generate hybrids of yeasts more tolerant than their parents (87,90). Del Castillo (91) has also developed a method of hybrid selection between homothallic wild-type and heterothallic strains and has shown that many genes control ethanol tolerance.

VII. CONCLUSIONS

From the foregoing discussions it is apparent that most of the studies in relation to ethanol are based on yeasts, specifically *Saccharomyces* species. This has mainly been because *Saccharomyces* has been traditionally employed for ethanol production. A current biotechnological endeavor is production of ethanol by yeasts that are able to metabolize pentose sugars, a major component in the breakdown of lignocellulosic waste (100,101). However, studies have also indicated that exposure to ethanol leads to the synthesis of stress proteins similar to heat shock proteins, and subsequent to synthesis of these proteins fungi acquire the ability to tolerate heat as well as ethanol stress. Synthesis of these proteins under ethanol stress in *C. albicans*, *S. cerevisiae*, and *N. crassa* suggests that it is a general physiological response of fungi under ethanol stress. Further, the changes in morphological form under ethanol stress are a fascinating and intriguing fungal response to chemical stress. However, it does seem clear from a number of studies that the ability of ethanol to disrupt various membrane properties of fungal cells provides the physiological basis of ethanol-induced inhibition of growth. Nevertheless, there is a great need to study ethanol tolerance of various fungi so that the underlying mechanisms of adaptation to ethanol can be clarified further.

ACKNOWLEDGMENTS

I acknowledge Prof. P. Ghosh and Prof. S. Chand, Biochemical Engineering Research Centre, Indian Institute of Technology, Delhi, for providing help and encouragement.

REFERENCES

1. Ingram, L. O. (1986). Microbial tolerance to alcohols: role of the cell membrane, *Trends Biotechnol. 4*: 40.
2. Ingram, L. O. (1976). Adaptation of membrane lipids to alcohols, *J. Bacteriol. 125*: 670.
3. Jimenez, J., and Benitez, T. (1987). Adaptation of yeast cell membranes to ethanol, *Appl. Environ. Microbiol. 53*: 1196.
4. Hoek, J. B., and Taraschi, T. F. (1988). Cellular adaptation to ethanol, *Trends Biochem. 13*: 269.
5. Walker-Caprioglio, H. M., Casey, W. M., and Parks, L. W. (1990). *Saccharomyces cerevisiae* membrane sterol modifications in response to growth in presence of ethanol, *Appl. Environ. Microbiol. 56*: 2853.
6. Jones, R. P., and Greenfield, P. F. (1987). Ethanol and the fluidity of the yeast plasma membrane, *Yeast 3*: 223.
7. Ingram, L. O., and Buttke, T. M. (1984). Effects of alcohols on microorganisms, *Adv. Microb. Physiol. 25*: 253.
8. Casey, G. P., and Ingledew, W. M. (1986). Ethanol tolerance in yeasts, *CRC Crit. Rev. Microbiol. 13*: 219.
9. Zeuthen, M. L., Dabrowa, N., Aniebo, C. M., and Howard, D. H. (1988). Ethanol tolerance and the induction of stress proteins by ethanol in *Candida albicans*, *J. Gen. Microbiol. 134*: 1375.
10. Tani, Y., Yamada, Y., and Kamihara, T. (1979). Morphological changes in *Candida tropicalis* pk 233 caused by ethanol and its prevention by myo-inositol, *Biochem. Biophys. Res. Commun. 91*: 351.
11. Uejima, Y., Koga, T., and Kamihara, T. (1987). Enhanced metabolism of phosphatidylinositol in *Candida tropicalis* in association with filamentous growth caused by ethanol, *FEBS Lett, 214*: 127.
12. Roychowdhury, H. B., and Kapoor, M. (1988). Ethanol and carbon source starvation enhance the accumulation of HSP 80 in *Neurospora crassa*, *Can. J. Microbiol. 34*: 162.
13. Beavan, M. J., Charpentier, C., and Rose, A. H. (1982). Production and tolerance of ethanol in relation to phospholipid fatty-acyl composition in *Saccharomyces cerevisiae* NCYC 431, *J. Gen. Microbiol. 128*: 1447.
14. Pamment, N. B. (1989). Overall kinetics and mathematical modeling of ethanol inhibition in yeasts, in *Alcohol Toxicity in Yeasts and Bacteria* (N. van Uden, ed.), CRC Press, Orlando, FL, p. 1.

15. Ghosh, T. K., and Tyagi, R. D. (1979). Rapid ethanol fermentation of cellulose hydrolysate. 1. Batch versus continuous systems, *Biotechnol. Bioeng. 21*: 1387.
16. Holzberg, I., Finn, R. K., and Steinkraus, K. H. (1967). A kinetic study of the alcoholic fermentation of grape juice, *Biotechnol. Bioeng. 9*: 413.
17. Toda, K., Asakura, T., and Ohtake, H. (1987). Inhibitory effect of ethanol on ethanol fermentation, *J. Gen. Appl. Microbiol. 33*: 421.
18. Rose, A. H. (1987). Responses to the chemical environment, in *The Yeast: Yeast and the Environment*, Vol. 2 (A. H. Rose and J. S. Harrison, eds.), Academic Press, London, p. 5.
19. Mishra, P., and Kaur, S. (1991). Lipids as modulators of ethanol tolerance in yeast, *Appl. Microbiol. Biotechnology 34*: 697.
20. Inoue, T., Takaoka, Y., and Hata, S. (1962). Studies on sake yeasts. V. Conditions which affect the tolerance of yeast to alcohol, *J. Ferment. Technol. 40*: 511.
21. Rose, A. H. (1980). Recent research on industrially important strains of *Saccharomyces cerevisiae*, in *Biology and Activity of Yeast* (F. A. Skinner, S. M. Passmore, and R. R. Davenport, eds.), Academic Press, London, p. 103.
22. Thomas, S. D., Hossack, J. A., and Rose, A. H. (1978). Plasma-membrane lipid composition and ethanol tolerance in *Saccharomyces cerevisiae*, *Arch. Microbiol. 117*: 239.
23. Pollack, J. H., and Hashimoto, T. (1985). Ethanol-induced germ tube formation in *Candida albicans, J. Gen. Microbiol. 131*: 3303.
24. Mishra, P., and Prasad R. (1990). An overview of lipids of *Candida albicans*, *Prog. Lipid Res. 29*: 65.
25. Mishra, P., and Prasad, R. (1991). Lipids of *Candida albicans*, in Candida albicans: *Cellular and Molecular Biology* (R. Prasad, ed.), Springer-Verlag, Berlin, Heidelberg, p. 128.
26. Berrah G., and Konetzka, W. A. (1962). Selective and reversible inhibition of the synthesis of bacterial deoxyribonucleic acid by phenethyl alcohols, *J. Bacteriol. 83*: 738.
27. Lester, G. (1965). Inhibition of growth, synthesis and permeability of *Neurospora crassa* by phenethyl alcohols, *J. Bacteriol. 90*: 29.
28. Kojima, M., Hamada, H., and Morita, T. (1989). Effect of alcohols on mitochondrial functions, in *Alcohol Toxicity in Yeasts and Bacteria* (N. Van Uden, ed.), CRC Press, Orlando, FL, p. 109.
29. Lingappa, B. T., Prasad, M., Lingappa, Y., Hunt, D. F., and Biemann, K. (1969). Phenethyl alcohol and tryptophol: auto antibiotics produced by fungus *Candida albicans*, *Science 163*: 192.
30. Thomas, S. D., and Rose, A. H. (1979). Inhibitory effect of ethanol on growth and solute accumulation by *Saccharomyces cerevisiae* as affected by plasma-membrane lipid composition, *Arch. Microbiol. 122*: 49.
31. Leão, C., and Van Uden, N. (1982). Effects of ethanol and other alkanols on the glucose transport system of *Saccharomyces cerevisiae*, *Biotechnol. Bioeng. 24*: 2601.
32. Loureiro-Dias, M. C., and Peinado, J. M. (1982). Effect of ethanol and other alkanols on the maltose transport system of *Saccharomyces cerevisiae*, *Biotechnol. Lett. 4*: 721.

33. Leão, C., and Van Uden, N. (1984). Effect of ethanol and other alkanols on general amino acid permease of *Saccharomyces cerevisiae*, *Biotechnol. Bioeng.* *26*: 403.
34. Leão, C., and Van Uden, N. (1983). Effect of ethanol and other alkanols on the ammonium transport system of *Saccharomyces cerevisiae*, *Biotechnol. Bioeng.* *25*: 2085.
35. Mishra, P., and Prasad, R. (1988). Role of phospholipid head groups in ethanol tolerance of *Saccharomyces cerevisiae*, *J. Gen. Microbiol.* *134*: 3205.
36. Mishra, P., and Prasad, R. (1989). Relationship between ethanol tolerance and fatty acyl composition of *Saccharomyces cerevisiae*, *Appl. Microbiol. Biotechnol.* *30*: 294.
37. Eddy, A. A. (1982). Mechanisms of solute transport into selected eukaryotic microorganisms, *Adv. Microb. Physiol.* *23*: 1.
38. Borst-Pauwels, G. W. F. H. (1981). Ion transport in yeast, *Biochim. Biophys. Acta* *650*: 88.
39. Serrano, R. (1977). Energy requirements for maltose transport in yeast, *Eur. J. Biochem.* *80*: 97.
40. Leão, C., and Van Uden, N. (1984). Effects of ethanol and other alkanols on passive proton influx in the yeast *Saccharomyces cerevisiae*, *Biochim. Biophys. Acta* *774*: 43.
41. Cartwright, C. P., Juroszek, J-R., Beavan, M. J., Ruby, F. M. S., De Morais, S. F. M., and Rose, A. H. (1986). Ethanol dissipates the proton-motive force across the plasma membrane of *Saccharomyces cerevisiae*, *J. Gen. Microbiol.* *132*: 369.
42. Pascual, C., Alonso, A., Garcia, I., Romay, C., and Kotyk, A. (1988). Effect of ethanol on glucose transport, key glycolytic enzymes, and proton extrusion in *Saccharomyces cerevisiae*, *Biotechnol. Bioeng.* *32*: 374.
43. Cartwright, C. P., Veazey, F. J., and Rose, A. H. (1987). Effect of ethanol on activity of the plasma membrane ATPase in, and accumulation of glycine by, *Saccharomyces cerevisiae*, *J. Gen. Microbiol.* *133*: 857.
44. Malpartida, F., and Serrano, R. (1981). Proton translocation catalyzed by the purified yeast plasma membrane ATPase reconstituted in liposomes, *FEBS Lett.* *131*: 351.
45. Dufour, J-P., Goffeau, A., and Tsong, T. Y. (1982). Active proton uptake in lipid vesicles reconstituted with a purified yeast plasma membrane ATPase, *J. Biol. Chem.* *257*: 9365.
46. Serrano, R. (1984). Plasma membrane ATPase of fungi and plants as a novel type of proton pump, *Curr. Top. Cell Regul.* *23*: 87.
47. Petrov, V. V., and Okorokov, L. A. (1990). Increase of the anion and proton permeability of *Saccharomyces carlsbergensis* plasmalemma by *n*-alcohols as a possible cause of its deenergization, *Yeast* *6*: 311.
48. Ferreras, J. M., Iglesias, R., and Girbes, T. (1989). Effect of the chronic ethanol action on the activity of the general amino acid permease from *Saccharomyces cerevisiae* var. *ellipsoideus*, *Biochim. Biophys. Acta* *979*: 375.
49. Iglesias, R., Ferreras, J. M., Arias, F. J., Munoz, R., and Girbes, T. (1991). Effect of continued exposition to ethanol on the activity of the ammonium and

fructose systems in *Saccharomyces cerevisiae* var *ellipsoideus*, *Biotechnol. Bioeng. 37*: 389.

50. Pampulha, M. E., and Loureiro-Dias, M. C. (1989). Combined effect of acetic acid, pH and ethanol on intracellular pH of fermenting yeast, *Appl. Microbiol. Biotechnol. 31*: 547.
51. Loureiro-Dias, M. C., and Santos, H. (1990). Effects of ethanol on *Saccharomyces cerevisiae* as monitored by in vivo 31-P and 13-C nuclear magnetic resonance, *Arch. Microbiol. 153*: 384.
52. Jain, M. K., and Wu, N. M. (1977). Effect of small molecules on the dipalmitoyllecithin liposomal bilayers: phase transition in lipid bilayer, *J. Memb. Biol. 34*: 157.
53. Walker-Caprioglio, H. M., Rodriguez, R. J., and Parks, L. W. (1985). Recovery of *Saccharomyces cerevisiae* from ethanol induced growth inhibition, *Appl. Environ. Microbiol. 50*: 685.
54. Koukou, A. I., Tsoukatos, D., and Drainas, C. (1990). Effect of ethanol on the phospholipid and fatty acid content of *Schizosaccharomyces pombe* membranes, *J. Gen. Microbiol. 136*: 1271.
55. Pelham, H. R. B. (1987). Heat shock proteins coming in from the cold, *Nature 332*: 776.
56. Hubbard, T. J. P., and Sander, C. (1991). The role of heat shock and chaperone proteins in protein folding: possible molecular mechanisms, *Protein Eng. 4*: 711.
57. Gething, M-J., and Sambrook, J. (1992). Protein folding in the cell, *Nature 355*: 33.
58. Plesset, J., Palm, C., and McLaughlin, C. S. (1982). Induction of heat shock proteins and thermotolerance by ethanol in *Saccharomyces cerevisiae*, *Biochem. Biophys. Res. Commun. 108*: 1340.
59. Watson, K., and Cavicchioli, R. (1983). Acquisition of ethanol tolerance in yeast cells by heat shock, *Biotechnol. Lett. 5*: 683.
60. Aguilera, A., and Benitez, T. (1985). Role of mitochondria in ethanol tolerance of *Saccharomyces cerevisiae*, *Arch. Microbiol. 142*: 389.
61. Watson, K., Dunlop, G., and Cavicchioli, R. (1984). Mitochondrial and cytoplasmic protein syntheses are not required for heat shock acquisition of ethanol and thermotolerance in yeast, *FEBS Lett. 172:* 299.
62. Ashburner, M., and Bonner, J. J. (1979). The induction of gene activity in *Drosophila* by heat shock, *Cell 17*: 241.
63. Ingolia, T. D., Slater, M. R., and Craig, E. A. (1982). *Saccharomyces cerevisiae* contains a complex multigene family related to the major heat-inducible gene of *Drosophila*, *Mol. Cell. Biol. 2*: 1388.
64. Kelly, P. M., and Schlessinger, M. J. (1982). Antibodies to two major heat-shock proteins cross-react with similar proteins in widely divergent species, *Mol. Cell. Biol. 2*: 267.
65. Dabrowa, N., and Howard, D. H. (1984). Heat shock and heat stroke proteins observed during germination of the blastoconidia of *Candida albicans*, *Infect. Immun. 44*: 537.

66. Plesofsky-Vig, N., and Brambl, R. (1985). Heat shock response in *Neurospora crassa*: protein synthesis and induced thermotolerance, *J. Bacteriol. 162*: 1083.
67. Wanner, R., Forster, H., Mendgen, K., and Staples, R. C. (1985). Synthesis of differentiation-specific proteins in germlings of the wheat stem rust fungus after heat shock, *Exp. Mycol. 9*: 279.
68. Lejohn, H. B., and Braithwaite, C. E. (1984). Heat and nutritional shock-induced proteins of the fungus *Achlya* are different and under independent transcriptional control, *Can. J. Biochem. Cell Biol. 62*: 837.
69. Li, G. C., Shiu, E. C., and Hahn, G. M. (1980). Similarities in cellular inactivation by hyperthermia or ethanol, *Rad. Res. 82*: 257.
70. Iida, H. (1988). Multistress resistance of *Saccharomyces cerevisiae* is generated by insertion of retrotransposons (Ty) into the 5′ coding region of adenylate cyclase gene, *Mol. Cell. Biol. 8*: 5555.
71. Panaretou, B., and Piper, P. W. (1990). Plasma-membrane ATPase action affects several stress tolerances of *Saccharomyces cerevisiae* and *Schizosaccharomyces pombe* as well as the extent and duration of the heat shock response, *J. Gen. Microbiol. 136*: 1763.
72. Aoyama, Y., Okikawa, T., and Yoshida, Y. (1981). Evidence for the presence of cytochrome P-450 functional in lanosterol 14 α-demethylation in microsomes of aerobically grown respiring yeast, *Biochim. Biophys. Acta 665*: 596.
73. Aoyama, Y., Yoshida, Y., and Sato, R. (1984). Yeast cytochrome P-450 catalyzing lanosterol 14 α-demethylation II Lanosterol metabolism by purified P-45014DM and by intact microsomes, *J. Biol. Chem. 259*: 1661.
74. Wiseman, A., Lim, T. K., and Woods, L. F. J. (1978). Regulation of the biosynthesis of cytochrome P-450 in brewer's yeast: role of cyclic AMP, *Biochim. Biophys. Acta 544*: 615.
75. Morita, T., and Mifuchi, I. (1984). Ethanol enhancement of cytochrome P-450 in yeast *Saccharomyces cerevisiae* D7, *Chem. Pharmacol. Bull. 32*: 1624.
76. Del Carratore, R., Morganti, C., Galli, A., and Bronzetta, G. (1984). Cytochrome P-450 inducibility by ethanol and 7-ethoxycoumarin O-deethylation in *Saccharomyces cerevisiae*, *Biochem. Biophys. Res. Commun. 123*: 186.
77. Blatiak, A., King, D. J., Wiseman, A., and Winkler, M. A. (1987). Cytochrome P-450 accumulation and loss as controlled by growth phase of *Saccharomyces cerevisiae*: relationship to oxygen, glucose and ethanol concentrations, *J. Gen. Microbiol. 133*: 1053.
78. Hayashida, S., Feng, D. D., and Hongo, M. (1974). Function of the high concentration alcohol-producing factor, *Agric. Biol. Chem. 38*: 2001.
79. Hayashida, S., Feng, D. D., and Hongo, M. (1975). Physiological properties of yeast cells grown in the proteolipid-supplemented media, *Agric. Biol. Chem. 39*: 1025.
80. Hayashida, S., Feng, D. D., Ohta, K., Chaitiumvong, S., and Hongo, M. (1976). Composition and role of *Aspergillus oryzae*-proteolipid as a high concentration alcohol-producing factor, *Agric. Biol. Chem. 40*: 73.
81. Hayashida, S., and Ohta, K. (1978). Cell structure of yeasts grown anaerobically in *Aspergillus oryzae*-proteolipid supplemented media, *Agric. Biol. Chem. 42*: 1139.

82. Hayashida, S., and Ohta, K. (1980). Effects of phosphatidylcholine or ergosteryl oleate on physiological properties of *Saccharomyces sake, Agric. Biol. Chem. 44*: 2561.
83. Jin, C. K., Chiang, H. L., and Wang, S. S. (1981). Steady state analysis of the enhancement in ethanol productivity of a continuous fermentation process enhancing protein-phospholipid complex as protective agent, *Enzyme Microbiol. Technol. 3*: 249.
84. Hossack, J. A., and Rose, A. H. (1976). Fragility of plasma membranes, in *Saccharomyces cerevisiae* enriched with different sterols, *J. Bacteriol. 127*: 67.
85. Larue, F., Lafon-Lafourcade, S., and Ribereau-Gayon, P. (1980). Relationship between the sterol content of yeast cells and their fermentation activity in grape must, *Appl. Environ. Microbiol. 39*: 808.
86. Lees, N. D., Lofton, S. L., Woods, R. A., and Bard, M. (1980). The effect of varied energy source and detergent on the growth of sterol mutants of *Saccharomyces cerevisiae, J. Gen. Microbiol. 118*: 209.
87. Ismail, A. A., and Ali, M. M. (1971). Selection of high ethanol yielding *Saccharomyces*. I. Ethanol tolerance and the effect of training in *Saccharomyces cerevisiae* Hansen, *Folia Microbiol. 16*: 346.
88. Brown, S. W., and Oliver, S. G. (1983). Isolation of ethanol tolerant mutants of yeasts by continuous selection, *Eur. J. Appl. Microbiol. Biotechnol. 16*: 116.
89. Aguilera, A., and Benitez, T. (1986). Ethanol sensitive mutants of *Saccharomyces cerevisiae, Arch. Microbiol. 143*: 337.
90. Jimenez, J., and Benitez, T. (1987). Genetic analysis of highly ethanol tolerant wine yeasts, *Curr. Genet. 12*: 421.
91. Del Castillo, L. (1985). Genetic aspects of ethanol tolerance and production by *Saccharomyces cerevisiae, Curr. Microbiol. 12*: 41.
92. Van de Mortel, J. B. J., Mulders, D., Korthout, H., Theuvenet, A. P. R., and Borst-Pauwels, G. W. F. H. (1988). Transient hyperpolarization of yeast by glucose and ethanol, *Biochim. Biophys. Acta 936*: 421.
93. Salgueiro, S. P., Sa-Correia, I., and Novais, J. M. (1988). Ethanol-induced leakage in *Saccharomyces cerevisiae*: kinetics and relationship of yeast ethanol tolerance and alcohol fermentation productivity, *Appl. Environ. Microbiol. 54*: 903.
94. Glaser, H-U., and Höfer, M. (1990). Effect of ethanol on cell growth, metabolism and cation fluxes in yeast *Metschnikowia reukaufii, J. Gen. Microbiol. 136*: 1265.
95. Ghareib, M., Youssef, K. A., and Khalil, A. A. (1988). Ethanol tolerance of *Saccharomyces cerevisiae* and its relationship to lipid content and composition, *Folia Microbiol. 33*: 447.
96. Janssens, J. H., Burris, N., Woodward, A., and Baily, R. B. (1983). Lipid-enhanced ethanol production by *Kluyveromyces fragilis, Appl. Environ. Microbiol. 45*: 598.
97. Munoz, E., and Ingledew, M. W. (1989). Effect of yeast hulls on stuck and sluggish wine fermentations: importance of lipid component, *Appl. Environ. Microbiol. 55*: 1560.

98. Ohta, K., and Hayashida, S. (1983). Role of Tween 80 and monolein in a lipid sterol-protein complex which enhances ethanol tolerance of sake yeasts, *Appl. Environ. Microbiol. 46*: 821.
99. Panchal, C. J., and Stewart, G. G. (1981). Regulatory factors in alcohol fermentations, in *Current Developments in Yeast Research* (G. G. Stewart and Russell, eds.), Pergamon Press, Canada, p. 9.
100. Prior, B. A., Killian, S. G., and du Preez, J. C. (1989). Fermentation of D-xylose by the yeasts *Candida shehatae* and *Pichia stipitis*—prospects and problems, *Process Biochem. 24*: 21.
101. Schneider, H. (1989). Conversion of pentoses to ethanol by yeasts and fungi, *CRC Crit. Rev. Biotechnol. 9*: 1.

9

Tolerance of Fungi to NaCl

Anders Blomberg and Lennart Adler

University of Göteborg
Göteborg, Sweden

I. INTRODUCTION

Like all microorganisms, fungi have to come to their own terms with the external aqueous environment. In freshwater habitats, where the suspending medium is more dilute than the internal milieu, inflow of water is prevented by the development of a counteracting turgor pressure. In more saline environments, the organisms encounter combined stresses due to both increased electrolyte concentration and lowered water potential. Fungi adapt to this situation primarily by excluding the ions from the cytoplasm and by employing alternative osmotica to restore the cell volume and turgor required for growth to proceed. It is hardly surprising that salt-tolerant fungi have developed very efficient mechanisms to meet these demands. In this chapter we focus on the physiology of salt tolerance in fungi with emphasis on the mechanisms used to counter salt stress. One of the virtues of fungi is the availability of genetically accessible species and the relative ease by which mutants can be generated. Thus, there is a growing interest for molecular biological initiatives to better illuminate the nature of the adaptive mechanisms. A major focus for this interest will be the identification of genes involved in determining salt tolerance, how these genes are regulated, and how changed salinity is sensed and the signal transduced to the relevant genes. Uncovering of such fundamental molecular mechanisms for salt tolerance in fungi may be relevant also for the understanding of the more complex processes in green plants. The widespread problem of salinization of soils in irrigated agriculture clearly adds an applied interest to such studies.

II. FUNGAL OCCURRENCE AND GROWTH IN RELATION TO SALINITY

Sodium and chloride ions prevail in almost all natural environments. From an ecological point of view, the sea might be the most obvious example; however, even in freshwater systems these ions are abundant. About 50,000 fungal species are known from terrestrial habitats, but less than 500 have been described from oceans and estuaries, including 209 species of higher filamentous fungi and 177 species of yeasts, whereas the lower marine fungi comprise probably less than 100 species (1). Marine fungi occur as parasites on plants and animals, as symbionts in lichenoid association with algae, and as saprobes on dead organic matter of plant or animal origin. Marine fungi classified as obligate are those species that grow and sporulate exclusively in a marine or estuarine habitat.

Obligate marine yeasts such as *Pichia spartinae* and *Metschnikowia bicuspidata* var. *australis* have been reported; however, most marine-occurring yeasts are facultative and grow in freshwater or terrestrial milieus in addition to the marine environment (1). Even if the predominance of NaCl in the sea sets the physicochemical framework for the physiology of marine organisms, other factors might contribute in selecting for successful competitors. For instance, the geographical distribution of higher marine fungi is controlled mainly by temperature and not by differences in salinities (2). *Wallemia sebi*, being the main cause of fungal spoilage of dried salted fish, was rarely isolated from dried fish in the tropical countries because of its relatively low maximal temperature of growth (3). Yeast species isolated from a Swedish estuary were tested for their temperature-related growth response, and most species were found to grow well below 10 °C. Noteworthy are species of *Cryptococcus*, which exhibited high growth rates even at 3 °C. On the contrary, isolated strains of *Saccharomyces cerevisae* were not capable of growth below 10 °C and were thus classified as probably nonmarine (4).

The alkaline pH of the sea (≈ 8) is an additional selective factor, reflected in the capacity of the marine yeast *Debaryomyces hansenii* to better select for potassium against sodium in alkaline media (5). In a study on the influence of nutrient quality on the salt tolerance of *S. cerevisiae* (6), it was found that the tolerance was conditioned by the quality of the carbon source. Acetate sustained growth up to 0.45 M NaCl, whereas in glucose medium this strain of *S. cerevisiae* grew at 1.5 M NaCl. The implications of salinity and related cellular strategies for osmoregulation in setting the limits for growth and survival in the marine oligotrophic environment have recently been reviewed (7).

Few fungal species require sodium ions for growth. However, the zoosporic fungi, including the genera *Althornia* and *Thraustochytrium*, demands Na^+ at concentrations that give the element the status of a macronutrient (8). An obligate requirement of Na^+ as a micronutrient for growth is practically difficult to demonstrate because it is almost impossible to completely eliminate sodium ions from media. However, if there is an absolute requirement, the threshold level is presumably low and *Neurospora crassa* exhibits maximal growth even at 30 μM external Na^+ (9). On the contrary, there seems to be an absolute requirement for potassium, and the growth rate of *S. cerevisiae* responds strongly to external levels of this ion in the range 0.2–0.4 mM. At these low levels of potassium addition, sodium also displayed a stimulatory effect on cellular growth (10). Potassium is believed to possess specific functions as cofactors for enzymes (11) as well as more general roles as a determinant of ionic and osmotic strength of the cell (7). Maybe sodium to some extent functionally substitutes for potassium under such low availability of this usually prevalent cation.

Food can be preserved by reduction in the available water content, either by dehydration or by addition of high concentrations of sugar or salt. Much of the available information about fungal resistance to extreme environments stems from studies on the food spoilage potential of fungi (12,13). Fungi isolated from salted or dried fish have been examined for their halotolerance, and *Basipetospora halophila, Exophiala werneckii,* and *Aspergillus penicilloides* were all found capable of growth at NaCl-saturated media. All these species were also found to exhibit their growth optimum at salinities corresponding to about −15 MPa (2.5 M NaCl). However, growth was sustained even in more dilute media. A somewhat halophilic, and not just halotolerant, character was displayed by *B. halophila, E. werneckii,* and *Polypaecilum pisce* since their growth rate in saline media exceeded that recorded in media supplemented with glucose/fructose or glycerol (14). Of the species examined for their growth response in NaCl medium, nearly all adhere to a generalized growth response curve with a distinct growth rate optimum in relation to the NaCl concentration (7). At higher and lower salinities the rate of growth is diminished; however, even at saturated levels of NaCl *B. halophila* retained roughly 50% of its maximum radial growth rate (14). To our knowledge and for unknown reasons, fungi have not been isolated from the hypersaline lakes despite the above-mentioned record of fungal species capable of growth at saturated levels of NaCl. Probably other environmental factors at these locations, such as high concentrations of magnesium or alkaline pH, are inhibitory (15).

As soil dries, bacterial activity rapidly declines, and the respiratory activity measured in soil at water potentials below −2 MPa is believed to be brought about solely by fungal activity (16). At lower water potentials (< −15 MPa), species of the genera *Aspergillus* and *Penicillium* dominate the active flora. These genera contain many representatives of species generally tolerant to low water availability (12). However, among the Basidiomycetes, no representatives are known to grow below water potentials of −7 MPa (12). An applied aspect concerns the commercially important fungus *Agaricus bisporus*, for which it has been shown that roughly 30 mM NaCl (−0.1 MPa) in the watering solution causes a 10% reduction in sporophore yield and at about 1 M NaCl (−5 MPa), there is no growth (17).

If a fungal species is to grow indefinitely in a specific environment, it requires the capacity to complete its life cycle within that niche. Spores of a number of halotolerant fungi have been examined for ability to germinate at various concentrations of NaCl. Generally, tested species, including *W. sallemi* (3) and *B. halophila* (14), germinated at NaCl concentrations compatible to those allowing subsequent growth. A means of avoiding germination at high salinity seems available to spores of *N. crassa* and other filamentous fungi. Charlang and Horowitz (18) presented evidence for the existence of a germination essential substance in conidia of *N. crassa*, which was lost from the cells at low water potential. This substance was identified as a siderophore (19), i.e., a high-affinity chelating compound that solubilizes ferric iron and transports it into the cell. It was subsequently demonstrated that conidia of *Aspergillus nidulans* and *Penicillium chrysogenum* also lost cellular siderophores at high salinity and became dependent on siderophore addition for germination (20).

III. WHY IS POTASSIUM BUT NOT SODIUM A PREFERRED INTRACELLULAR CATION?

It is one of the puzzles in biology why K^+ is a preferred intracellular ion, particularly in contrast with Na^+. There is only a small difference between the two ions in cationic size, and their effects on the native state of proteins are very similar, as reflected in their close position in the Hofmeister series, which ranks ions according to their stabilizing or destabilizing effects on protein conformation (21). For anyone dissolving high concentrations of salt in water, it is evident, however, that the addition of KCl makes the solution cold whereas NaCl does not. The heat absorbed by the KCl solution is used to force water molecules into a more active state, and the hydration interactions of potassium ions are significantly weaker than those of sodium ions (22). There is little to indicate biological effects of these differences in hydration properties, but they were recently implicated in explaining the different rates of temperature-dependent unfolding of malate dehydrogenase from *Halobacterium marismortui* in solutions of KCl and NaCl (23).

An interesting aspect of the K^+/Na^+ paradox was discussed by Wiggins (24), who proposed that cells select in favor of K^+ in order to minimize formation of "extreme" water structures. Water adjacent to polyelectrolytes such as nucleic acids and proteins carries counterions and therefore contains a higher concentration of solutes than water in nearby regions. Since the chemical potential of water must be the same throughout the system, water in regions away from the surfaces compensates by volume expansion. This enhances the ability of the water molecules to make strong, straight hydrogen bonds with each other, and such water becomes "bonded" and less reactive. The implication is that cell water contains populations of water molecules with different activities and densities and therefore different solvent properties. Since K^+ has less affinity than Na^+ for the dense water, K^+ ions are more likely than Na^+ ions to associate with the negative charges of the macromolecules and thereby decrease the osmolality, density, and reactivity of the water surrounding the polyelectrolytes. By equilibration with the rest of the water of the cell, induction of extreme forms of water is diminished. Furthermore, the formation of low-density, bonded water is also promoted by hydrophobic surfaces in enzyme clefts. Influx of K^+ converts such inert water to highly reactive water (25), which might explain why some enzymes display higher molecular activity in the presence of KCl than NaCl (26), and even why some cellular processes, such as protein synthesis (27), specifically require K^+.

IV. TOLERANCE OF INTRACELLULAR ENZYMES TO IONS

In vitro studies of enzymes from plants have shown that the enzyme activity is sensitive to salinity and in general inhibited by NaCl concentrations above 100–200 mM (28,29). A comparative study of the marine fungus *Asteromyces cruciatus* and the terrestrial *Chaetomium globosum* showed that malate dehydrogenase of both fungi was as sensitive to NaCl as the enzyme from higher plants (28). Paton and Jennings (30) observed that the same enzyme from the marine fungus *Dendryphiella salina* was inhibited by salt concentrations above 50 mM, and that the salt tolerance of the enzyme did not increase when the organism was cultured in medium adjusted to seawater salinity. Similarly, the kinetics and water relations of isocitrate dehydrogenase of the salt-tolerant *Zygosaccharomyces rouxii* and the less tolerant *S. cerevisiae* did not reveal any significant differences (31), indicating no correlation between cellular and enzymatic salt tolerance. A similar conclusion was reached by Onishi (32), who noted that glucose consumption and glycerol production by cell free extracts of the salt-tolerant yeast *Pichia miso* were completely inhibited by an NaCl concentration that did not inhibit the activity of intact cells. It thus appears from the limited number of studies cited that the generalized response of fungal enzymes to high ionic strength does not display exceptional tolerance and is principally similar to that of their green plant counterparts, although exceptions cannot be excluded (see, for example, Ref. 33). Extrapolations from in vitro studies should, however, be performed with caution. Not only does the in vivo protein concentration differ by several orders of magnitude from in vitro conditions [the protein concentration in a yeast cell is about 150 mg/ml; (34)], but also pH and substrate concentration may deviate importantly from the physiological conditions. An interesting illustration to the importance of the anionic environment for enzyme tolerance to high ionic strength comes from the studies of the NAD-dependent glycerol 3-phosphate dehydrogenase of *Debaryomyces hansenii* (35). This enzyme is strongly inhibited by potassium or sodium chloride at high ionic strength, while in potassium or sodium glutamate the enzyme maintained an about twofold stimulated activity even at 0.9 M ionic strength. The main counteranions of the cell are to be found among amino acids of which glutamate is predominant (36), carboxy acids, phosphorylated intermediates, and negatively charged macromolecules. The observed salt response of glycerol 3-phosphate dehydrogenase may indicate that the effects of raised intracellular ionic strength following upon cell dehydration can be well tolerated owing to the benign nature of cellular counteranions.

V. ADAPTATION TO HIGH CONCENTRATIONS OF SALT: NaCl EXCLUSION

It was shown by Scott in 1957 that the consequences of solute stress on microbial activity were largely related to the reduction in water availability whereas solute-specific effects were less important (37). The central theme of this unifying concept is cellular exclusion of the stress solute and selective production and intracellular accumulation of alternative osmotica compatible with cellular functions (7,15). However, solute-specific impacts on fungi clearly exist, as exemplified by a collection of strains of the highly sugar-tolerant yeast *Z. rouxii*, which exhibited dramatically different responses to NaCl (38). While most of these strains were clearly osmotolerant in that they grew at high sugar or NaCl concentrations, some displayed impaired growth at a NaCl concentration of 150 mM. Although the precise cellular nature of the salt sensitivity of these strains is not known, tentative candidates for impaired cell functions would conceivably be found among the mechanisms for NaCl exclusion described below (Section V.C).

A. Intracellular Sodium and Chloride Concentrations

During growth in basal medium, fungal cells exhibit low intracellular levels of sodium ions (in the range 5–30 mM), as noted for *S. cerevisiae* (39,40), *N. crassa* (9), *D. hansenii* (5), and *A. nidulans* (41). Substantial increase in the steady-state cellular sodium level was not observed for *N. crassa* until the external NaCl concentration reached about 100 mM (9), while *S. cerevisiae* responded to slight external increases (5). During growth in medium containing 1 M NaCl, the intracellular sodium level of *S. cerevisiae* increased to about 150 mM (40), indicating that at higher salinities sodium becomes one of the main cations in fungi comparable in amount to the prevalent potassium. Accordingly, the intracellular K^+/Na^+ ratio was determined to be about equal to unity in both *D. hansenii* and *S. cerevisiae* during adaptation to high external concentrations of NaCl (42). However, the intracellular sodium ion level is related to both the external potassium ion concentration (9) and the hydrogen ion concentration (5).

In contrast to the extensive knowledge on sodium accumulation, less is reported for the cellular levels of the externally prevalent chloride anion. *S. cerevisiae* seems to exclude this anion (39), while other fungi admit it to varying extent (43,44). During saline growth of *D. salina* the cytoplasmic Cl^- concentration was comparable to that of Na^+ (45).

B. Influx of Sodium Ions

A general alkali cation carrier mediates major Na^+ influx in *S. cerevisiae* (46). This carrier system is saturable by monovalent cations with the specificity sequence $K^+ > Rb^+ > Cs^+ > Na^+ > Li^+$ (47). Both growth and assay conditions have been found to influence the characteristics of the carrier. Cells grown at K^+ concentrations in the micromolar range displayed a high-affinity system with a K_m for K^+ of 24 μM, while cells proliferating at millimolar concentrations of K^+ exhibited a K_m of 2 mM (48). The two modes of transport could also be distinguished on the basis of temperature dependence, CCCP sensitivity, and response to ATP depletion. The existence of at least two independent transport systems has been further substantiated by the isolation of the corresponding genes *TRK1* and *TRK2* (49,50). Both *trk1Δ* and *trk1Δtrk2Δ* mutants are more sensitive to NaCl compared to wild type, and growth is inhibited on YPD agar at 0.8 and 0.4 M NaCl, respectively. However, this sensitivity is linked to the requirement for K^+, since the NaCl inhibition of the mutants can be remedied by the inclusion of 100 mM K^+ (L. Adler and A. Blomberg, unpublished result).

Mechanosensitive ion channels have been identified in the plasma membrane of *S. cerevisiae* by using patch-clamp electrical recording technique on spheroplasts (51). These channels were permeable to both cations and anions. Furthermore, uptake of phosphate by *S. cerevisiae* is mediated by two systems with different affinity, the high-affinity system being sodium dependent. Two different sodium sites with affinity constants of 0.04 and 29 mM were reported (52). Thus, the existence of a sodium-phosphate cotransport system has been proposed. Similarly, phosphate uptake in a biflagellate marine fungus exhibited a specific requirement for sodium (53). Even though most solute transport in fungi is mediated by proton antiporters, a sodium-dependent symport mechanism for glycerol transport in *D. hansenii* and *Z. rouxii* has recently been documented (see Section VI.B).

C. Efflux of Sodium Ions

It was demonstrated by Conway et al. in 1954 (54) that *S. cerevisiae* was able to export Na^+ against the Na^+ concentration gradient across the membrane. It was later observed that the rate of Na^+ efflux was a function of the intracellular Na^+ content and that the discrimination for sodium compared to potassium increased at high intracellular levels of Na^+ (55). A Na^+/H^+ antiporter of the plasma membrane of *S. cerevisiae* and *N. crassa* was proposed to mediate the cellular sodium efflux (56–58). Furthermore, the systems for sodium influx and efflux were reported to be independent in *S. cerevisiae*, experimentally verified by the effects of *p*-chloromercuribenzenesulfonate (PCMBS), which exclusively inhibited efflux without affecting the influx (57). For *N. crassa* it seems that there are two different systems for Na^+ efflux, one of which being induced in sodium-grown cells (58). Genetic, physiological, and molecular data to demonstrate an electroneutral Na^+/H^+ antiporter in the plasma membrane of the fission yeast *Schizosaccharomyces pombe* have recently been presented (59). The screen used to isolate Na^+ efflux mutants employed lithium selection, since this ion is (a) transported by the same system as Na^+ and (b) toxic to the cell at low concentrations. The use of lithium thus eliminates selection for mutations in genes involved in general osmotolerance. A lithium-resistant mutant, termed *sod2*, was isolated, which displayed a sodium tolerance that had increased from 0.6 to 1.2 M Na^+. The changed phenotype was brought about by gene amplification, and consequently overexpression of the cloned *SOD2* gene behind a strong promoter also conferred enhanced tolerance. Interestingly, the *sod2* transcript was not induced by NaCl in the wild type. The *SOD2* gene disruption mutant was not capable of exporting sodium, indicating that this gene codes for an exclusive Na^+ efflux system in *S. pombe*, and the mutants were unable to grow at NaCl concentrations above 125 mM.

In animal cells the efflux of Na^+ is mediated by Na^+, K^+-ATPase, and the Na^+ gradient generated plays an essential role in the transport of cellular substrates as well as in pH homeostasis (60). On the other hand, eukaryotic cells with cell walls generate their membrane potential by H^+-ATPase-mediated proton pumping, and their secondary transport systems are coupled to the proton gradient (see Section V.D). However, recently, a novel Na^+ pumping transport system, a Na^+-ATPase termed *ENA1*, was identified in *S. cerevisiae* (61). The nucleotide sequence of the cloned DNA fragment exhibited two almost identical genes in tandem. Disruption of both genes produced a strain with defects in the efflux of sodium and lithium, as well as increased sensitivity to Na^+. From sequence data, it was concluded that the gene had previously been cloned by hybridization to two of the conserved regions of P-type ATPases, the gene at that time designated *PMR2* (62). The sodium sensitivity of the disruption mutant was pH dependent, and sodium toxicity was particularly pronounced at alkaline pH values. Sodium was, however, less toxic at acidic pH values, and a slight but significant efflux was recorded. It thus seems likely that *S. cerevisiae* contains two independent systems for sodium extrusion, one Na^+/H^+ antiporter mainly functional at acidic pH values and a P-type Na^+-ATPase responsible for Na^+ efflux at more alkaline conditions. It has been proposed that the failure of *S. pombe* to grow at high pH values might reflect the absence of a Na^+-ATPase system such as *ENA1* in that yeast (59).

D. Plasma Membrane H^+-ATPase and Salt Tolerance

The proton gradient generated by the fungal plasma membrane H^+-ATPase seems to be the driving force for active transport of nutrients by mainly symport mechanisms (63). The corresponding gene in *S. cerevisiae, PMA1*, has been cloned and was shown to be essential since a deletion mutation was lethal in haploid cells (64). By homologous recombination, mutations were introduced into the *PMA1* promoter, and the resulting strains displayed about 20% of the wild-type ATPase content (65). These mutants exhibited growth defects below pH 5.5, but at higher pH values their growth characteristics were indistinguishable from those of the wild type. Even addition of 1.15 M NaCl at pH 6 affected the mutants and the wild type to a similar extent. It was thus concluded that the *PMA1* gene product is not the factor limiting salt tolerance in *S. cerevisiae*. Tolerance to NaCl was not, however, tested under more acidic conditions or at higher NaCl concentrations. A second H^+-ATPase gene, *PMA2*, exists, which is not transcribed during normal growth and thus does not normally compensate for the *PMA1* deletion phenotype. *PMA2* exhibits 90% homology to the *PMA1* structural gene but shows less conservation in the promoter region, indicating that its expression might be initiated only under specific conditions (66). Constructed double mutants for *PMA1* and *PMA2* were, however, not tested for their salt tolerance, which might be a fruitful pursuit.

Partially purified membranes isolated from *D. hansenii* grown under nonsaline conditions contained an ATPase activity that possessed a pH optimum at 6–6.5, which is roughly similar to the pH optimum of the enzyme from *S. cerevisiae* (67). However, an additional alkaline ATPase optimum (pH ≈ 8) was reported for *D. hansenii* when 1.5 M NaCl was present in the growth medium. This alkaline ATPase activity showed high in vitro tolerance to NaCl and proposed to promote tolerance of *D. hansenii* to high NaCl concentrations under alkaline conditions. The plasma membrane ATPase activity of *Z. rouxii* was recently reported to be NaCl induced, but with no apparent change in its pH response (optimum at pH 7) (68).

E. Vacuolar Involvement in Salt Tolerance

Vacuoles are dominant organelles in the cells of higher plants, algae, and fungi. In plants it is well documented that the large central vacuole plays an important role in regulating the cellular turgor pressure and the cytoplasmic ionic content. Although not as extensively studied, it is conceivable that the fungal vacuole might have similar osmoregulatory functions.

The vacuole occupies 10–20% of the total cell volume in fungal cells (69–71), and it is well established that basic amino acids as well as phosphate are accumulated in this compartment (72). Less is known about the selective confinement of sodium or chloride ions to the vacuole in fungi exposed to saline environments. Jennings and colleagues (45) addressed the question of ion compartmentation in hyphae of the filamentous marine fungus *D. salina* by applying X-ray microanalysis. The concentration of Na^+ and Cl^- as determined in cytoplasm and vacuoles did, however, not indicate any preferential vacuolar accumulation of these ions in hypha subjected to enhanced external salinity. It was thus concluded that the vacuole of *D. salina* makes a relatively small contribution in cytoplasmic ionic adjustment. During short-time adaptation to sodium-rich media, *S. cerevisiae* preferentially accumulates Na^+ in the vacuole (73). To our knowledge, data for the vacuolar content of chloride ions are lacking. It is noteworthy, however, that the tonoplast contains a specific Cl^- transport system (72), which might indicate a selective sequestering of chloride ions to this compartment. The question of selective compartmentation of sodium and chloride ions in fungal vacuoles during saline growth merits further investigation.

Despite the lack of substantial evidence for a preferential NaCl accumulation in the fungal vacuole, data on the involvement of vacuoles in the fungal halotolerance have come from studies of mutants defective in the sorting of vacuolar proteins. Banta et al. (71) found that some of these mutants of *S. cerevisiae* [*vpt* mutants later renamed *vps* mutants (72)] were unable to grow on 1.5 M NaCl medium. All mutants that entirely lacked visible vacuoles (class C mutants) displayed this osmosensitivity, while among the other vacuolar mutant classes only certain representatives exhibited the same phenotype. Interestingly, all the osmosensitive mutants displayed common morphological features such as vesicles and Berkley bodies, regardless of vacuolar appearance. In a recent study roughly 100 salt-sensitive mutants, comprising 17 complementation groups, were analyzed for their vacuolar morphology. All the osmosensitive mutants were defective in vacuole biogenesis and thus termed salt-sensitive vacuolar (*ssv*) mutants (74). The identity of two of the corresponding *SSV* genes is known: (a) *SSV1* (identical to *VPS1*) encodes an 80-kDa GTP-binding protein that appears to be involved in microtubule bundling, and to play an essential role in vacuolar protein sorting, and (b) *SSV15* (identical to *OSM2*) encodes a chorismate mutase, the involvement of which in osmoregulation and vacuolar biogenesis remains obscure. In addition, the nucleotide sequence of the *VPS15* gene, whose corresponding mutant is osmosensitive(72), revealed significant similarity to the serine-threonine family of protein kinases, indicating phosphorylation reactions being involved in the vacuolar protein sorting (72). All the identified *SSV* or *VPS* genes are involved in processes that could simultaneously affect diverse and just remotely related cellular functions. In accordance, many of the vacuolar mutants displayed additional phenotypes such as temperature sensitivity, poor growth on nonfermentable carbon sources, and low frequency of DNA transformation (71).

Vacuole targeting of components of the osmoregulatory system would explain the osmosensitivity of these mutants. However, no evidence for a preferential vacuolar location of any of the enzymes involved in osmoregulation has, to our knowledge, been presented; on the contrary, one of the salt-induced enzymes in *S. cerevisiae*, glycerol 3-phosphate phosphatase (see Section VI), was not detected in purified vacuoles (75). Thus, whether the osmoregulatory defect is a consequence of the aberrant vacuole biogenesis or a parallel independent event is at present difficult to ascertain.

VI. ADAPTATION TO HIGH CONCENTRATIONS OF SALT: POLYOL ACCUMULATION

The constrained admission of NaCl into fungi requires compensatory osmotic adjustments such that turgor, cell volume, and growth can be maintained in saline environments. Achievement of this option involves adjustment of the intracellular osmotic potential by use of a compatible solute (76).

Polyols are produced by most fungi (77,78), and exposure of fungi to lowered water potential characteristically leads to increased production and intracellular accumulation of polyols (7,15). As is clear from batch culture as well as chemostat studies of well over 20 different yeasts and filamentous fungi, glycerol appears to be the primary osmotic adjuster in growing cells subjected to NaCl stress (36,41,79–86), while glycerol decreases and finally disappears when the organism enters the stationary phase (36,79,87). During this period the content of a higher polyol may transiently increase, but not to the extent that it fully compensates for the decrease in glycerol (36,84,87). Normally, the higher polyols present in fungi, such as erythritol, arabinitol (arabitol), and mannitol, respond little to salinity (however, see below), but can be kept at relatively high concentrations in the cells, independent of extracellular salt concentration. Cultured in chemostat at various growth rates in the salinity range 0–1.4 M NaCl, the marine yeast *D. hansenii* maintained an intracellular arabinitol concentration in the range 0.25–0.5 M, whereas intracellular glycerol increased from below detection in basal medium to 1.1–1.5 M at the highest salinity (82). Data on the extent to which glycerol and higher polyols contribute to the osmotic potential in fungi growing at high salinity are summarized in Table 1. In cases where both polyols and inorganic ions have been analyzed and internal concentrations calculated on the basis of intracellular volume determinations, the total osmotic contribution from the osmoresponsive glycerol pool, plus the contribution from the other pools, added up to and exceeded that of the external osmotic pressure, thereby maintaining the salt-stressed cells in a turgid state (82,86,88). It thus seems clear that glycerol serves as a growth-associated compatible solute in fungi, and that its probable function is turgor/volume regulation under solute stress. The higher polyols that are more or less constitutively present in the cells may serve to buffer cells growing at high water potential (low salinity) against dehydration upon sudden solute stresses. They may also offer protection to the structure of key macromolecules (see Section VI.A) on dehydration and other stressful conditions experienced during nongrowing conditions.

The accumulation of glycerol during growth at high salinity should preferably be carried out without too great a loss to the surroundings. Changes in the fatty acyl composition of membrane phospholipids as demonstrated for salt-tolerant yeasts grown at elevated NaCl concentrations may reduce the membrane permeability to glycerol (89). However, the observed leakage of glycerol from various yeast species grown at high salinity might indicate a higher membrane permeability to glycerol than that of green algae of the genus *Dunaliella*, which uses glycerol as a compatible solute and very effectively retains it. By NMR spectroscopy Brown et al. (90) demonstrated an extremely low permeability coefficient for glycerol of the plasma membrane of *Dunaliella salina*. In the salt-tolerant yeast *D. hansenii*, maintenance of a

Table 1 Percent Contribution of Polyols to Counteract the External Osmotic Potential of Some Fungi Growing at Elevated Concentrations of NaCl

Fungus	Salt added to growth medium	Polyols identified	% contribution to counteract external osmotic potential from: Total polyols	Glycerol	References
Dendryphiella salina	0.5 M NaCl	Mannitol, arabinitol, glycerol, erythritol	59	26	(81,88)
Saccharomyces cerevisiae	0.5 M NaCl	Glycerol	36[b]	36	(86)
	0.9 M NaCl		67[b]	67	
Penicillium chrysogenum	0.7 M NaCl	Mannitol, arabinitol, erythritol, glycerol	30[c]	17	(80)
Aspergillus niger	0.7 M NaCl	Mannitol, arabinitol, erythritol, glycerol	50[c]	38	(80)
Debaryomyces hansenii	0.7 M NaCl	Arabinitol, ribitol,[a] glycerol	75[b]	30	(82)
	1.4 M NaCl		75[b]	53	
Zygosaccharomyces rouxii	1.2 m NaCl	Arabinitol, glycerol	58[b]	51	(85)

[a]From Ref. 83.
[b]Average for cells cultured in chemostat at different dilution rates.
[c]From an assumed fresh weight:dry weight ratio of 2.5.

high intracellular glycerol concentration is partly achieved by a constitutive uptake system for glycerol (91), which is driven by the sodium gradient through a glycerol-sodium symport (92). Thus, the cells take help from the sodium gradient across the membrane to recapture glycerol lost to the medium. A similar sodium-dependent transport system for glycerol was described for *Z. rouxii* (93), and genetic evidence has indicated a glycerol transport system also in the salt-tolerant hyphal fungus *A. nidulans* (94). It is likely that such transport capacity is common among the highly salt-tolerant fungi. The observation by Ushio et al. (87) that mutants of *Z. rouxii* with alterations of the endogenously produced sterols leaked both glycerol and arabinitol and showed a marked decrease in NaCl tolerance underlines that the effectiveness by which the polyol pool is maintained is important in determining the growth-limiting NaCl concentration. This was demonstrated in a more direct way by André et al. (95). A NaCl-sensitive mutant of *D. hansenii* that displayed a decreased glycerol-producing ability regained its capacity to grow at high salinity when provided with low concentrations (1 mM) of extracellular glycerol. The exogenously added glycerol was used by the salt-stressed mutant to increase its internal glycerol to wild-type levels.

Thus, there is mounting evidence that glycerol serves as the major compatible solute in fungi subjected to NaCl stress, but is glycerol a stringently required osmotic adjuster? Among the Oomycetes and in the marine fungus *Thraustochytrium*, no or few polyols are produced (43,77), and instead it is the content of proline that increases when these organisms are subjected to osmotic stress (43,96). In *Geotrichum candidum*, arabinitol appears to serve as a major osmoticum in salt-stressed cells (97), and for *Neocosmospora vasinfecta*, growth at increased KCl concentration is also associated with a large increase in arabinitol (98). It is furthermore reported that increased external salinity induces not only increased accumulation of glycerol in *A. nidulans, Aspergillus niger,* and *P. chrysogenum*, but also increased levels of erythritol (41,80). Among the Basidiomycetes, which are relatively intolerant to water stress (99), mannitol is a predominant polyol (77) and growth of *Agaricus bisporus* in casing soil with added NaCl resulted in elevated concentrations of mannitol in the fruiting bodies (100). In view of the above observations, it is clear that the fungal response to NaCl stress can involve increased accumulation of solutes other than glycerol. However, there is a great need for further information to substantiate that the higher polyols function as true osmotic adjusters, i.e., that they respond primarily to changes of the external osmolality rather than to other environmental or physiological conditions.

A. Polyols as Compatible Solutes

A number of studies have demonstrated that various enzymes, although inhibited by low concentrations of electrolytes, maintain activity in molar concentrations of glycerol, providing a functional explanation for the selective accumulation of glycerol in water-stressed fungi (7,15,101). The molecular mechanisms behind the compatible nature of glycerol and other polyols have not been elucidated in detail. While monohydroxy alcohols destabilize proteins against heat denaturation, polyhydroxy alcohols stabilize the native conformation of the protein, and the protective effect increases with the concentration and the number of hydroxymethyl groups on the polyols (102, 103). From equilibrium dialysis studies of proteins in glycerol-water mixtures, Gekko and Timasheff (104) concluded that the protein-stabilizing effect of glycerol in aqueous solvent is related to the exclusion of glycerol from the hydration volume around the protein. This entropically unfavorable distribution would tend to stabilize the protein by minimizing its surface area. By a similar technique Bull and Breese (105) showed that protein hydration in the presence of various solutes increased through the series ethylene glycol, glycerol, adonitol, erythritol, mannitol, and sucrose. It has been argued by Brown (15) that glycerol serves as an exceptionally good compatible solute because of its intermediate stabilizing effects, by which a moderate stabilization can be combined with sufficient conformational flexibility to maintain biological activity. According to Wiggins (24), compatible solutes accumulate selectively in the low-density regions of the intracellular water population whereby the densities of the two water populations in the cell tend to converge and more normal reactivities and solvent properties of intracellular water are restored. From studies on the effects of compatible solutes on the solvent properties of water in membranes (106), it appears that glycine betaine, which is used as a compatible solute by bacteria and plants, fulfills such a role, whereas glycerol showed no or little such effects.

B. Regulation of Glycerol Accumulation

The common pathway for glycerol production involves outflow from the glycolytic flux through reduction of dihydroxyacetone phosphate to glycerol 3-phosphate by cytoplasmic glycerol 3-phosphate dehydrogenase (GPD) and subsequent dephosphorylation of glycerol 3-phosphate to glycerol. The dissimilation of glycerol occurs normally via glycerol kinase and mitochondrial GPD to produce dihydroxyacetone phosphate, but in *Schizosaccharomyces pombe* there is genetic and biochemical evidence for this dissimilation to occur via glycerol dehydrogenase and dihydroxyacetone kinase (for references, see 7).

Following transfer of *D. hansenii* to a highly saline medium, there is an instant admission of Na^+ into the cell followed by a period of sodium expulsion and glycerol accumulation (107). Obviously, the reactions involved in glycerol formation must be resistant to these conditions, and the high sodium ion tolerance of GPD in the presence of glutamate (see Section IV) may be necessary to promote glycerol production under such circumstances. The increased glycerol production of salt-adapted *D. hansenii* is not achieved by any marked changes of the glycerol-producing enzyme activities. An approximately twofold increase of the specific activities of GPD (91) and glycerol 3-phosphatase (108) was observed when the salinity of the growth medium was increased to 1.4 M. In *S. cerevisiae*, on the other hand, growth at 1.1 M NaCl increased the activity of GPD about sevenfold (109) and that of glycerol 3-phosphatase about fivefold (N. Akhtar, unpublished observations). The response of the *S. cerevisiae* GPD activity required protein synthesis since it was prevented by cycloheximide, a translation inhibitor in eukaryotes (110). Western blot analysis has confirmed that the salt-induced increase of GPD activity is due to de novo synthesis of the enzyme (109), suggesting that the control is exerted on gene level. Since the glycerol-dissimilating enzymes are repressed by glucose in *S. cerevisiae*, salt-induced glycerol formation is regulated solely through control of the producing pathway in high glucose media.

In *Z. rouxii* the glycerol production rate does not change much with external salinity; instead the cells adjust the proportion retained within the cell. Accordingly, there are only slight changes in GPD activity with changing salinity (111). Thus, in *Z. rouxii* the cytoplasmic pool appears to be controlled mainly by the balance between the outward flux via membrane leakage and the inward flux via the sodium-glycerol cotransport system (112), which increases its affinity for glycerol as the external salinity is increased. In the strongly salt-tolerant *D. hansenii* the uptake via the sodium-glycerol symport has a higher affinity for glycerol than that of *Z. rouxii*, but its affinity is not affected by medium salinity. In *S. cerevisiae* there is no evidence of active transport of glycerol across the membrane (113). Experiments with cycloheximide-treated cells showed, however, that salt-shocked cells retain the produced glycerol better than nonstressed cells (110). This may result from changed permeability properties of the plasma membrane of the shrunken cells or the presence of regulated glycerol specific pathways, as discussed by André et al. (109).

Little is yet known about how fungi sense osmotic stress and how the signal is transduced to the glycerol system. Potassium uptake is induced by osmotic stress in bacteria and serves as second messenger that triggers accumulation of trehalose and betaine (114). However, in *S. cerevisiae* glycerol accumulation in response to NaCl stress appears to occur normally whether or not cells are starved for potassium (115; I. Tunblad-Johansson, unpublished). According to an alternative concept, a strain sensor in the membrane or its associated cytoskeleton senses changes in cell volume or turgor, which are then signaled to the osmoregulatory process. Disassembly of the *S. cerevisiae* microtubuli by treatment with the drug nocodazole or disruption of actin structures by use of a temperature-sensitive *S. cerevisiae* actin mutant (116) did, however, not affect the salt-induced glycerol accumulation (L. Adler and K. Larsson, unpublished), indicating a noninvolvement of these structures in the glycerol response. However, observations from studies of the germinating spore of *Phycomyces blakesleeanus* suggest that protein phosphorylation participates in the regulation of glycerol synthesis. During germination, this spore produces large amounts of glycerol to mediate swelling (117), and one of the enzymes in glycerol production, glycerol 3-phosphatase, is transiently about 10-fold activated during early germination. Interestingly, the activation of glycerol 3-phosphatase could be reproduced in vitro by the addition of cAMP or bovine cAMP-dependent protein kinase (118), indicating that cAMP-dependent protein phosphorylation boosts glycerol production in these spores. Recently, osmosensitive mutants of *S. cerevisiae* were isolated that displayed both reduced induction of GPD and defective growth arrest when subjected to salt stress. A gene complementing one of the mutants, *HOG1*, was cloned and sequenced, and its deduced amino acid sequence was 49% identical to that of a protein kinase involved the pheromone-induced growth arrest of *S. cerevisiae* (119). These interesting results strongly indicate a role for protein phosphorylation in the cellular response to osmotic stress of *S. cerevisiae.* Examination of signal-transferring networks, directed by inositol 1,4,5-trisphosphate, diacylglycerol, and Ca^{2+}, as potential transducers for osmotic signaling has to our knowledge not yet been conducted.

VII. CONCLUDING REMARKS

The response of a fungus to NaCl stress involves the functioning of many diverse capacities of the organism. The impact of the external salinity on the intracellular physiology is lessened primarily by exclusion of Na^+ and by a compensatory accumulation of polyols, mainly glycerol. This control of the composition of the cytoplasm to combine appropriate turgor with an internal milieu that is conducive to enzyme activity is exerted by a regulated functioning of transport as well as biosynthetic systems. The outcome of the process such as reflected in the K^+-Na^+ interplay and the balance between ions and polyols in the cytoplasm seems to vary among fungi and depend on genetic as well as environmental factors. However, from the rather dramatic effects on salt tolerance in yeast by mutations affecting sodium efflux systems (59,61) compared to the less severe effects by defective polyol accumulation (87,95,120), it appears that capacity for sodium export is a highly decisive factor in fungal salt tolerance. Investigations of fungal tolerance to NaCl at the physiological level have recently initiated a growing interest in the molecular processes. With the experimental tractability and developed genetics of many fungal species we now can foresee a rapidly expanding understanding of features underlying fungal tolerance to sodium chloride.

REFERENCES

1. Kohlmeyer, J., and Kohlmeyer, E. (1979). *Marine Mycology. The Higher Fungi*, Academic Press, New York.
2. Kohlmeyer, J. (1983). Geography of marine fungi, *Aust. J. Bot. Suppl. Ser. 10*: 67.

3. Wheeler, K. A., Hocking, A. D., and Pitt, J. I. (1988). Effects of temperature and water activity on germination and growth of *Wallemia sebi, Trans. Br. Mycol. Soc. 90*: 365.
4. Norkrans, B. (1966). Studies on marine occuring yeasts: growth related to pH, NaCl concentration and temperature, *Arch. Mikrobiol. 54*: 374.
5. Hobot, J. A., and Jennings, D. H. (1981). Growth of *Debaryomyces hansenii* and *Saccharomyces cerevisiae* in relation to pH and salinity, *Exp. Mycol. 5*: 217.
6. Bellinger, Y., and Larher, F. (1990). Nutritional aspects of salt tolerance of the yeast *Saccharomyces cerevisiae, CR Acad. Sci. Paris 310*: 351.
7. Blomberg, A., and Adler, L. (1992). Physiology of osmotolerance in fungi, *Adv. Microbial Physiol. 33*: 145.
8. Jennings, D. H. (1986). Fungal growth in the sea, in *The Biology of Marine Fungi* (S. T. Moss, ed.), Cambridge University Press, Cambridge, p. 1.
9. Slayman, C. W., and Tatum, E. L. (1964). Potassium transport in *Neurospora*. I. Intracellular sodium and potassium concentrations, and cation requirements for growth, *Biochim. Biophys. Acta 88*: 578.
10. Camacho, M., Ramos, J., and Rodriguéz-Navarro, A. (1981). Potassium requirement of *Saccharomyces cerevisiae, Curr. Microbiol. 6*: 295.
11. Bostian, K. A., and Betts, G. F. (1978). Kinetics and reaction mechanism of potassium-activated aldehyde dehydrogenase from *Saccharomyces cerevisiae, Biochem. J. 173*: 787.
12. Pitt, J. I. (1975). Xerophilic fungi and the spoilage of foods of plant origin, in *Water Relations of Foods* (R. B. Duckworth, ed.), Academic Press, London, p. 273.
13. Tilbury, R. H. (1980). Xerotolerant (osmophilic) yeasts, in *Biology and Activities of Yeasts* (F. A. Skinner, S. M. Passmore, and R. R. Davenport, eds.), Academic Press, London, p. 153.
14. Andrews, S., and Pitt, J. I. (1987). Further studies on the water relation of xerophilic fungi, including some halophiles, *J. Gen. Microbiol. 133*: 233.
15. Brown, A. D. (1990). *Microbial Water Stress Physiology. Principles and Perspectives.* Wiley, Chichester.
16. Griffin, D. M., and Luard, E. J. (1979). Water stress and microbial ecology, in *Strategies of Microbial Life in Extreme Environments* (M. Shilo, ed.), Verlag Chemie, Weinheim, p. 49.
17. Awad, A. S., and Nair, N. G. (1989). Salt tolerance of *Agaricus bisporous* in relation to water stress and toxicity of sodium ions, *Ann. Appl. Biol. 115*: 215.
18. Charlang, G. W., and Horowitz, N. H. (1971). Germination and growth of *Neurospora* at low water activities, *Proc. Natl. Acad. Sci. USA 68*: 260.
19. Horowitz, N. H., Charlang, G., Horn, G., and Williams, N. P. (1976). Isolation and identification of the conidial germination factor of *Neurospora crassa, J. Bacteriol. 127*: 135.
20. Charlang, G., Ng, B., Horowitz, N. H., and Horowitz, R. M. (1981). Cellular and extracellular siderophores of *Aspergillus nidulans* and *Penicillium chrysogenum, Mol. Cell. Biol. 1*: 94.

21. Yancey, P. H., Clark, M. E., Hand, S. C., Bowlus, R. D., and Somero, G. N. (1982). Living with water stress: evolution of osmolyte systems, *Science 217*: 1214.
22. Erlander, S. R. (1970). The structure of water, *Sci. J. 5A*: 60.
23. Zaccai, G., and Eisenberg, H. (1990). Halophilic proteins and the influence of solvent on protein stabilization, *Trends. Biochem. Sci. 15*: 333.
24. Wiggins, P. M. (1990). Role of water in some biological processes, *Microbiol. Rev. 54*: 432.
25. Wiggins, P. M. (1988). Water structure in polymer membranes, *Prog. Polym. Sci. 13*: 1.
26. Wiggins, P. M., and MacClement, B. A. (1987). Two states of water found in hydrophobic clefts: their possible contribution to mechanisms of cation pumps and other enzymes, *Int. Rev. Cytol. 108*: 249.
27. Wyn Jones, R. G. (1984). Physiochemical aspects of osmotic adaption, in *Recent Advances in Phytochemistry* (B. N. Timmermann, C. Steelink, and F. A. Loewus, eds.), Plenum Press, New York, p. 55.
28. Flowers, T. J., Troke, P. F., and Yeo, A. R. (1977). The mechanism of salt tolerance in halophytes, *Annu. Rev. Plant Physiol. 28*: 89.
29. Pollard, A., and Wyn Jones, R. G. (1979). Enzyme activities in concentrated solutions of glycinebetaine and other solutes, *Planta 144*: 291.
30. Paton, F. M., and Jennings, D. H. (1988). Effect of sodium and potassium chloride and polyols on malate and glucose 6-phosphate dehydrogenases from the marine fungus *Dendryphiella salina, Trans. Br. Mycol. Soc. 91*: 205.
31. Brown, A. D. (1976). Microbial water stress, *Bacteriol. Rev. 40*: 803.
32. Onishi, H. (1963). Osmophilic yeasts, *Adv. Food Res. 12*: 53.
33. Adler, L. (1978). Properties of alkaline phosphatase of the halotolerant yeast *Debaryomyces hansenii, Biochim. Biophys. Acta 522*: 113.
34. Hess, B., Boiteux, A., and Krüger, J. (1969). Cooperation of glycolytic enzymes, *Adv. Enz. Regul. 7*: 149.
35. Nilsson, A., and Adler, L. (1990). Purification and characterization of glycerol-3-phosphate dehydrogenase (NAD^+) in the salt tolerant yeast *Debaryomyces hansenii, Biochim. Biophys. Acta 1034*: 180.
36. Adler, L., and Gustafsson, L. (1980). Polyhydric alcohol production and intracellular amino acid pool in relation to halotolerance of the yeast *Debaryomyces hansenii, Arch. Microbiol. 124*: 123.
37. Scott, W. J. (1957). Water relations of food spoilage microorganisms, *Adv. Food Res. 7*: 83.
38. Rodriguéz-Navarro, A. (1971). Inhibition by sodium and lithium in osmophilic yeasts, *Ant. van Leeuwenh. 37*: 225.
39. Conway, E. J., and Armstrong, W. M. (1961). The total intracellular concentration of solutes in yeast and other plant cells and the distensibility of the plant-cell wall, *Biochem. J. 81*: 631.
40. Watson, T. G. (1970). Effects of sodium chloride on steady-state growth and metabolism of *Saccharomyces cerevisiae, J. Gen. Microbiol. 64*: 91.
41. Beever, R. E., and Laracy, E. P. (1986). Osmotic adjustment in the filamentous fungus *Aspergillus nidulans, J. Bacteriol. 168*: 1358.

42. Norkrans, B., and Kylin, A. (1969). Regulation of the potassium to sodium ratio and of the osmotic potential in relation to salt tolerance in yeasts, *J. Bacteriol. 100*: 836.
43. Wethered, J. M., and Jennings, D. H. (1985). Major solutes contributing to the solute potential of *Thraustochytrium aureum* and *T. roseum* after growth in media of different salinities, *Trans. Br. Mycol. Soc. 85*: 439.
44. Luard, E. J. (1982). Accumulation of intracellular solutes by two filamentous fungi in response to growth at low steady state osmotic potential, *J. Gen. Microbiol. 128*: 2563.
45. Clipson, N. J. W., Hajbagheri, M. A., and Jennings, D. H. (1990). Ion compartmentation in the marine fungus *Dendryphiella salina* in response to salinity; X-ray microanalysis, *J. Exp. Bot. 41*: 190.
46. Borst-Pauwels, G. W. F. H. (1981). Ion transport in yeast, *Biochim. Biophys. Acta 650*: 88.
47. Armstrong, W. M., and Rothstein, A. (1964). Discrimination between alkali metal cations by yeast. I. Effect of pH on uptake, *J. Gen. Physiol. 48*: 61.
48. Rodriguéz-Navarro, A., and Ramos, J. (1984). Dual system for potassium transport in *Saccharomyces cerevisiae, J. Bacteriol. 159*: 9440.
49. Gaber, R. F., Styles, C. A., and Fink, G. R. (1988). TRK1 encodes a plasma membrane protein for high-affinity potassium transport in *Saccharomyces cerevisiae, Mol. Cell. Biol. 8*: 2848.
50. Ko, C. H., and Gaber, R. F. (1991). TRK1 and TRK2 encodes structurally related K^+ transporters in *Saccharomyces cerevisiae, Mol. Cell Biol. 11*: 4266.
51. Gustin, M. C., Zhou, X-L., Marutinac, B., and Kung, C. (1988). A mechanosensitive ion channel in the yeast plasma membrane, *Science 242*: 762.
52. Roomans, G. M., Blasco, F., and Borst-Pauwels, G. W. F. H. (1977). Cotransport of phosphate and sodium by yeast, *Biochim. Biophys. Acta 467*: 65.
53. Belsky, M. M., Goldstein, S., and Menna, M. (1970). Factors affecting phosphate uptake in the marine fungus *Dermocystidium* sp., *J. Gen. Microbiol. 62*: 399.
54. Conway, E. J., Ryan, H., and Carton, E. (1954). Active transport of sodium ions from the yeast cell, *Biochem. J. 58*: 158.
55. Rothstein, A. (1974). Relationship of cation influxes and effluxes in yeast, *J. Gen. Physiol. 64*: 608.
56. Rodriguéz-Navarro, A., Sancho, E. D., and Pérez-Lloveres, C. (1981). Energy source for lithium efflux in yeast, *Biochim. Biophys. Acta 640*: 352.
57. Rodriguéz-Navarro, A., and Ortega, D. (1982). The mechanism of sodium efflux in yeast, *FEBS Lett. 138*: 205.
58. Ortega, M. D., and Rodriguéz-Navarro, A. (1986). Sodium ion transport in *Neurospora crassa, Physiol. Plant. 66*: 705.
59. Jia, Z.-P., McCullough, N., Martel, R., Hemmingsen, S., and Young, P. G. (1992). Gene amplification at a locus encoding a putative Na^+/H^+ antiporter confers sodium and lithium tolerance in fission yeast, *EMBO J. 11*: 1631.
60. Krulwich, T. A. (1983). Na^+/H^+ antiporters, *Biochim. Biophys. Acta 726*: 245.
61. Haro, R., Garciadeblas, B., and Rodriguéz-Navarro, A. (1991). A novel P-type ATPase from yeast involved in sodium transport, *FEBS Lett. 291*: 189.

62. Rudolph, H. K., Antebi, A., Fink, G. R., Buckley, C. M., Dorman, T. E., LeVitre, J., Davidow, L. S., Mao, J., and Moir, D. T. (1989). The yeast secretory pathway is perturbed by mutations in PMR1, a member of a Ca^{2+} ATPase family, *Cell 58*: 133.
63. Eddy, A. A. (1982). Mechanisms of solute transport in selected eukaryotic microorganisms, *Adv. Microbial Physiol. 23*: 1.
64. Serrano, R., Kielland-Brandt, M. C., and Fink, G. R. (1986). Yeast plasma membrane ATPase is essential for growth and has homology with (Na^+ + K^+), K^+- and Ca^{2+}-ATPases, *Nature 319*: 689.
65. Vallejo, C. G., and Serrano, R. (1989). Physiology of mutants with reduced expression of plasma membrane H^+-ATPase, *Yeast 5*: 307.
66. Schlesser, A., Ulaszewski, S., Ghislain, M., and Gouffeau, A. (1988). A second transport ATPase gene in *Saccharomyces cerevisiae, J. Biol. Chem. 263*: 19480.
67. Comerford, J. G., Spencer-Phillips, P. T. H., and Jennings, D. H. (1985). Membrane-bound ATPase activity, properties of which are altered by growth in saline conditions, isolated from the marine yeast *Debaryomyces hansenii, Trans. Br. Mycol. Soc. 85*: 421.
68. Watanabe, Y., Yamaguchi, M., Sanemitsu, Y., and Tamai, Y. (1991). Characterization of plasma membrane proton-ATPase from salt-tolerant yeast *Zygosaccharomyces rouxii* cells, *Yeast 7*: 599.
69. Lindman, B., and Norkrans, B. (1982). Membraneous interrelationships—an ultrastructural study of the halotolerant yeast *Debaryomyces hansenii, Protoplasma 110*: 66.
70. Clipson, N. J. W., Jennings, D. H., and Smith, J. L. (1989). The response to salinity at the microscopic level of the marine fungus *Dendryphiella salina* Nicot and Pugh as investigated stereologically, *New Phytol. 113*: 21.
71. Banta, L. M., Robinson, J. S., Klionsky, D. J., and Emr, S. D. (1988). Organelle assembly in yeast: characterization of yeast mutants defective in vacuolar biogenesis and protein sorting, *J. Cell Biol. 107*: 1369.
72. Klionsky, D. J., Herman, P. K., and Emr, S. D. (1990). The fungal vacuole: composition, function, and biogenesis, *Microbiol. Rev. 54*: 266.
73. Ortega, M. D. (1988). Distribucion del potasio y sodio en vacuola y citoplasma de *Saccharomyces cerevisiae, Microbiologia SEM 4*: 61.
74. Latterich, M., and Watson, M. D. (1991). Isolation and characerization of osmosensitive vacuolar mutants of *Saccharomyces cerevisiae, Mol. Microbiol. 5*: 2417.
75. Wiemken, A., Schellenberg, M., and Urech, K. (1979). Vacuoles: the sole compartments of digestive enzymes in yeast (*Saccharomyces cerevisiae*)? *Arch. Microbiol. 123*: 23.
76. Brown, A. D., and Simpson, J. R. (1972). Water relation of sugar-tolerant yeasts: the role of intracellular polyols, *J. Gen. Microbiol. 72*: 589.
77. Pfyffer, G. E., Pfyffer, B. U., and Rast, D. M. (1986). The polyol pattern, chemotaxonomy, and phylogeny of the fungi, *Sydowia 39*: 160.
78. Pfyffer, G. E., Boraschi-Gaia, C., Weber, B., Hoesch, L., Orpin, C. G., and Rast, D. M. (1990). A further report on the occurrence of acyclic sugar alcohols in fungi, *Mycol. Res. 94*: 219.

79. Hocking, A. D. (1986). Effects of water activity and culture age on the glycerol accumulation patterns of five fungi, *J. Gen. Microbiol. 132*: 269.
80. Adler, L., Pedersen, A., and Tunblad-Johansson, I. (1982). Polyol accumulation by two filamentous fungi grown at different concentrations of NaCl, *Physiol. Plant 56*: 139.
81. Wethered, J. M., Metcalf, E. C., and Jennings, D. H. (1985). Carbohydrate metabolism in the fungus *Dendryphiella salina*. VIII. The contribution of polyols and ions to the mycelial solute potential in relation to the external osmoticum, *New Phytol. 101*: 631.
82. Larsson, C., Morales, C., Gustafsson, L., and Adler, L. (1990). Osmoregulation of the salt-tolerant yeast *Debaryomyces hansenii* grown in a chemostat at different salinities, *J. Bacteriol. 172*: 1769.
83. Meikle, A. J., Chudek, J. A., Reed, R. H., and Gadd, G. M. (1991). Natural abundance ^{13}C nuclear magnetic resonance spectroscopic analysis of acyclic polyol and trehalose accumulation by several yeast species in response to salt stress, *FEMS Microbiol. Lett. 82*: 163.
84. Van Eck, J. H., Prior, B. A., and Brandt, V. E. (1989). Accumulation of polyhydroxy-alcohols by *Hansenula anomala* in response to water stress, *J. Gen. Microbiol. 135*: 3505.
85. Van Zyl, P. J., and Prior, B. A. (1990). Water relations of polyol accumulation by *Zygosaccharomyces rouxii* in continuous culture, *Appl. Microbiol. Biotechnol. 33*: 12.
86. Ölz, R., Larsson, K., Adler, L., and Gustafsson, L. (1993). Energy flow and osmoregulation of *Saccharomyces cerevisiae* grown in a chemostat under NaCl stress, *J. Bacteriol.* (accepted).
87. Ushio, K., Sawatani, M., and Nakata, Y. (1991). Effect of sterol composition of the salt tolerance of *Zygosaccharomyces rouxii, J. Ferment. Bioeng. 71*: 390.
88. Jennings, D. H., and Burke, R. M. (1990). Compatible solutes—the mycological dimension and their role as physiological buffering agents, *New Phytol. 116*: 277.
89. Tunblad-Johansson, I., and Adler, L. (1987). Effects of sodium chloride concentrations on phospholipid fatty acid composition of yeasts differing in osmotolerance, *FEMS Microbiol. Lett. 43*: 275.
90. Brown, F. F., Sussman, I., Avron, M., and Degani, H. (1982). NMR studies of glycerol permeability in lipid vesicles, erythrocytes and the alga *Dunaliella, Biochim. Biophys. Acta 690*: 165.
91. Adler, L., Blomberg, A., and Nilsson, A. (1985). Glycerol metabolism and osmoregulation in the salt-tolerant yeast *Debaryomyces hansenii, J. Bacteriol. 162*: 300.
92. Lucas, C., da Costa, M., and van Uden, N. (1990). Osmoregulatory active sodium-glycerol co-transport in the halotolerant yeast *Debaryomyces hansenii, Yeast 6*: 187.
93. Van Zyl, P. J., Kilian, S. G., and Prior, B. A. (1990). The role of an active transport mechanism in glycerol accumulation during osmoregulation by *Zygosaccharomyces rouxii, Appl. Microbiol. Biotechnol. 34*: 231.

94. Visser, J., Van Rooijen, R., Dijkema, C., Swart, K., and Sealy-Lewis, H. M. (1988). Glycerol uptake mutants of the hyphal fungus *Aspergillus nidulans, J. Gen. Microbiol. 134*: 655.
95. André, L., Nilsson, A., and Adler, L. (1988). The role of glycerol in osmotolerance of the yeast *Debaryomyces hansenii, J. Gen. Microbiol. 134*: 669.
96. Luard, E. J. (1982). Growth and accumulation of solutes by *Phytophtora cinnamomi* and other lower fungi in response to changes in external osmotic potential, *J. Gen. Microbiol. 128*: 2583.
97. da Costa, M. S., and Niederpruem, D. J. (1982). Arabitol accumulation in *Geotrichum candidum, Arch. Microbiol. 131*: 283.
98. Kelly, D. J. A., and Budd, K. (1991). Polyol metabolism and osmotic adjustment in the mycelial ascomocete *Neocosmospora vasinfecta* (E. F. Smith), *Exp. Mycol. 15*: 55.
99. Pitt, J. I. (1989). Water activity, the neglected parameter, *Aust. Microbiol. 10*: 93.
100. Kalberer, P. P. (1990). Influence of the water potential of the casing soil on crop yield and on dry-matter content, osmotic potential and mannitol content of the fruit bodies of *Agaricus bisporus, J. Horticult. Sci. 65*: 573.
101. Brown, A. D. (1978). Compatible solute and extreme water stress in eukaryotic microorganisms, *Adv. Microbial Physiol. 17*: 181.
102. Gerlsma, S. Y. (1968). Reversible denaturation of ribonuclease in aqueous solutions as influenced by polyhydric alcohols and some other additives, *J. Biol. Chem. 243*: 957.
103. Gerlsma, S. Y., and Stuur, E. R. (1972). The effect of polyhydric and monohydric alcohols on the heat induced reversible denaturation of lysozyme and ribonuclease, *Int. J. Pept. Prot. Res. 4*: 377.
104. Gekko, K., and Timasheff, S. N. (1981). Mechanism of protein stabilization by glycerol: preferential hydration in glycerol-water mixtures, *Biochemistry 20*: 4667.
105. Bull, H. B., and Breese, K. (1978). Interaction of alcohols with proteins, *Biopolymers 17*: 2121.
106. Wiggins, P. M., and van Ryn, R. T. (1986). The solvent properties of water in desalination membranes, *J. Macromol. Sci. Chem. A23*: 875.
107. Burke, R. M., and Jennings, D. H. (1990). Effect of sodium chloride on growth characteristics of the marine yeast *Debaryomyces hansenii* in batch and continuous culture under carbon and potassium limitation, *Mycol. Res. 94*: 378.
108. Nilsson, A. (1990). Glycerol metabolism and salt tolerance of the yeast *Debaryomyces hansenii*, Ph.D. thesis, University of Göteborg, Sweden.
109. André, L., Hemming, A., and Adler, L. (1991). Osmoregulation in *Saccharomyces cerevisiae.* Studies on the osmotic induction of glycerol production and glycerol 3-phosphate dehydrogenase (NAD^+), *FEBS Lett. 286*: 13.
110. Blomberg, A., and Adler, L. (1989). Roles of glycerol and glycerol-3-phosphate dehydrogenase (NAD^+) in acquired osmotolerance of *Saccharomyces cerevisiae, J. Bacteriol. 171*: 1087.

111. Brown, A. D., and Edgley, M. (1980). Osmoregulation in yeast, in *Symposium on Genetic Engineering of Osmoregulation* (D. W. Rains, R. C. Valentine, and A. Hollander, eds.), Plenum Press, New York, p. 75.
112. Van Zyl, P. J., Kilian, S. G., and Prior, B. A. (1990). The role of an active transport mechanism in glycerol accumulation during osmoregulation by *Zygosaccharomyces rouxii, Appl. Microbiol. Biotechnol. 34*: 231.
113. Brown, A. D. (1974). Microbial water relations: features of the intracellular composition of sugar-tolerant yeasts, *J. Bacteriol. 118*: 769.
114. Higgins, C. F., Cairney, J., Stirling, D. A., Sutherland, L., and Booth, I. R. (1987). Osmotic regulation of gene expression: ionic strength as an intracellular signal? *Trends Biochem. Sci. 12*: 339.
115. Meikle, A. J., Reed, R. H., and Gadd, G. M. (1991). The osmotic responses of *Saccharomyces cerevisiae* in K^+-depleted medium, *FEMS Microbiol. Lett. 78*: 89.
116. Novick, P., and Botstein, D. (1985). Phenotypic analysis of temperature-sensitive yeast actin mutants, *Cell 40*: 405.
117. Van Laere, A. J., and Hulsman, E. (1987). Water potential, glycerol synthesis, and water content of germinating *Phycomyces* spores, *Arch. Microbiol. 147*: 257.
118. Van Schaftingen, E., and Van Laere, A. J. (1985). Glycerol formation after breaking of dormancy of *Phycomyces blakesleeanus* spores. Role of an interconvertible glycerol-3-phosphatase, *Eur. J. Biochem. 48*: 399.
119. Brewster, J. L., de Valoir, T., Dwyer, N. D., Allen, D. M., and Gustin, M. C. (1991). HOG1, a FUS3 homologue implicated in the growth control and osmoregulatory response to osmotic stress. Yeast Cell Biology Meeting Abstracts, Cold Spring Harbor Laboratory, Cold Spring Harbor, NY.
120. Yagi, T., and Tada, K. (1988). Isolation and characterization of the salt-sensitive mutants of a salt-tolerant yeast *Zygosaccharomyces rouxii, FEMS Microbiol. Lett. 49*: 317.

10

Responses of Xerophilic Fungi to Changes in Water Activity

Ailsa D. Hocking

Division of Food Science and Technology
Commonwealth Scientific and Industrial Research Organisation
North Ryde, New South Wales, Australia

I. INTRODUCTION

Most microorganisms have mechanisms for dealing with water stress, but relatively few have evolved with physiological adaptations that enable them to grow well in low-water-activity (a_w) environments, i.e., 0.85 a_w or less. Many terms have been used to describe these microorganisms: halophilic, osmophilic, osmotolerant, xerophytic, xerotolerant, xerophilic, and so forth. The term "halophilic" is correctly applied to the specialized group of bacteria that require high concentrations of salt for growth. Of the other terms, xerophilic (from the Greek = dry-loving) is perhaps most appropriate, as it describes the preference or requirement for reduced water activity (not for increased osmotic pressure, as is implied by the term "osmophilic").

Microorganisms from most of the major groups are represented in environments of reduced water activity: cyanobacteria, eubacteria, archaebacteria, algae, yeasts, and filamentous fungi. Species from each of these groups can be found in highly saline situations, but only the yeasts and filamentous fungi have mastered the art of growth in environments containing high sugar concentrations.

The strategies employed by different microorganisms to enable them to survive and grow at reduced a_w vary, but all rely on a common approach.

This is the intracellular accumulation of a solute or solutes to balance the a_w inside the cell against the external a_w, thus preventing movement of water out of the cell. At low a_w, these solutes must be accumulated at very high concentrations, but at the same time, they must not inhibit the cell's normal functions, such as protein synthesis or DNA replication. Consequently, they are known as compatible solutes (1).

II. PHYSIOLOGY OF XEROPHILIC YEASTS

Yeasts that are able to grow at reduced a_w are widespread and important in the food industry, both as fermentation organisms in soy sauce and miso production (2) and as spoilage organisms in glucose syrups, sweetened condensed milk, honey, jams, dried fruits, fruit juice concentrates, confectionery and bakery products, curing brines, and salted meats (3). The most important genera are *Debaryomyces, Hansenula, Pichia, Zygosaccharomyces,* and *Torulopsis*. Osmoregulation has so far been studied only in the genera *Saccharomyces, Zygosaccharomyces, Debaryomyces,* and *Hansenula.*

A. Production of Polyhydric Alcohols by Yeasts

It was established as early as 1945 that osmophilic yeasts produced polyhydric alcohols from sugar fermentation. Nickersen and Carroll (4) found that *Zygosaccharomyces acidifaciens* (= *Zygosaccharomyces bailii*) produced a substance they tentatively identified as glycerol, in yields of about 20% of the weight of sugar fermented. Spencer and Sallans (5) surveyed 79 strains of osmophilic yeasts for polyol production and found that about half of them (the faster-growing strains) produced mainly glycerol and D-arabitol. The slower-growing strains produced mainly glycerol and erythritol.

These early workers did not speculate on the reasons for such high accumulations of polyols in osmophilic yeasts. Despite substantial reviews by Scott (6), Ingram (7), and Onishi (2), there was no information about the physiological or biochemical bases of ability of some yeasts to grow in concentrated solutions and to produce and accumulate polyols.

Anand and Brown (8) investigated the growth rate patterns of a number of xerophilic and nonxerophilic yeasts in sugar solutions and solutions of polyethylene glycol. They found that the nonxerophilic yeasts had much more sharply defined water activity optima for growth than the xerophilic yeasts, and that the limiting a_w for growth was affected by the type of controlling solute. Both types of yeast were less tolerant of low a_w in the presence of polyethylene glycol than in the presence of sugars.

B. Comparative Studies on *Saccharomyces cerevisiae* and *Zygosaccharomyces rouxii*

1. Polyol Accumulation

Early studies investigated the differences in biochemical and physiological behavior of the xerophilic species *Zygosaccharomyces rouxii* and the closely related but nonxerophilic *Saccharomyces cerevisiae* in response to growth at reduced a_w (1,9–13). These studies have shed considerable light on the mechanisms that enable *Z. rouxii* to grow at reduced a_w, and the reasons why *S. cerevisiae* is much less successful.

A physiological function for the large accumulations of polyols in osmophilic yeasts was proposed by Brown and Simpson (1). They concluded that the primary function of the polyol was to act as a compatible solute, enabling the metabolic processes of the cell to be maintained, while also functioning to balance the external osmotic pressure. They showed that the activity of a crude preparation of NADP-specific isocitrate dehydrogenase was affected only slightly when glycerol was used to lower a_w, but when sucrose was the controlling solute, enzyme activity dropped to about 10% of the maximum rate, indicating that glycerol was more "compatible" with cellular activities than sucrose.

Brown (9) observed that *Z. rouxii* accumulated large quantities of arabitol in response to lowered a_w (0.92), whereas *S. cerevisiae* accumulated much less arabitol and smaller quantities of glycerol and a hexitol (probably mannitol). Amphotericin B caused rapid leakage of arabitol from *Z. rouxii*, indicating that membrane integrity is a major requirement for the retention of intracellular polyols. There are fundamental differences in the responses of the respiratory apparatus of these two yeasts to high concentrations of glucose, with respiratory quotient and NADH oxidase activity of *S. cerevisiae* greatly reduced by 36% (w/v) glucose, whereas these parameters are only affected to a minor degree in *Z. rouxii* (10).

Later studies (11,14) showed that glycerol, not arabitol, was accumulated by *Z. rouxii* in response to lowered a_w, while the intracellular concentration of arabitol remained constant. The total amount of glycerol produced by *Z. rouxii* remains nearly constant over a range of a_w, but the proportion retained by the cell rises as a_w is reduced. The response of *S. cerevisiae* to reduced a_w is the synthesis of higher levels of glycerol, much of which is rapidly leaked to the growth medium. Thus, the response of *Z. rouxii* to reduced a_w is at the level of glycerol permeation and transport, whereas that of *S. cerevisiae* is metabolic. Although arabitol can act as a "second line" compatible solute, cushioning the yeasts against initial osmotic stress (14), the important compatible solute and major osmoregulatory substance in *Z. rouxii* is glycerol, as indicated by its enzymological effects (1,14,15).

An obligate xerophilic mutant of *Z. rouxii* has been isolated (16,17), which is unable to grow at water activities corresponding to 20% (w/v) sucrose (a_w 0.98) or less. The composition of the cell envelope of the mutant yeast differs significantly from the parent strain in glucose:mannose ratio, hexosamine content, and lipid composition: lipids in the mutant contain more unsaturated fatty acids than the parent strain.

2. The Trigger for Glycerol Synthesis

In *S. cerevisiae*, glycerol synthesis in respone to a change in a_w is dependent on glycolysis and NADH generation, whereas in *Z. rouxii* NADPH generated by the pentose phosphate cycle is important. The metabolic control of both these pathways has been detailed (12), but the nature of the initial osmoregulatory signal that triggers the increased production or retention of glycerol remains to be elucidated. In *S. cerevisiae* at least, the apparent immediate response to the osmotic stress caused by an increase in salt concentration was the redirection of carbohydrate metabolism to glycerol production (12). In *S. cerevisiae*, glycerol is produced by a glycerol 3-phosphate dehydrogenase reduction of dihydroxyacetone phosphate to glycerol 3-phosphate, which is then dephosphorylated by a phosphatase to glycerol (18). G3P dehydrogenase levels increase with decreasing a_w of the growth medium, irrespective of controlling solute. This increase in glycerol 3-phosphate dehydrogenase levels is due to de novo synthesis, not to activation of a preexisting enzyme (18). Similarly, in *Z. rouxii*, protein synthesis has been demonstrated to be necessary for osmotic adaptation and glycerol accumulation (19).

The trigger for this response may be K^+ depletion, since transition from high to low a_w media causes a transient depletion of cellular K^+ in *S. cerevisiae* and *Z. rouxii*. K^+ transport in yeasts is dependent on ATP and is coupled to a proton pump that is responsible for the creation of an electrochemical potential across the plasma membrane (20). In *Z. rouxii* and *S. cerevisiae* a 5:1 coupling exists between proton efflux and K^+ uptake, a coupling destroyed by solute stress. After a lag period (perhaps while protein synthesis recovers), the second response of *S. cerevisiae* is an increase in the activity of G3P dehydrogenase, which, as suggested, may be induced by increased intracellular glycerol concentration (12).

However, exogenous K^+ does not play a significant role in glycerol synthesis in osmotically stressed *S. cerevisiae*, since this yeast can still synthesize and accumulate similar amounts of glycerol in K^+-replete and K^+-depleted media containing 2.28% NaCl (21).

3. Exclusion of Na^+

The exclusion of Na^+ from the cells is an important factor in survival for yeasts growing in high salt concentrations. In cells of *Z. rouxii*, the existence has been demonstrated of an ATPase, assumed to be Na^+/K^+ activated and Mg^{2+} dependent, which could serve as a sodium pump protecting the cells in a high salt environment (22). This ATPase is inhibited by ouabain, a cardiac glycoside that specifically inhibits cation transport. The ouabain-inhibited ATPase was found to be more active in cells grown in 18% NaCl than in cells grown without added salt.

4. Responses to Osmotic Shock

Salt stress affects phosphofructokinase and *sn*-glycerol 3-phosphate dehydrogenase, two enzymes in the glycerol synthetic pathway in *Z. rouxii* and *S. cerevisiae* (13). When cells of the two species were transferred from high-a_w media (0.997) to media containing 10% (w/v) NaCl, the two enzymes reacted differently. In *Z. rouxii*, the activity of glycerol 3-phosphate dehydrogenase remained steady, and the activity of phosphofructokinase decreased by about half. In *S. cerevisiae*, glycerol 3-phosphate dehydrogenase activity increased dramatically, though only after a lag period of about 40 hr, and phosphofructokinase activity doubled. The viability of *Z. rouxii* did not decrease on transfer from basal medium to 10% NaCl. Glycerol synthesis continued a steady rate, and a higher proportion was retained in the medium of lower a_w. *S. cerevisiae*, in contrast, showed a dramatic decrease in viability on transfer; those cells which survived took 140 hr to fully adapt to growth in 10% NaCl, compared with less than 6 hr for *Z. rouxii*. Glycerol was produced at a much higher rate, but a large proportion was lost to the growth medium (13).

In *Z. rouxii*, changes in intracellular levels of glycerol and inorganic ions (Na^+, K^+, and Cl^-), were observed when cells were subjected to hypo- and hyperosmotic shock by transfer from 1 M NaCl to medium without added salt, or 2 M NaCl (23). The cell volume increased and decreased under conditions of hypo-osmotic and hyperosmotic stress, respectively, indicating gross movement of water into or out of the cells. The intracellular concentrations of solute, calculated from the total content of glycerol or inorganic ions and the cell volume, became almost equivalent to the external osmotic pressure after the shift to hypo-osmotic or hyperosmotic conditions (23). Gross hypo-osmotic shock (transfer from 2.5 M NaCl to 100 mM sodium carbonate buffer) caused the release of large amounts of protein and ultraviolet-absorbing material from *Z. rouxii* (24). Cells subjected to this alkaline hypo-osmotic shock could no longer grow on media containing high levels of NaCl.

C. Solute Accumulation in Other Xerophilic Yeasts

1. Glycerol Accumulation by *Debaryomyces hansenii* and the Role of K^+

Although polyols, particularly glycerol, apparently play a major role in osmoregulation in yeasts, other factors must also be involved. In *Debaryomyces hansenii*, a highly salt-tolerant yeast, Na^+ is excluded, and K^+ accumulated, so that the internal $K^+:Na^+$ ratio is much higher than that of the media. *S. cerevisiae* is also able to accumulate some K^+, but it is not nearly as efficient as *D. hansenii* under the same conditions (25). Although the K^+ probably contributes toward maintaining the internal osmotic balance in *D. hansenii*, glycerol is also produced and accumulated as the salinity of the culture medium increases (26,27). The latter effect is probably more important.

The intracellular levels of polyols in *D. hansenii* are markedly enhanced by high salinity, the dominant solutes being glycerol in the log phase (27) and arabitol in stationary-phase cells (28). In a low-salt medium (4 mM NaCl) arabitol was the main intracellular solute throughout the growth cycle. The arabitol accumulated by *D. hansenii* during the stationary phase is probably more important as an energy store than as an osmoregulator. Arabitol accumulation is independent of glucose concentration in the growth medium (at least between 1% and 9%) and is rapidly depleted during regrowth in fresh glucose medium (29). Significant amounts of glycerol were detected in the growth medium when *D. hansenii* was grown in 2.7 M NaCl, although it appeared to be reabsorbed and metabolized later in the growth cycle (28). This contrasts with the lack of glycerol detected in the growth medium of *Z. rouxii* (13). However, a salt-tolerant isolate of *Z. rouxii* from soy sauce was reported to excrete glycerol into the medium when grown in 18% NaCl (30). *D. hansenii* releases considerable quantities of glycerol into the growth medium if it is suddenly transferred to a medium of lower salinity (27).

Glycerol uptake and metabolism in *D. hansenii* has been studied by comparing a wild-type strain with a mutant incapable of utilizing glycerol (31). Glycerol was taken up by active transport, and uptake was similar in both strains. However, the mutant was unable to reutilize glycerol produced for osmoregulation because of a deficiency in mitochondrial G3P dehydrogenase. Alder et al. (31) concluded that dissimilation of glycerol in *D. hansenii* proceeds via glycerol kinase and mitochondrial G3P dehydrogenase, although an alternative pathway via glycerol dehydrogenase and dihydroxyacetone phosphate kinase (the DHA pathway) also exists.

2. Polyol Accumulation in *Hansenula anomala*

Hansenula anomala is a xerophilic ascomycetous yeast that can cause spoilage in fruit concentrates, maple syrup, condensed milk, molasses, and salt beans (3,32). It has been investigated for industrial production of glycerol (33). Like other yeasts studied, *H. anomala* produces glycerol as a compatible solute when grown at reduced a_w (34,35). Accumulation occurs during exponential growth, and the intra:extracellular concentration ratio may be as high as 10,000-fold (34). Arabitol is accumulated also, particularly during late exponential phase (35) and stationary phase (34), but at a much lower concentration ratio, and the amount of intracellular glycerol decreases. When *H. anomala* was grown in 2.25 M NaCl with acetate as the sole carbon source, arabitol was the only polyol detected by ^{13}C nuclear magnetic resonance (NMR) (35).

D. Conclusion

The ability of xerophilic yeasts to grow in high solute concentrations appears to be based primarily on the intracellular accumulation of compatible solutes, particularly glycerol. Active transport mechanisms for glycerol appear to exist in both *Z. rouxii* (36) and *D. hansenii* (31). Regulation of the glycerol permeability of the membrane is important, so that the metabolic energy expended in the synthesis of glycerol is not wasted by leakage from the cell. Maintenance of a low intracellular concentration of Na^+ is important for enzyme function, and K^+ and H^+ fluxes appear to be important in this respect. Such fluxes may also prove to be the initial osmoregulatory signal for glycerol production and accumulation.

III. PHYSIOLOGY OF XEROPHILIC FILAMENTOUS FUNGI

A. Water Relations of Xerophilic Filamentous Fungi

1. The Xerophilic Fungi

Pitt (37) defined a xerophilic fungus as "a fungus which is capable of growth, under at least one set of environmental conditions, at a water activity below 0.85." Compared with the published works on xerophilic yeasts, there is little information on the osmoregulatory mechanisms of xerophilic filamentous fungi. It had been assumed (14) that these microorganisms used polyols, probably mainly glycerol, as compatible solutes in the same way as yeasts and halophilic algae, and this has since been shown to be the case (38–43).

Despite the lack of studies on the physiological bases of xerophilism in these fungi, a great deal of empirical information on growth of filamentous fungi at low a_w exists. Snow (44) determined the minimum humidities under which spores of various xerophilic fungal species would germinate. The most xerophilic was *Aspergillus echinulatus*, occasional spores of which germinated at 0.62 a_w after 2 years. Other members of the *Aspergillus glaucus* group (*Eurotium* species) had minima nearer 0.70 a_w. Snow postulated that the thick walls of the ascospores might enable them to survive long periods of desiccation, but their ability to germinate and grow at such low water activities suggested that "the osmotic pressure of their sap must be much higher than that of the majority of other moulds," perhaps the first inference that these fungi accumulated intracellular solutes to balance the external a_w.

An extensive review (37) listed the known xerophiles and their reported a_w limits, drawing on published data (44–47). Of the 44 fungi listed, only three were yeasts. The filamentous fungi were from a limited number of genera: *Aspergillus* (including the ascomycetous states *Emericella* and *Eurotium*), *Chrysosporium, Eremascus, Paecilomyces, Penicillium, Wallemia,* and *Xeromyces.* More recent studies have added a few more species to the list, but the number of genera containing xerophilic species is still small (48–51).

2. Effects of Type of Solute in the Growth Medium

Type of solute, pH, temperature, and nutrient may affect the ability of a xerophilic fungus to grow at low a_w (37,44–46,52–56), but none has a specific solute requirement such as is found in halophilic bacteria. *Wallemia sebi,* considered by Frank and Hess (57) to be an obligate halophile, also grows well on media containing high sugar (58) or high glycerol concentrations (52). Two fungi commonly found on salted dried fish in tropical regions, *Polypaecilum pisce* and *Basipetospora halophila* (59), grow significantly better on media containing salt, but they do not have a salt requirement (49,55).

The more fastidious xerophiles such as *Xeromyces bisporus* and *Chrysosporium fastidium*, normally found in high sugar environments, are intolerant of salts such as NaCl and KCl irrespective of pH, but will grow in glycerol, although poorly (52,60). Under favorable conditions, such fungi demonstrate supreme adaptation to their low-a_w environment, growing rapidly at water activities between 0.75 and 0.70. Indeed, *X. bisporus* is the most xerophilic of all known microorganisms, being capable of growth near 0.60 a_w (45,52).

The less specialized xerophiles such as *Eurotium, Aspergillus,* and *Penicillium* species are able to grow well in a wider range of solutes, but they generally cannot grow at such low water activities as the extreme xerophiles (48,52,53,60).

3. Effects of a_w on Sporulation

As a_w of the growth medium decreases, the ability of a fungus to produce reproductive structures is affected. This is particularly noticeable in teleomorphic xerophilic fungi such as *Eurotium* species and *Xeromyces bisporus.*

These fungi are able to form their anamorphic state at a lower a_w than that at which the teleomorph is formed (45). Production of some *Eurotium* teleomorphs ceases at 0.84 a_w, but may be produced as low as 0.77 a_w in other species. Most species of *Eurotium* can form their *Aspergillus* state at 0.02 a_w units below that at which the teleomorph is produced. The differential between the lowest a_w value at which the teleomorph is formed and that at which only the *Aspergillus* state can be produced varies from 0.02 to 0.09 a_w units (45). This effect of reduced a_w on production of *Eurotium* cleistothecia was also noted by Kushner et al. (61), although these authors did not record the a_w values at which the teleomorph was no longer formed. Even in anamorphic fungi, reduced a_w can have an effect on the types of spores produced: *Paecilomyces farinosus* produced more blastospores per unit biomass at 0.958 a_w than at higher water activities (62).

X. bisporus (Fig. 1) is remarkable in that it is able to form mature ascospores at 0.67 a_w, much lower than any of the common *Eurotium* species can even grow (45). In *X. bisporus*, the entire sexual life cycle can occur in

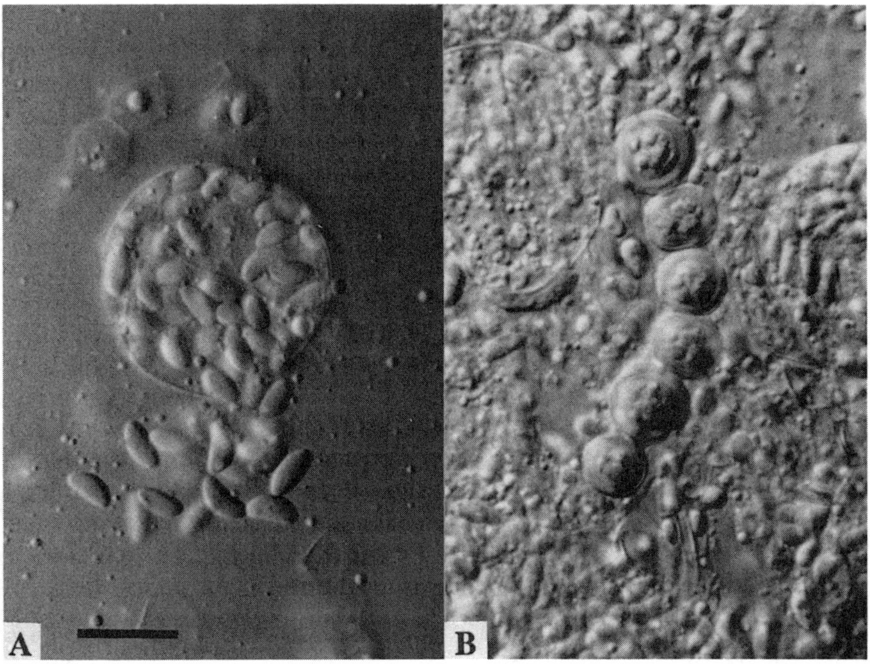

Figure 1 *Xeromyces bisporus*, an extremely xerophilic species, produces cleistothecia and ascospores (A) at water activities as low as 0.67. At very high (0.96–0.93) and very low (0.67–0.62) a_w, this fungus produces only thick-walled aleurioconidia (B).

an extremely concentrated environment, equivalent to syrup of 77.5% (w/w) fructose of 65% (w/w) glycerol. The mechanisms that enable DNA to be transcribed, or meiosis to occur, under such extreme conditions are unknown. *X. bisporus* can still grow very slowly and produce aleurioconidia at 0.62 a_w, equivalent to 80% (w/w) fructose (45).

4. Effects of High a_w on Xerophilic Filamentous Fungi

Most xerophilic fungi exhibit maximum growth rates at an a_w below 0.997; i.e., their optimum a_w for growth is below that of normal bacteriological media (6,46,52–54,60). This effect is more pronounced in extreme xerophiles, some of which are quite unable to grow at high a_w. *X. bisporus*, under the most favorable conditions of pH, nutrient, temperature, and solute type, is unable to grow above 0.96 a_w (52).

The optimum a_w values for growth of most *Eurotium* species lie between about 0.96 and 0.90 (49–52) (Fig. 2). An exception is the extremely xero-

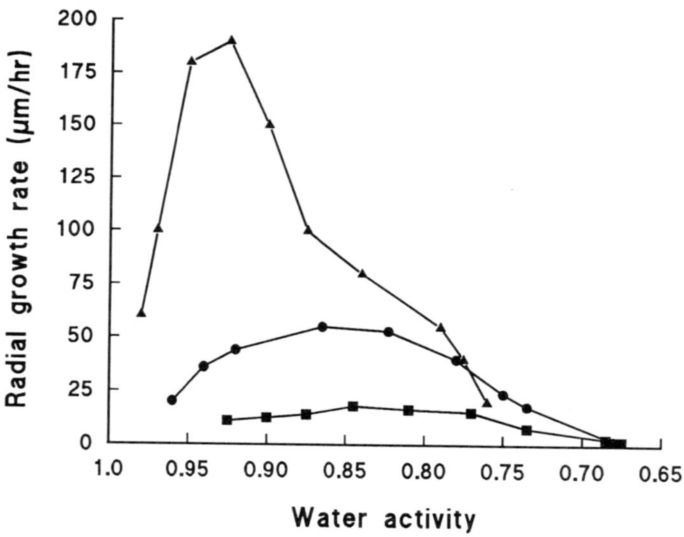

Figure 2 The effect of a_w on growth rates of three xerophilic fungi on glucose/fructose-based media at 25°C, showing optimal growth rates well below the a_w of normal bacteriological media (0.997). (▲) *Eurotium amstelodami*, representative of most *Eurotium* species, has a fairly sharply defined growth optimum between 0.96 and 0.90. (●) *Xeromyces bisporus* has a less well-defined growth optimum between about 0.92 and 0.80. (■) *E. halophilicum* is an extremely xerophilic species, which grows at a similar rate between about 0.90 and 0.74 a_w. (Data from Refs. 49, 51, and 52.)

philic species *Eurotium halophilicum*, which has a maximum a_w for growth at 0.94 and grows optimally between 0.88 and 0.74 a_w (49) (Fig. 2). *X. bisporus* grows best between 0.92 and 0.80 a_w (52) (Fig. 2).

The halophilic xerophiles, *Polypaecilum pisce* and *Basipetospora halophila*, grow optimally over a wider a_w range on media containing NaCl rather than glucose as the controlling solute (49,55), with the optimal a_w range for *P. pisce* being 0.96–0.85 and for *B. halophila* 0.92–0.83 at 30°C (55) (Fig. 3).

The physiological limitations that inhibit or prevent growth of xerophilic fungi at high a_w values are not understood. Studies on *X. bisporus* indicate that this species leaks large quantities of glycerol, particularly at high (0.95) a_w (63). It is possible that the energy efficiency of *X. bisporus* is very poor, so this species may have a requirement for high concentrations of sugars, to provide both energy for growth and a reduced a_w environment.

Studies on the effects of high a_w on growth and metabolism of *Eurotium rubrum* (as *Aspergillus sejunctus*) (64) found that the fungus had a lower protein-synthesizing capacity at higher water activities. It is possible that high a_w affects the permeability of the cell membrane, limiting the rate of nutrient uptake, and this could be a rate-limiting step affecting growth and protein synthesis (64).

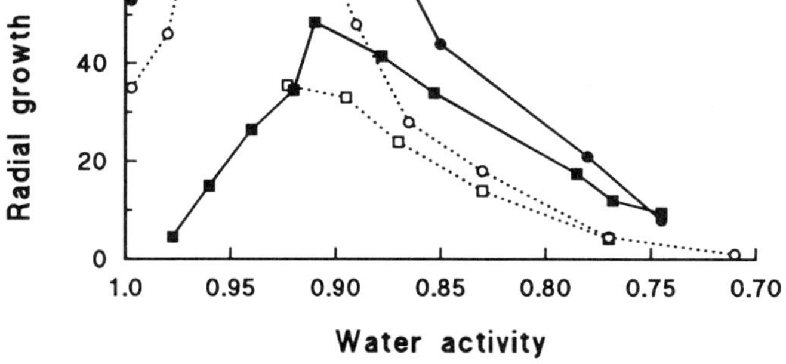

Figure 3 Growth of two halophilic xerophiles, *Polypaecilum pisce* (●, ○) and *Basipetospora halophila* (■, □), at 30°C on media where the a_w is adjusted with NaCl (solid symbols) or glucose/fructose (open symbols). These fungi grow optimally over a wider a_w range in NaCl than in glucose/fructose-based media. (Data from Ref. 55.)

5. Interactions and Competitive Inhibition Between Species of Xerophilic Filamentous Fungi

The combination of reduced a_w and type of solute exerts strong selective pressures on the fungi present in an ecosystem. This is particularly evident in systems containing high concentrations of NaCl (Fig. 4). Although there are no known obligately halophilic fungi, some species grow much faster in the presence of salt and are thus highly adapted and highly competitive in saline environments. In salt-based media below 0.88 a_w, *Polypacilum pisce* and *Basipetospora halophila* will both outgrow *Eurotium* species such as *E. rubrum* at 30°C (65,66), but below 0.78 a_w, *B. halophila* is dominant. At 30°C in NaCl, *Wallemia sebi* appears to compete poorly, irrespective of a_w, if growth rate is the only criterion considered (Fig. 4).

However, competitiveness and relative growth rates of xerophilic fungi on low-a_w foods, such as dried fish or very high sugar substrates, are also influenced by other, less easily defined factors, such as the composition and structure of the food itself (66). Growth rate alone is not the only factor

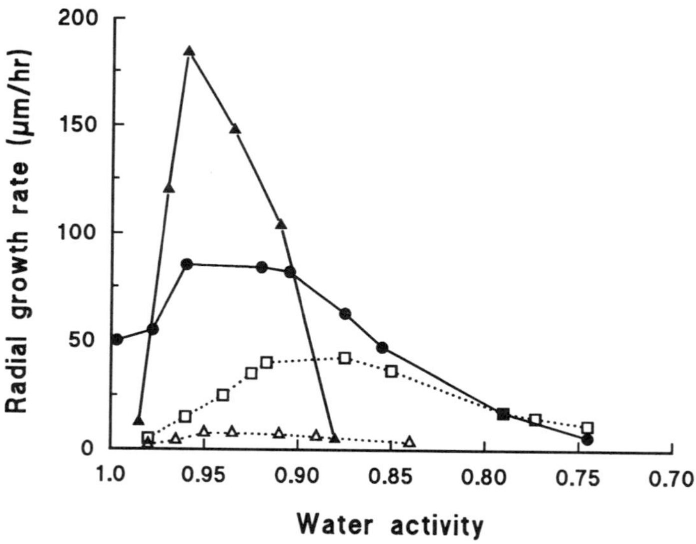

Figure 4 Comparison of growth rates over a range of a_w values can give an indication of the competitiveness of fungal species in mixed culture. At 30°C, in media with NaCl as the controlling solute, *Eurotium rubrum* (▲) is dominant between 0.98 and 0.88 a_w. Below 0.88 a_w, *Polypaecilum pisce* (●) and *Basipetospora halophila* (□) will outgrow common *Eurotium* species, with *B. halophila* dominating below 0.78 a_w. *Wallemia sebi* (△) grows very slowly over its a_w range at 30°C, and competes poorly. (Data from Refs. 56 and 66.)

that determines dominance in mixed populations. Interactions between the fungi, and the number of spores, the rapidity and efficiency with which they can be produced, and their longevity, will also determine how effectively a fungus can compete.

B. The Function of Polyols in Xerophilic Filamentous Fungi

1. Distribution of Polyols

The distribution and function of polyols in filamentous fungi has been reviewed by Lewis and Smith (67). However, these authors specifically omitted glycerol from their review. The most commonly occurring polyol in higher fungi was mannitol; other polyols, such as arabitol, xylitol, dulcitol, and sorbitol, also occurred. Mannitol and other polyols are frequently regarded as storage products of carbohydrate metabolism. Lewis and Smith (67) mentioned in passing that mannitol "may be important in maintaining high osmotic pressures in mycelium during periods of water stress," but there was no mention of polyols functioning primarily as osmoregulatory compounds. Conidia of fungi such as *Penicillium* and *Aspergillus* species may contain large amounts of trehalose, mannitol, and other polyols. It is thought that these compounds may be used as the energy source for endogenous respiration and for germination (68,69). Polyols may, in addition to their role as storage carbohydrates, perform the function of storage of reducing power. Polyols may also have a role in coenzyme regulation, e.g., in the regulation of NADPH/NADP ratios (67).

2. Polyol Accumulation in Response to Reduced a_w

Before 1980, accumulated polyols in fungi were assumed to have a storage function (70,71). However, Luard (38,39) reported that glycerol was accumulated in response to decreased a_w in the xerophilic species *Chrysosporium fastidium* and in the "facultative" xerophile *Penicillium chrysogenum*. Mannitol was also present in both species, but decreased in concentration as the a_w was reduced. Arabitol and erythritol were also detected in low concentrations; the levels of arabitol remained constant irrespective of a_w, and erythritol concentrations appeared to be related to growth rate rather than a_w. Luard suggested that mannitol may function as a carbohydrate source or energy reserve for glycerol production, arabitol may act as a minor osmoregulatory compounds, and erythritol may function as a reducing power reserve. She assigned to glycerol the role of primary osmoregulatory compound in these fungi. Luard (38,72) also reported that the amino acid proline is the main osmoregulatory substance in the lower fungus *Phytophthora cinnamomi*. Polyols, apart from a trace of arabitol, were not detected. Potassium was the predominant cation in the three species studied. *C. fastidium* was able to accumulate K^+ and to exclude Na^+ at low a_w (38).

Natural abundance ^{13}C NMR has been used to show that glycerol was accumulated by four xerophilic fungi and one nonxerophilic species in response to reduced a_w when grown on sugar-based or sorbitol-based media (73). Erythritol, arabitol, mannitol, and trehalose were also detected, but at lower concentrations. Culture age is important when assessing the glycerol content of fungal mycelium, as glycerol is rapidly depleted during the production of conidia, especially when cultures are grown on salt-based media with a restricted energy source (74). Thus, although glycerol obviously has an osmoregulatory function, it may also serve as an energy reserve for sporulation.

Natural abundance ^{13}C NMR has also been used to demonstrate that glycerol is important in maintaining the osmotic balance in the copper-tolerant fungus *Penicillium ochro-chloron* when it is grown in elevated concentrations of Na^+ or Cu^{2+} (0.523 M NaCl or 1.129 M $CuSO_4$) (75). A 15-fold increase in glycerol concentration was observed between *P. ochro-chloron* cultures grown in the control medium (0.1 g/L NaCl) and those grown in a medium containing 0.5 M NaCl (approx. 0.96 a_w). Concentrations of other low-molecular-weight organic compounds did not rise significantly, although cell K^+ rose slightly.

As in yeasts, intracellular glycerol accumulated in response to reduced a_w is quickly lost if fungi are transferred to growth media of higher a_w. Transfer of three *Aspergillus* species, *A. niger, A. ochraceus,* and *A. tamarii* from media containing 12.8% NaCl to media without added NaCl resulted in a decrease in intracellular glycerol, and a concomitant increase in glycerol in the growth media (43).

3. Polyols and Maintenance of a Positive Turgor Pressure

Hyphal tip extension in growing fungi requires the maintenance of a positive turgor pressure relative to the external environment. When the hyphal osmotic potential of various fungi growing over a range of water activities was measured by psychrometry, all fungi maintained a positive turgor potential with respect to the growth medium (60). For each species, the difference in turgor potential between the hyphae and the growth medium was reasonably constant, irrespective of the a_w (or water potential) of the medium. However, lower fungi such as *P. cinnamomi* and *Phellinus noxius*, which are intolerant of reduced a_w, only maintained their hyphae at about 0.9–1.0 MPa above the growth medium, whereas the xerophilic species (*Eurotium, Aspergillus, Penicillium, Chrysosporium,* and *Xeromyces*) maintained a positive difference of 2.5–3.3 MPa between hyphal turgor potential and the water potential of the growth medium (60).

IV. GENERAL CONSIDERATIONS

A. Solute Accumulation and Limiting a_w for Growth

Growth of microorganisms at reduced a_w depends, in part at least, on the accumulation of a compatible solute to balance the solute concentration in the external environment. In a review on solute accumulation and the regulation of cell water activity, Borowitzka (76) drew together convincing evidence that glycerol is the most effective compatible solute (Table 1). Those microorganisms which are able to produce and retain glycerol are the most xerophilic, e.g., the alga *Dunaliella*, the yeasts *Debaryomyces hansenii* and *Zygosaccharomyces rouxii*, and a number of xerophilic fungi. Halophilic bacteria that accumulate K^+ are an exception, but represent a special case. Accumulation of glycerol permits growth below 0.75 a_w. The next most effective compatible solute appears to be betaine, either alone or in combination with glutamate or K^+. Organisms accumulating betaine are capable of growth as low as 0.90 a_w. Accumulation of α-glycosylglycerol in cyanobacteria allows growth to 0.92 a_w, and proline and/or glutamate accumulation allows growth to 0.94–0.96 a_w in bacteria. Many microorganisms can grow as low as 0.96 a_w by accumulating a variety of solutes such as mannitol, floridoside,

Table 1 Water Activity Regulating Solutes of Microorganisms and Their Limiting a_w for Growth

Microorganism	Main solute(s) accumulated	Approx. minimum a_w for growth
Cyanobacteria, freshwater	Sucrose	0.98
Bacteria, most species	Glycine, betaine, proline, glutamate	0.97–0.90
Algae, most species	Proline, mannitol	0.96
Bacteria, marine	Proline, glutamate	0.95
Cyanobacteria, marine	α-glucosylglycerol	0.92
Cyanobacteria, salt lakes	Betaines	0.90–0.75
Halophilic bacteria	K^+	0.75
Dunaliella spp. (salt lake algae)	Glycerol	0.75
Xerophilic yeasts (e.g., *S. rouxii*, *D. hansenii*)	Glycerol	0.83–0.62
Xerophilic filamentous fungi (e.g., *X. bisporus, W. sebi*, *Eurotium* species)	Glycerol	0.72–0.61

Source: Adapted from Ref. 76.

isofloridoside, sucrose, trehalose, and mixtures of amino acids. Although the accumulation of a particular solute does not rigidly determine the limiting a_w for growth, nevertheless, a good correlation exists between the type of solute accumulated and the minimum a_w at which a particular microorganism can grow (Table 1) (76).

B. Membrane Structure, Lipid Composition, and Growth at Reduced a_w

Little information has been published on the mechanisms that enable the maintenance of positive turgor potential of fungal hyphae, particularly at low a_w against strong concentration gradients. Small, uncharged molecules such as polyols easily pass through biological membranes, so there must be some changes in membrane structure and/or function at reduced a_w that render the plasma membrane less permeable to glycerol. Either xerophile membranes are intrinsically different, or reduced a_w induces changes that alter membrane permeability.

One possible mechanism is a change in composition of the membrane phospholipids, in particular the degree of unsaturation of the fatty acids, as the degree of unsaturation affects fluidity, and hence permeability, of the membrane (77). De Gier et al. (78) showed that the introduction of double bonds into the paraffin chains of lecithin liposomes caused a marked increase in their permeability to glycerol and erythritol.

Similarly, incorporation of branched-chain or unsaturated fatty acids or fatty acids of reduced chain length into the membranes of the mycoplasma *Acholeplasma laidlawii* increased the permeability to glycerol and erythritol (79). The incorporation of cholesterol into the membrane reduced the nonelectrolyte permeability of both the plasma membrane of *A. laidlawii* and liposomes derived from it (79,80). A separate study showed that the [^{14}C] erythritol equilibrium flux for *A. laidlawii* cells grown on different fatty acids decreased as the degree of saturation increased, with cells grown on linoleic acid (C18:2) much more permeable to erythritol than those grown on stearic acid (C18:0) (80).

Microorganisms adjust the fatty acid composition of their membrane phospholipids to compensate for changes in growth temperature. At lower temperatures, the degree of unsaturation increases, allowing the membrane to become more fluid, and thus more permeable. This occurs in bacteria (77,81,82), in yeasts (83,84), and in fungi (85,86), and also in protozoa (87). However, no significant differences were observed in the fatty acid composition of *S. cerevisiae* cells grown at 30°C and 15°C (88).

In comparison with temperature effects, little has been published on the effects of a_w on the fatty acid composition of biological membranes, especially in eukaryotes. A review on fungal lipids (85) cited a Russian reference that reported that "the ratio of saturated to unsaturated fatty acids produced by [the fungus] *Blakeslea trispora* varied inversely with sugar concentration."

In *Candida albicans*, NaCl concentration affects the lipid content and fatty acid composition (89). NaCl in excess of 1% decreases the relative amounts of total unsaturated fatty acids. Cells harvested from media that contained 3.5% NaCl or higher concentrations contained higher levels of saturated fatty acids, while the levels of total unsaturated fatty acids decreased (89). Increasing concentrations of glucose and galactose in the growth medium led to a decrease in the content of unsaturated fatty acids in *S. cerevisiae* (90). Such changes might be expected to cause a decrease in membrane fluidity and permeability, based on the "fluid mosaic" model of Singer and Nicolson (91).

Although the effects of a_w on the lipid composition of fungi have been little studied, a sizable pool of data exists on the lipid and fatty acid composition of fungi (85). The fatty acid composition varies greatly according to species and growth conditions. In hyphomycetes and ascomycetes, the most commonly occurring fatty acids are 16:0 (palmitic), 18:0 (stearic), 18:1 (oleic), and 18:2 (linoleic), with smaller amounts of 18:3 (linolenic) and some other C14, C20, and C22 fatty acids.

There is little information on whether the fatty acids of xerophilic species differ from those of closely related nonxerophilic species, or whether the fatty acid composition in xerophiles is less affected by changes in a_w than in nonxerophiles. Differences have been reported in fatty acid composition of a two *Penicillium* species, a xerophile, *P. janczewskii*, and a non-xerophile, *P. digitatum*, with the lipids of *P. digitatum* being more unsaturated because of the presence of up to 10% α-linolenic acid (C18:3ω3) (63). Such a high degree of unsaturation could affect membrane fluidity and account for the inability of *P. digitatum* to retain accumulated glycerol intracellularly (63). Increased salinity in the growth media of three nonxerophilic fungi (*Alternaria phragmospora, A. chlamydospora,* and *Ulocladium chartarum*) caused a decrease in the major fatty acid (C18:1), but the unsaturated fatty acids were greater than the saturated fatty acids (92).

Sterol composition is also important in membrane permeability, and a number of reports show that incorporation of cholesterol into membranes renders them less permeable to small organic molecules (93). The nonelectrolyte permeability of *A. laidlawii* membranes is reduced by incorporation of cholesterol (79,80). The presence of a 3-β-hydroxyl group and a planar configuration of the sterol molecule may be essential requirements for permeability to be reduced (80).

C. Active Transport Mechanisms

The possibility of the existence of active transport mechanisms is another major consideration in the permeability of membranes to molecules such as glycerol. There is good evidence for such a transport system in the xerophilic yeasts *D. hansenii* (31) and *Z. rouxii* (36), and it seems logical that similar systems should also exist in xerophilic filamentous fungi. However, these systems remain to be investigated.

D. Genetic Controls

The genetic control of the a_w-regulating response in eukaryotic xerophiles and extremely halophilic bacteria has yet to be studied; however, some clues to possible mechanisms and their control may be gained from studies on the bacteria *Escherichia coli* and *Salmonella typhimurium*. *E. coli*, when confronted with reduced a_w, can activate or induce transport systems for several a_w-lowering solutes, such as proline and glycine betaine, and precursors such as choline (94). Osmoresponsive operons encoding for inducible transport systems for K^+ and for proline have been identified in *E. coli* (95), and a gene encoding a transport system for betaine has been identified in *S. typhimurium* (96). Expression of these gene rises if the a_w of the growth medium is reduced (95,96). There is some evidence that cytomplasmic ionic strength plays an important regulatory role in this osmotic response, modulating gene expression and enzyme function in response to changes in extracellular a_w (97).

Investigations into the genetic control of osmoregulation continue in bacteria (98–100), but there appears to be very little progress in working toward an understanding of this process in filamentous fungi.

REFERENCES

1. Brown, A. D., and Simpson, J. R. (1972). Water relations of sugar tolerant yeasts: the role of intracellular polyols, *J. Gen. Microbiol. 72*: 589.
2. Onishi, H. (1963). Osmophilic yeasts, *Adv. Food Res. 12*: 53.
3. Tilbury, R. H. (1978). The microbial stability of intermediate moisture foods with respect to yeasts, in *Intermediate Moisture Foods* (R. Davies, G. G. Birch, and K. J. Parker, eds.), Applied Science Publishers, London, p. 138.
4. Nickerson, W. J., and Carroll, W. R. (1945). On the metabolism of *Zygosaccharomyces, Arch. Biochem. 7*: 257.
5. Spencer, J. F. T., and Sallans, H. R. (1956). Production of polyhydric alcohols by osmophilic yeasts, *Can. J. Bot. 2*: 72.
6. Scott, W. J. (1957). Water relations of food spoilage microorganisms, *Adv. Food Res. 7*: 83.

7. Ingram, M. (1957). Micro-organisms resisting high concentrations of sugars or salts, in *VIIth Symposium of the Society for General Microbiology: Microbial Ecology*, Cambridge University Press, Cambridge, p. 90.
8. Anand, J. C., and Brown, A. D. (1968). Growth patterns of the so-called osmophilic and nonosmophilic yeasts in solutions of polyethylene glycol, *J. Gen. Microbiol. 52*: 205.
9. Brown, A. D. (1974). Microbial water relations: features of the intracellular composition of sugar tolerant yeasts, *J. Bacteriol. 118*: 769.
10. Brown, A. D. (1975). Microbial water relations. Effects of solute concentration on the respiratory activity of sugar tolerant and non-tolerant yeasts, *J. Gen. Microbiol. 86*: 241.
11. Edgley, M., and Brown, A. D. (1978). Response of xerotolerant and non-tolerant yeasts to water stress, *J. Gen. Microbiol. 104*: 434.
12. Brown, A. D., and Edgley, M. (1980). Osmoregulation in yeast, in *Genetic Engineering of Osmoregulation* (D. W. Rains, R. D. Valentine, and A. Hollander, eds.), Plenum Press, New York, p. 75.
13. Edgley, M., and Brown, A. D. (1983). Yeast water relations: physiological changes induced by solute stress in *Saccharomyces cerevisiae* and *Saccharomyces rouxii, J. Gen. Microbiol. 129*: 3453.
14. Brown, A. D. (1978). Compatible solutes and extreme water stress in eukaryotic microorganisms, *Adv. Microbial Physiol. 17*: 181.
15. Simpson, J. R. (1976). Water relations of the sugar-tolerant yeast *Saccharomyces rouxii,* Ph.D. thesis, University of New South Wales, Australia.
16. Koh, T. Y. (1975). Studies on the "osmophilic" yeast *Saccharomyces rouxii* and an obligate osmophilic mutant, *J. Gen. Microbiol. 88*: 101.
17. Koh, T. Y. (1975). The isolation of obligate osmophilic mutants of the yeast *Saccharomyces rouxii, J. Gen. Microbiol. 88*: 184.
18. André, L., Hemming, A., and Adler, L. (1991). Osmoregulation in *Saccharomyces cerevisiae*. Studies on the osmotic induction of glycerol production and glyc;erol 3-phosphate dehydrogenase (NAD^+), *FEBS Lett. 286*: 13.
19. Jacobus van Zyl, P., Prior, B. A., and Kilian, S. G. (1991). Regulation of glycerol metabolism in *Zygosaccharomyces rouxii* in response to osmotic stress, *Appl. Microbiol. Biotechnol. 36*: 369.
20. Pena, A. (1975). Studies on the mechanism of K^+ transport in yeast, *Arch. Biochem. Biophys. 167*: 397.
21. Meikle, A. J., Reed, R. H., and Gadd, G. M. (1991). The osmotic responses of *Saccharomyces cerevisiae* in potassium-depleted medium, *FEMS Microbiol. Lett. 78*: 89.
22. Steinkraus, K. H., Ayres, R., Olek, A., and Farr, D. (1985). Factors influencing the permeability of the cell membrane of the osmotolerant yeast *Saccharomyces rouxii* grown in the presence and absence of 18% NaCl. I. Na^+/K^+-activated $Mg2^+$-dependent ATPase, *Int. J. Food Microbiol. 1*: 291.
23. Yagi, T. (1991). Effects of increases and decreases in the external salinity on the intracellular glycerol and inorganic ion content in the salt-tolerant yeasts *Zygosaccharomyces rouxii, Microbios 68*: 109.

24. Hosono, K. (1991). The release of proteins and UV-absorbing materials from salt-tolerant yeast, *Zygosaccharomyces rouxii*, by osmotic shock, *J. Ferment. Bioeng. 72*: 445.
25. Norkrans, B., and Kylin, A. (1969). Regulation of the potassium to sodium ratio and the osmotic potential in relation to salt tolerance in yeasts, *J. Bacteriol. 100*: 836.
26. Gustafsson, L., and Norkrans, B. (1976). On the mechanisms of salt tolerance. Production of glycerol and heat during the growth of *Debaryomyces hansenii, Arch. Microbiol. 110*: 177.
27. André, L., Nilsson, A., and Adler, L. (1988). The role of glycerol in osmotolerance of the yeast *Debaryomyces hansenii, J. Gen. Microbiol. 134*: 669.
28. Adler, L., and Gustafsson, L. (1980). Polyhydric alcohol production and intracellular amino acid pool in relation to halotolerance of the yeast *Debaryomyces hansenii, Arch. Microbiol. 124*: 123.
29. Nobre, M. F., and Da Costa, M. S. (1985). Factors influencing the accumulation of arabinitol in the yeast *Debaryomyces hansenii, Can. J. Microbiol. 31*: 467.
30. Onishi, H., and Shiromaru, Y. (1984). Physiological changes induced by salt stress in a salt-tolerant soy-yeast *Saccharomyces rouxii, FEMS Microbiol. Lett. 25*: 175.
31. Adler, L., Blomberg, A., and Nilsson, A. (1985). Glycerol metabolism and osmoregulation in the salt-tolerant yeast *Debaryomyces hansenii, J. Bacteriol. 162*: 300.
32. Barnett, J. A., Payne, R. W., and Yarrow, D. (1990). *Yeasts. Characteristics and Identification*, 2nd ed., Cambridge University Press, Cambridge.
33. Parekh, S. K., and Pandey, N. K. (1985). Production of glycerol by *Hansenula anomala, Biotechnol. Bioeng. 27*: 1089.
34. Van Eck, J. H., Prior, B. A., and Brandt, E. V. (1989). Accumulation of polyhydroxy alcohols by *Hansenula anomala* in response to water stress, *J. Gen. Microbiol. 135*: 3505.
35. Bellinger, Y., and Larher, F. (1991). Salt tolerance and osmolyte composition of the yeast *Hansenula anomala* grown in the presence of fermentable or nonfermentable sources of carbon, *Sci. Aliments 11*: 37.
36. Jacobus van Zyl, P., Kilian, S. G., and Prior, B. A. (1990). The role of an active transport mechanism in glycerol accumulation during osmoregulation by *Zygosaccharomyces rouxii, Appl. Microbiol. Biotechnol. 34*: 231.
37. Pitt, J. I. (1975). Xerophilic fungi and the spoilage of foods of plant origin, in *Water Relations of Foods* (R. B. Duckworth, ed.), Academic Press, London, p. 273.
38. Luard, E. J. (1980). Water relations of fungi with particular reference to xerophytic species, Ph.D. thesis, Australian National University, Canberra, Australia.
39. Luard, E. J. (1982). Accumulation of intracellular solutes by two filamentous fungi in response to growth at low steady state osmotic potential, *J. Gen. Microbiol. 128*: 2563.

40. Hocking, A. D., and Norton, R. S. (1983). Natural-abundance ^{13}C nuclear magnetic resonance studies on the internal solutes of xerophilic fungi, *J. Gen. Microbiol. 129*: 2915.
41. Gadd, G. M., Chudek, J. A., Foster, R., and Reed, R. H. (1984). The osmotic responses of *Penicillium ochrochloron*: changes in internal solute levels in response to copper and salt stress, *J. Gen. Microbiol. 130*: 1969.
42. Hocking, A. D. (1986). Effects of water activity and culture age on the glycerol accumulation patterns of five fungi, *J. Gen. Microbiol. 132*: 269.
43. Zidan, M. A., and Abdel-Mallek, A. Y. (1987). Effect of NaCl on the accumulation of glycerol by three *Aspergillus* species, *J. Basic Microbiol. 27*: 393.
44. Snow, D. (1949). The germination of mould spores at controlled humidities, *Ann. Appl. Biol. 36*: 1.
45. Pitt, J. I., and Christian, J. H. B. (1968). Water relations of xerophilic fungi isolated from prunes, *Appl. Microbiol. 16*: 1853.
46. Ayerst, G. (1969). The effects of moisture and temperature on growth and spore germination in some fungi, *J. Stored Prod. Res. 5*: 127.
47. Mislivec, P. B., and Tuite, J. (1970). Temperature and relative humidity requirements of species of *Penicillium* isolated from yellow dent corn kernels, *Mycologia 62*: 75.
48. Hocking, A. D., and Pitt, J. I. (1979). Water relations of some *Penicillium* species at 25°C, *Trans. Br. Mycol. Soc. 73*: 141.
49. Andrews, S., and Pitt, J. I. (1987). Further studies on the water relations of xerophilic fungi, including the characterisation of some halophiles, *J. Gen. Microbiol. 133*: 233.
50. Wheeler, K. A., Hocking, A. D., and Pitt, J. I. (1988). Water relations of some *Aspergillus* species isolated from dried fish, *Trans. Br. Mycol. Soc. 91*: 631.
51. Wheeler, K. A., and Hocking, A. D. (1988). Water relations of *Paecilomyces variotii, Eurotium amstelodami, Aspergillus candidus* and *Aspergillus sydowii*, xerophilic fungi isolated from dried fish, *Int. J. Food Microbiol. 7*: 73.
52. Pitt, J. I., and Hocking, A. D. (1977). Influence of solute and hydrogen ion concentration on the water relations of some xerophilic fungi, *J. Gen. Microbiol. 101*: 35.
53. Avari, G. P., and Allsopp, D. (1983). The combined effect of pH, solutes and water activity (a_w) on the growth of some xerophilic *Aspergillus* species, *Biodeterioration 5*: 548.
54. Magan, N., and Lacey, J. (1984). Effect of temperature and pH on water relations of field and storage fungi, *Trans. Br. Mycol. Soc. 82*: 71.
55. Wheeler, K. A., Hocking, A. D., and Pitt, J. I. (1988). Influence of temperature on the water relations of *Polypaecilum pisce* and *Basipetospora halophila*, two halophilic fungi, *J. Gen. Microbiol. 134*: 2255.
56. Wheeler, K. A., Hocking, A. D., and Pitt, J. I. (1988). Effects of temperature and water activity of germination and growth of *Wallemia sebi, Trans. Br. Mycol. Soc. 90*: 365.
57. Frank, M., and Hess, E. (1941). Studies on salt fish. V. Studies on *Sporendonema epizoum* from "dun" salt fish, *J. Fish. Res. Board Can. 5*: 276.

58. Vaisey, E. B. (1954). Osmophilism of *Sporandonema epizoum, J. Fish. Res. Board Can. 11*: 901.
59. Pitt, J. I., and Hocking, A. D. (1985a). New species of fungi from Indonesian dried fish, *Mycotaxon 22*: 197.
60. Luard, E. J., and Griffin, D. M. (1981). Effect of water potential on fungal growth and turgor, *Trans. Br. Mycol. Soc. 76*: 33.
61. Kushner, L., Rozenweig, W. D., and Stotzky, G. (1979). Effects of salts, sugars and salt-sugar combinations on growth and sporulation of an isolate of *Eurotium rubrum* from pancake syrup, *J. Food Prot. 41*: 706.
62. Inch, J. M. M., and Trinci, A. P. J. (1987). Effects of water activity on growth and sporulation of *Paecilomyces farinosus* in liquid and solid media, *J. Gen. Microbiol. 133*: 247.
63. Hocking, A. D. (1986). Some physiological responses of fungi growing at reduced water activities, Ph.D. thesis, University of New South Wales, Australia.
64. Stevens, L., Dix, N. J., and Thompstone, A. (1983). Effects of high water activity on growth and metabolism of *Aspergillus sejunctus, Trans. Br. Mycol. Soc. 80*: 527.
65. Hocking, A. D. (1991). Xerophilic fungi in intermediate and low moisture foods, in *Handbook of Applied Mycology. Vol. 3: Foods and Feeds* (D. K. Arora, K. G. Mukerji, and E. H. Marth, eds.), Marcel Dekker, New York, p. 69.
66. Wheeler, K. A., and Hocking, A. D. (1993). Interactions among xerophilic fungi associated with dried fish, *J. Appl. Bacteriol. 74*: 164.
67. Lewis, D. H., and Smith, D. C. (1967). Sugar alcohols (polyols) in fungi and green plants. I. Distribution, physiology and metabolism, *New Phytol. 66*: 143.
68. Ballio, A., Vittorio, V., and Russi, S. (1964). The isolation of trehalose and polyols from the conidia of *Penicillium chrysogenum* Thom, *Arch. Biochem. Biophys. 107*: 177.
69. Horikoshi, K., Iida, S., and Ikeda, Y. (1965). Mannitol, and mannitol dehydrogenases in conidia of *Aspergillus oryzae, J. Bacteriol. 89*: 326.
70. Corina, D. L., and Munday, K. A. (1971). Studies on polyol function in *Aspergillus clavatus*: a role for mannitol and ribitol, *J. Gen. Microbiol. 69*: 221.
71. Holligan, P. M., and Lewis, D. H. (1973). The soluble carbohydrates of *Aspergillus clavatus, J. Gen. Microbiol. 75*: 155.
72. Luard, E. J. (1982). Growth and accumulation of solutes by *Phytophthora cinnamomi* and other lower fungi in response to changes in external osmotic potential, *J. Gen. Microbiol. 128*: 2583.
73. Hocking, A. D., and Norton, R. S.)1983). Natural-abundance ^{13}C nuclear magnetic resonance studies on the internal solutes of xerophilic fungi, *J. Gen. Microbiol. 129*: 2915.
74. Hocking, A. D. (1986). Effects of water activity and culture age on the glycerol accumulation patterns of five fungi, *J. Gen. Microbiol. 132*: 269.
75. Gadd, G. M., Chudek, J. A., Foster, R., and Reed, R. H. (1984). The osmotic responses of *Penicillium ochro-chloron*: changes in internal solute levels in response to copper and salt stress, *J. Gen. Microbiol. 130*: 1969.

76. Borowitzka, L. J. (1981). Solute accumulation and regulation of cell water activity, in *Physiology and Biochemistry of Drought Resistance in Plants* (L. G. Paleg and D. Aspinall, eds.), Academic Press, Sydney, p. 97.
77. Cronan, J. E., and Gelmann, E. P. (1975). Physical properties of membrane lipids: biological relevance and regulation, *Bacteriol. Rev. 39*: 232.
78. De Gier, J., Mandersloot, J. G., and van Deenan, L. L. M. (1968). Lipid composition and permeability of liposomes, *Biochim. Biophys. Acta 150*: 666.
79. McElhaney, R. N., de Gier, J., and van der Neut-Kok, E. C. M. (1973). The effect of alterations in fatty acid composition and cholesterol content on the nonelectrolyte permeability of *Acholeplasma laidlawii* B cells and derived liposomes, *Biochim. Biophys. Acta 298*: 500.
80. De Kruyff, B., de Greef, W. J., van Eyk, R. V. W., Demel, R. A., and van Deenan, L. L. M. (1973). The effect of different fatty acid and sterol composition on the erythritol flux through the cell membrane of *Acholesplasma laidlawii, Biochim. Biophys. Acta 298*: 479.
81. Ohno, Y., Yano, I., and Masui, M. (1979). Effect of NaCl concentration and temperature on the phospholipid and fatty acid composition of a moderately halophilic bacterium, *Pseudomonas halosaccharolytica, J. Biochem. 85*: 413–421.
82. Miller, K. J. (1985). Effects of temperature and sodium chloride concentration on the phospholipid and fatty acid compositions of a halotolerant *Planococcus* sp., *J. Bacteriol. 162*: 263.
83. Kates, M., and Paradis, M. (1973). Phospholipid desaturation in *Candida lipolytica* as a function of temperature and growth, *Can. J. Biochem. 51*: 184.
84. Prasad, R. (1985). Lipids in the structure and function of yeast membrane, *Adv. Lipid Res. 21*: 187.
85. Wassef, M. K. (1977). Fungal lipids, *Adv. Lipid Res. 15*: 159.
86. Martin, C. E., Siegel, D., and Aaronson, L. R. (1981). Effects of temperature acclimation on *Neurospora* phospholipids. Fatty acid desaturation appears to be a key element in modifying phospholipid fluid properties, *Biochim. Biophys. Acta 665*: 399.
87. Martin, C. E., Hiramitsu, K., Kitajima, Y., Nozawa, Y., Skriver, L., and Thompson, G. A. (1976). Molecular control of membrane properties during temperature acclimation. Fatty acid desaturase regulation of membrane fluidity in acclimating *Tetrahymena* cells, *Biochemistry 15*: 5218.
88. Hunter, K., and Rose, A. H. (1972). Lipid composition of *Saccharomyces cerevisiae* as influenced by growth temperature, *Biochim. Biophys. Acta 260*: 639.
89. Combs, T. J., Guarneri, J. J., and Pisaon, M. A. (1968). The effect of sodium chloride and the lipid content and fatty acid composition of *Candida albicans, Mycologia 60*: 1232.
90. Brown, C. M., and Johnson, B. (1970). Influence of the concentration of glucose and galactose on the physiology of *Saccharomyces cerevisiae* in continuous culture, *J. Gen. Microbiol. 64*: 279.
91. Singer, S. J., and Nicolson, G. L. (1972). The fluid mosaic model of the structure of cell membranes, *Science, New York 175*: 720.

92. Mulder, J. L., Ghannuom, M. A., Khamis, L., and Abu Elteen, K. (1989). Growth and lipid composition of some dematiaceous hyphomycete fungi grown at differeent salinities, *J. Gen. Microbiol. 135*: 3393.
93. Bloch, K. E. (1983). Sterol structure and membrane function, *CRC Crit. Rev. Biochem. 14*: 47.
94. Styrvold, O. B., Falkenberg, P., Landfald, B., Eshoo, M. W., Bjørnsen, T., and Strøm, A. R. (1986). Selection, mapping and characterization of osmoregulatory mutant of *Escherichia coli* blocked in the choline-glycine betaine pathway, *J. Bacteriol. 165*: 856.
95. Gowrishankar, J. (1985). Identification of osmoresponsive genes in *Esherichia coli*: evidence for participation of potassium and proline transport systems in osmoregulation, *J. Bacteriol. 164*: 434.
96. Cairney, J., Booth, I. R., and Higgins, C. F. (1985). *Salmonella typhimurium proP* gene encodes a transport system for the osmoprotectant betaine, *J. Bacteriol. 164*: 1218.
97. Higgins, C. F., Cairney, J., Stirling, D. A., Sutherland, L., and Booth, I. R. (1987). Osmotic regulation of gene expression: ionic strength as an intracellular signal? *Trends Biochem. Sci. 12*: 339.
98. Jovanovich, S. B., Martinell, M., Record, M. T., and Burgess, R. R. (1988). Rapid response to osmotic upshift by osmoregulated genes in *Escherichia coli* and *Salmonella typhimurium, J. Bacteriol. 170*: 534.
99. Lucht, J. M., and Brehmer, E. (1991). Characterization of mutations affecting osmoregulated *proU* promoter of *Escherichia coli* and identification of 5′ sequences required for high-level expression, *J. Bacteriol. 173*: 801.
100. Ramirez, R. M., and Villarejo, M. (1991). Osmotic signal transduction to *proU* is independent of DNA supercoiling in *Escherichia coli, J. Bacteriol. 173*: 879.

11

Anaerobic Fungi

Colin G. Orpin

Division of Tropical Crops and Pastures
Commonwealth Scientific and Industrial Research Organisation
St. Lucia, Queensland, Australia

I. INTRODUCTION

Anaerobic organisms similar in morphology and life cycle to Chytridiomycete fungi were first recognized as inhabitants of the rumen of sheep in 1975 (1). Proof of their identity as fungi was provided by the presence of chitin in the cell walls of the vegetative stages (2). Several genera of anaerobic fungi have now been described from the alimentary tracts and feces of ruminants and other, mostly large, herbivores (3–5). Despite extensive searches, anaerobic fungi have not been isolated from any habitat other than the alimentary tracts of herbivorous mammals (4).

The anaerobic fungi currently known (4,6,7) have life cycles consisting of an alternation of stages between a motile, flagellated zoospore stage, free-living in the gut contents, and a vegetative, reproductive stage, associated with plant fragments in the digesta. They were placed in the family Neocallimasticaceae within the Spizellomycetales of the Chytridiomycetes (8). The vegetative stage is either monocentric or polycentric, and the zoospores are generally mono- or polyflagellate, depending on the species. Sufficient evidence, particularly their variable morphology, uniform possession of hydrogenosomes, and novel fine structural features, has accumulated to warrant the erection of a new family, Anaeromycetales, to embrace the anaerobic fungi (9).

Anaerobic fungi from the rumens of domestic sheep and cattle have been studied more extensively than those from other herbivores, and much of which follows refers to isolates from these animals. Since the environmental conditions in the hindgut of large herbivores are similar in many respects to those of the rumen, and the fungi in both digestive systems are similar in morphology and in fine structure, it is likely that the metabolism and basic characters of the species of anaerobic fungi from the hindgut fermenters are similar to those of rumen species.

The most important stresses imposed by habitat conditions on anaerobic fungi can be summarized as a lack of molecular oxygen, low redox potential, scarcity of readily metabolizable compounds, high competition with other anaerobes and facultative anaerobes for readily assimilable nutrients, and the continual flow of digesta through the alimentary tract. Otherwise their habitat is rather friendly, and the provision of favorable conditions by the host animal includes a constant temperature, availability of organic substrates, removal of metabolites, periodic mixing, median to slightly acidic pH, and abundant inorganic nutrients and vitamins (10). The relationship of the anaerobic fungi with the host animal is one of symbiosis rather than parasitism, since the anaerobic fungi can apparently live in no other natural environment and contribute to the provision of nutrients, in the form of volatile fatty acids and microbial cells, to the host animal. The host animal depends on the activities of the rumen microorganisms for its nutritional requirements and thus its survival.

Transfer from animal to animal poses different stresses, such as exposure to oxygen and the essentiality of survival in air for finite periods to enable the organisms to be transmitted to new hosts.

II. ENVIRONMENTAL CONDITIONS IN THE HABITATS OF ANAEROBIC FUNGI

All anaerobic fungi so far discovered inhabit the alimentary tracts of herbivores, where they live in the reticulorumen of many ruminants and the hindgut of many large nonruminant herbivores, such as horses, elephants, and rhinoceri (3–5). Although anaerobic fungi have been isolated from all regions of the alimentary tracts of ruminants (11), they form large, growing populations only in the rumen. The environment of the hindgut is similar to that of the rumen, dominated by the scarcity of molecular oxygen, a low redox potential, abundant organic matter, and large numbers of other, often competitive, microorganisms (10,12).

The microbial population of the rumen is large, consisting of many species of anaerobic protozoa, anaerobic and facultatively anaerobic bacteria, anaerobic fungi, and other unidentified organisms (19). The population densities of microorganisms per gram of rumen contents are commonly as follows: protozoa, 10^6; bacteria, 10^{10}; anaerobic fungi, 10^5. Intense competition for readily metabolizable nutrients such as glucose occurs. The rumen fungi metabolize most soluble sugars, but are not dependent on them because they can utilize plant structural carbohydrates for growth (4,6,7).

The gas phase of rumen contents contains a mixture of carbon dioxide, methane, nitrogen, and low levels of hydrogen sulfide. Typical values for the ratios of these gases are 65:27:7:0.2:0.01:0.6, respectively (13). During eating and soon afterward, some oxygen is present in the gas phase. Because of high population densities of facultatively anaerobic bacteria and the oxygen-scavenging abilities of the facultative protozoa and fungi, oxygen tensions in rumen contents are generally low, less than 1.6 μM (14), and often unmeasurable. At these concentrations oxygen is not toxic to the rumen fungi (15), which have an active NADH oxidase system that rapidly reduces the oxygen to water. The oxygen-scavenging abilities of the anaerobic protozoa also contribute to protection of the fungi from oxygen (16).

The pH of rumen contents is usually within the range 5.7–7.0 (10). In much of this range, growth of the fungi is good, since their optimal pH for growth is in the range 6.0–7.0 (1,17,18). This redox potential may be as low as −240 mV (10).

The environmental temperature is tightly regulated and is near to or at the core temperature of the animal, usually in the range of 37–41 °C depending on species. In the rumen, optimal growth of the fungi is at or near the body temperature of the host animal (1,17,18). The temperature remains fairly constant particularly in the hindgut, but in the rumen some minor fluctuation of temperature occurs during feeding and drinking (10).

A major adaption of anaerobic fungi to growth in the alimentary tracts of herbivores is their ability to catabolize plant structural carbohydrates. The alimentary tracts of healthy, well-fed herbivores normally contain abundant organic matter, principally plant cell walls originating from the diet. Since much of the plant structural carbohydrates' cellulose and hemicelluloses' remain undigested during transit of the alimentary tract (10), substrates metabolizable by these fungi are present at all times. However, the anaerobic fungi, although present throughout the alimentary tract of ruminants, do not thrive in the hindgut or small intestine. In the abomasum and duodenum, the low pH would preclude their growth, as would the high pH values in the lower part of the small intestine. However, it is not known why they apparently do not grow actively in the hindgut of ruminants although they are undoubtedly present (11). Optical microscopy reveals an apparent absence of zoospores in the large intestine and cecum, and only an occasional sporangium may be observed.

Rumination, that is the regurgitation and rechewing of the ingested food, is an important part of digestion in the ruminant animal. The action effectively increases fracture sites and cut ends on the plant tissues, encouraging more extensive microbial digestion by increasing the surface area and sites of invasion. As will be described later, the rumen fungi invade plant tissues principally at such fracture sites (18,19).

The rumen is a highly cellulolytic environment, but one in which chitin is not extensively attacked (20). Many aerobic fungi contain cellulose or chitin in their cell walls. The possession of chitin as a major cell wall constituent of anaerobic fungi, rather than cellulose, ensures a greater likelihood of their survival in this highly cellulolytic habitat.

III. LIFE CYCLES, INTERHOST TRANSFER, AND INTERACTIONS WITH DIET

In order to transfer to new host animals, the fungi need to survive in air for finite periods. Unlike the anaerobic rumen bacteria, which in general do not produce a resistant survival stage and are transferred by direct contact or in aerosols traveling over short distances (10), some species of anaerobic fungi are known to produce oxygen-resistant forms that maintain viability in feces for several months (21). In this way inoculation of new animals is not dependent on close proximity of animals but may be achieved by way of fecal-contaminated food (6,7).

The life cycles of anaerobic fungi as presently described consist of an alternation of generations between a motile zoospore stage, free-living in gut contents, and a nonmotile, vegetative, reproductive stage associated with the digesta fragments. Clearly, in species that develop an oxygen-resistant stage, the situation is not that simple. Both normal vegetative stage and zoospores die readily in air. In addition, large numbers of anaerobic fungi can be isolated from the hindgut of sheep and cattle, yet few fungal cells can be detected microscopically, suggesting perhaps the possible existence of a small resistant stage not detected easily by microscopy (6). Resistant sporangia have been identified in one species of *Neocallimastix* (21), and similar structures may be formed by other rumen species surviving transit of the alimentary tract and can be isolated from dried feces (22). However, in the Eastern gray kangaroo (*Macropus giganteus*) anaerobic fungi occur in large numbers in the forestomach, but they have not been isolated from other regions of the alimentary tract or the feces (23), suggesting that resistant stages do not occur in species of anaerobic fungi from the kangaroo, or, if they do, that they do not survive transit through the alimentary tract. The viability of anaerobic fungi in the feces of hindgut-fermenting herbivores (horse, etc.) is not known.

Few ruminants practice coprophagy directly, but dietary contamination by feces is common and may be an important means for the transfer of fungi between individual ruminants. Although not examined exhaustively, microscopic evidence suggests that the common rumen isolates are similar in most ruminants, and inoculation of sheep with fungi from reindeer in vitro was successful (24), suggesting a lack of host-species specificity. Transfer of anaerobic fungi from infected to noninfected animals occurs readily when the animals are contained in the same animal house (24), but whether this is by fecal contamination of dust or by aerosol generated during rumination (26), or both, was not determined. Experimental inoculation of the rumen of a fungus-free animal with as little as 100 mg dry weight of feces was successful, confirming the feasibility of transfer via fecal material (25).

Anaerobic fungi readily colonize the prerumen of young milk-fed ruminants (27). All rumen species that are currently known can utilize lactose as a carbon source (24,28), which would enable them to survive in the alimentary tract of preruminants provided the redox potential and other nutrients provided from the milk diet were satisfactory. Rapid colonization of the prerumen by facultatively anaerobic bacteria quickly generates a low redox potential, enabling colonization by the fungi to occur. Anaerobic fungi can be found in the prerumen within a few days of birth, but population densities are rather unstable (27). It is clearly not essential for fibrous plant tissues to be present prior to colonization by the fungi, but their presence may stabilize the fungal population by minimizing outflow of fungal growth in the liquid phase of the stomach contents.

The high microbial population density in the rumen results in rapid loss of soluble carbohydrates from the diet (10). To compete successfully with the motile ciliate protozoa and bacteria, and to minimize the probability of washout of fungi from the rumen, the life cycle of the anaerobic fungi is tightly geared to the feeding behavior of the host animal. In animals fed once daily, a synchronized production of zoospores from mature sporangia occurs in response to the intake of heme-containing dietary material (29). Heme is an essential nutrient for *Neocallimastix* spp. (30). The released zoospores show a clear chemotactic response to glucose, sucrose, and fructose, soluble carbohydrates that occur commonly in fresh plant tissues and are released into rumen liquor soon after ingestion by the host animal (31). This behavior enables the zoospores to locate a suitable, nutrient-rich site for encystment and growth within minutes of the food entering the rumen and minimizes washout during the feeding period, when dilution of rumen contents by saliva and water is highest. Sites selected by the zoospores are principally cut ends of plant tissues, damaged regions, and stomata (7), where they have direct access to the interior regions of the tissue.

During digestion within the rumen, lignified tissues are least metabolized (10). After germination of the spore, the rhizoidal system penetrates the structure of the plant tissue and eventually anchors itself to a xylem bundle (32). When plant tissues are invaded at a cut end, the zoospores usually locate a xylem bundle for initial attachment. Precisely how this occurs is not known, but presumably the sucrose-glucose-fructose chemotaxis enables the cell to locate a fresh plant particle, but additional taxis forces must operate at the xylem bundle to induce the zoospore to attach and encyst at that site. Highly lignified tissues often support higher populations of anaerobic fungi than less lignified tissues (33,34).

IV. NUTRITION

The nutritional requirements of anaerobic fungi have received little attention. Only one species, *Neocallimastix patriciarum* (30), has been grown in a simple minimal medium. This species utilized ammonium ions as a nitrogen source and required heme and the vitamins D-biotin and thiamine for growth. Sulfur could be supplied as sulfide (present in rumen contents), and a wide range of carbohydrates supported growth. Growth was markedly stimulated by the provision of amino acids, other vitamins, volatile fatty acids, and long-chain fatty acids. It is likely that a similar nutrient requirement pattern occurs with other species of anaerobic fungi. A defined medium has been developed for *Neocallimastix hurleyensis* (35), which also supports the growth of many rumen species (11). Growth on the minimal medium was sparse, but in the rumen many compounds that stimulate growth of the fungi in vitro will be present in the diet.

V. RESPIRATION AND METABOLISM

A. Fermentative Metabolism

The lack of available oxygen in the gut compartments where microbial digestion of plant tissues occurs severely limits oxidative catabolism of the diet. Thus all microbial species that thrive in the rumen are strict anaerobes or facultative anaerobes, employing fermentative metabolism of the diet to generate energy for growth. All the anaerobic fungal species so far examined have a mixed acid fermentation; that is, they produce a mixture of volatile fatty acids and D(−)lactate as major products of fermentation of carbohydrates. Most species also produce ethanol and some produce succinate in small proportions. Carbon dioxide and hydrogen are gaseous fermentation products common to all species (28,36).

The pattern of fermentation products is markedly affected by the presence of methanogenic bacteria (37–39) that can utilize the hydrogen and formate produced by the fungus. Removal of these fermentation products often results in increased substrate utilization and increased growth rate of the fungus. In vivo, the close proximity of all microorganisms would ensure that much cross-feeding of fermentation products to other organisms would occur; hydrogen and formate would be rapidly utilized by the surrounding microflora, as would the D(−)lactate, succinate, and ethanol. Culture of anaerobic fungi with methanogens often increases the rate of degradation of plant cell wall structural carbohydrates (39).

Intermediary metabolism of glucose is via the glycolytic pathway (40). Malate generated from oxaloacetate enters the hydrogenosome, where it is converted to pyruvate, which is transported out of the hydrogenosome; the hydrogen removed is converted to molecular hydrogen and the pyruvate further metabolized to lactate, ethanol, and acetate by cytosolic enzymes. Some pyruvate is also converted to acetate in the hydrogenosome, but this is suppressed by high carbon dioxide levels (40).

B. Hydrogen Formation and Oxygen Sensitivity

Aerobic fungi possess mitochondria that are involved in oxidative energy generation, but no structures resembling mitochondria have been demonstrated in the anaerobic fungi. However, all species so far examined by electron microscopy possess hydrogenosomes. These organelles are present in both the zoospores and the vegetative stage (8,40–43). In the zoospores they are arranged in close proximity to the kinetosomes at the base of the flagella; in the vegetative stage they are found in the sporangium and in the rhizoids. Hydrogenosomes are believed to be involved in energy generation (44).

During anaerobic fermentation, products of both higher and lower oxidation states than the substrate are generated (45). One function of the hydrogenosomes is to remove reducing equivalents produced during fermentation, as molecular hydrogen, by means of the enzymes NADPH:ferredoxin reductase and hydrogenase (40).

The anaerobic fungi are regarded as strict anaerobes and they cannot be cultured in aerobic or microaerophilic media. At liquid-phase oxygen concentrations below 4 μM, *Neocallimastix patriciarum* is, however, able to consume oxygen (15) and remove it from the system. Hydrogen production is inhibited by molecular oxygen, with an apparent K_i (50% inhibition of hydrogen production) of 1.4 μM O_2. The significance of oxygen in the metabolism of *Neocallimastix* is not known, though low oxygen concentrations may enhance electron flow through the hydrogenosome in the presence of high levels of carbon dioxide, as has been proposed for the effect of low oxygen levels stimulating growth in the anaerobe *Trichomonas vaginalis* (37).

Exposure of *Neocallimastix* to air results in the generation of superoxide and peroxide; although the organism possesses a superoxide dismutase, it does not possess catalase (14,40). Peroxide radicals are therefore believed to be the cause of oxygen toxicity in these fungi, but owing to the low rumen oxygen concentrations and NADH oxidase, the fungi can survive the influx of air that occurs during feeding and drinking.

C. Cellular Constituents

The anaerobic environment produces some constraints on the biochemical abilities of anaerobic fungi. Since the catabolism of the aromatic rings of lignin usually requires the utilization of molecular oxygen (46), the lignin in plant tissues is not fermented extensively by anaerobic fungi (34) even though some lignin is released into the culture medium (34,47). The absence of oxygen, however, also precludes the synthesis of sterols, which normally require an oxygen-dependent step for cyclization of squalene in the sterol biosynthesis pathway (48). Thus, in *N. patriciarum* and *Piromyces communis*, squalene and tetrahymenol are incorporated into cell membranes instead of sterols (49).

VI. ENZYMOLOGY OF LIGNOCELLULOSE DEGRADATION BY ANAEROBIC FUNGI

A. Plant Cell Walls and Anaerobic Fungi

While all species of anaerobic fungi so far described will utilize glucose, xylose, and cellobiose as carbon sources, the survival of dietary soluble carbohydrates in the rumen is severely limited by the intense competition between the high population densities of rumen microorganisms for readily metabolized substrates. In organisms with a long life cycle, such as the anaerobic fungi, which may take up to 30 hr to complete a single generation (4,50), the ability to effectively utilize a slowly metabolized carbon source that remains within the rumen for this length of time is essential for their survival. Without the ability to attach to, and slowly utilize, plant cell walls, the anaerobic fungi would be rapidly washed from the rumen by dilution of the liquid phase of the rumen contents. Thus all species of anaerobic fungi attach to, grow on, and utilize chemical structures in the plant cell walls. To hydrolyze plant cell wall polysaccharides, anaerobic fungi produce extracellular enzymes with a wide range of activities (50–55), including cellulase, xylanase, and glycosidase activities.

B. Cellulases

Two types of enzyme activities are associated with cellulase, endo-β-1,4-glucanase and exo-β-1,4-glucanase (cellobiohydrolase). These enzymes, acting in concert, produce cellobiose and cellodextrins from cellulose; these are metabolized further by β-glucosidase and cellodextrinase to produce glucose, which can be used as a source of carbon and energy.

A number of native endo- and exo-acting glucanases with different molecular weights have been identified in the culture supernatant fluids of rumen fungi (56–59). It is thought that they are produced initially in a large, multienzyme complex, probably associated with a membrane fraction of the fungal cell wall and subsequently released into the culture fluid (60). In vivo, the multienzyme complex would be released from rhizoids penetrating the plant tissues, and reaction products would be directly available to the fungus and not to other rumen organisms.

Seven carboxymethylcellulases (endo-acting) and six avicelases (probably exocellulases) have been identified in supernatant fluids of cultures of *Neocallimastix frontalis* EB188 (59). Molecular weights range from 116 kDa to 19 kDa. The relationship between these cellulases is not clear since differing degrees of glycosylation (61) and possible proteolytic action by proteases secreted into the medium (62,63) may indicate a higher number of cellulases than are actually synthesized. What is clear, however, is that *Neocallimastix* spp. are capable of producing a cellulase complex of greater activity than several aerobic fungi (56,60) and that the production of cellulase activity is much higher when the fungus is grown in the presence of methanogens (as in the rumen) than in their absence.

Despite the normal enzymes being glycosylated, glycosylation is not necessary for activity but is apparently important for excretion. Nonglycosylated enzymes may have modified substrate specificities (61).

The ability of *Neocallimastix* cellulases to retain activity when not glycosylated has enabled a range of plant cell-wall hydrolyzing genes to be isolated from a cDNA library prepared from *N. patriciarum* mRNA, constructed in lambda ZAPII, and identified by their expression (64–67). Some of the recombinant genes isolated from this library encoded cellulases and xylanases containing more than one active site and some have active sites with multiple substrate specificities (65–67).

Five different cellulase genes have been cloned from *N. patriciarum* (64, 65). All of these genes express proteins in *Escherichia coli* that have cellodextrinase and endo-β-1,4-glucanase activities. Three of them (*CelA, CelD,* and *CelE*) also have significant cellobiohydrolase activity and release cellobiose from crystalline cellulose, while *CelD* also has xylanase and xylo-oligosaccharase activity (66). In addition, *CelB* and *CelD* have cellobiosidase activity. Interestingly, while *CelA, CelB,* and *CelC* are inducible by the presence of cellulose, *CelD* is produced constitutively. Constitutive cellulases are rare, and the enzyme expressed by *CelD* may be of major importance to the fungus during initial colonization of a plant fragment in the competitive environment of the rumen. During this process, which may occur as soon as 10 min after the host animal has started eating (18,19), the free zoospore locates a freshly ingested plant fragment and subsequently encysts and germinates. A penetration rhizoid grows into the plant tissue and anchors the young plant to its substratum by penetrating cell walls. When this occurs soon after the host animal is fed, some soluble carbohydrates, such as glucose, may be present in the rumen liquor at relatively high concentrations. Many cellulases are repressed by the presence of glucose or cellobiose (68), and if *N. patriciarum* cellulases were all subject to such repression, penetration of the plant particle would be delayed and the advantage gained by the rapid invasion of the particle by the zoospore in the presence of large numbers of competing microorganisms would be lost. In addition, the *CelD* enzyme is a true wide-specificity enzyme, with (in the recombinant enzyme) high specific activities for amorphous and crystalline cellulose and xylan (66). This would seem ideal for the rapid degradation of plant cell wall polysaccharides for rapid penetration of the cell wall by the fungus, removing a wait for the production of induced enzymes.

The cellobiohydrolases produced during culture of *N. patriciarum* occur in several forms, which release cellobiose from crystalline cellulose, in addition to hydrolyzing methylumbelifferylcellobioside (69). In the supernatant fluids of aged cultures, a form with the very low molecular weight of about 15 kDa occurs. In actively growing cultures, the molecular weights of isolated cellobiohydrolases range from 45 to 60 kDa, whereas a form with a molecular weight of 60 kDa has been isolated from vegetative growth. The relationships between these forms has yet to be determined, but the ultimate formation of a low-molecular-weight cellobiohydrolase may be beneficial for the deep penetration of enzyme into the structure of the plant cell wall and may be a special evolutionary feature of the fungus supporting its survival in the highly competitive environment of the rumen. The relationships of these native cellobiohydrolases to the recombinant cellulases with cellobiohydrolase activity, *CelA, CelD,* and *CelE,* are not yet known.

C. Cellodextrinase

Cellodextrinase activity (the ability to hydrolyze short-chain β-1,4-linked glucans) was detected with all the cellulase enzymes expressed from the recombinant genes isolated from CDNA libraries prepared from *N. patriciarum* (64–66).

D. Xylanases

Two different xylanases have been identified, *XylA* and *XylB* (67,70). While both also have xylo-oligosaccharase activities, neither has cellulase activity unlike many microbial xylanases (71), but *XylB* has some cellobiosidase activity. *Xyla* has substantial homology with prokaryote xylanases, suggesting horizontal gene transfer from prokaryotes at some time during the evolution of the fungus. *XylA* is a highly active endoxylanase, hydrolyzing about 90% of oat spelt xylan (67).

E. Glycosidases

For the complete degradation of plant cell wall polysaccharides, glycosidases are necessary. A range of glycosidase activities has been detected in culture supernatants and whole cells of several species of anaerobic fungi. To completely degrade cellulose to glucose, glucosidase is required. This enzyme has been detected in culture supernatant fluids of *N. patriciarum* (51) and of *N. frontalis* (53,55,72) and other rumen fungal cultures. Four different glucosidases, differentiated by molecular weight, were found in *N. frontalis* culture supernatant fluids (72), but their relationships were not determined. During the examination of five cloned cellulases from *N. patriciarum*, no glucosidase activity was detected, but cellobiosidase activity was present in two clones; one of these lacked cellobiohydrolase activity (64). Other glycosidases, including strong furanosidase, were detected in culture supernatants and cells of three different rumen isolates (51).

Cellobiosidase, which has no activity on cellulose, is produced by *Neocallimastix* sp. (51,55,73) and its activity detected in culture supernatant fluids of other species (51).

Xylosidase activity is also common in culture supernatants of anaerobic fungi, and the enzyme has been examined in more detail by Hebraud and Fevre (74).

F. Aryl Esterases

Plant cell walls contain lignin, which is bonded to the hemicellulose by ester and ether linkages. A *Neocallimastix* sp. contained both feruloyl esterase and *p*-coumaroyl esterase activities (75). These enzymes are believed to cleave feruloyl and *p*-coumaroyl ester bridges linking lignin to structural carbohydrates in the plant cell walls and could be very important in enabling the fungus to properly penetrate the cell wall for maximum accessibility to polysaccharides for growth. The molecular size of the *p*-coumaroyl esterase is about 11 kDa; this small size may assist the penetration of the enzyme deep within the tissues.

When compared with rumen bacteria, *Neocallimastix* sp. produced more feruoylesterase, and rumen bacteria contain no, or little, *p*-coumaroyl esterase (75). Thus the anaerobic fungus may cleave the aryl ester bonds between the plant cell wall polysaccharides and lignin more effectively than rumen bacteria.

G. Dissimilation of Plant Cell Walls

Thus *N. patriciarum* produces enzymes capable of the complete dissimilation of both cellulose and xylan, the major polysaccharides found in plant cell walls. In fact, *CelD* and *CelE*, a modified trimer of *CelD*, can do this individually. The only major enzymes lacking from the majority of the anaerobic fungi that are involved in degradation of the plant cell wall polysaccharides are pectinases. This is not surprising, since the action of pectinases would result in the separation of cells and destruction of the integrity of the plant tissue, with the fungus being separated from the digesta particles and washed from the rumen before their life cycle could be completed. Low levels of pectinase activity were detected by Williams and Orpin (52) in three rumen isolates but growth on pectin was not detected; however, a single strain of *Neocallimastix* has been shown to grow on pectin (58).

The production of novel multisubstrate, multiactive site enzymes together with a protease may give the anaerobic fungi an advantage over organisms producing less complex enzymes and allow them to penetrate the complex matrix of the plant cell wall with ease.

VII. CONCLUSIONS

The anaerobic fungi are well adapted to the environmental conditions existing in the rumen. They have a fermentative metabolism, do not require oxygen for synthesis of cellular components, and have evolved hydrogenosomes for the removal of reducing equivalents in the form of molecular hydrogen, and protection from low levels of molecular oxygen is achieved by removal of oxygen by NADH oxidase. The production of oxygen-tolerant survival stages by some species assists transfer to new host animals. They are nutritionally undemanding and most herbivore diets will supply their basic nutritional needs.

In the highly competitive rumen environment, the anaerobic fungi have evolved a comprehensive array to extracellular enzymes for the partial digestion of lignocellulose providing a continuing supply of nutrients; attachment to dietary particles ensures their survival in the rumen despite their slow growth. Integration of the life cycle with the feeding behavior of the host animal enables zoospores to invade freshly ingested plant tissues. The production of cellulase at low levels constitutively enables freshly germinated zoospores to penetrate plant tissues despite the frequent presence of dietary glucose, which, in many other organisms, completely suppresses cellulase production.

REFERENCES

1. Orpin, C. G. (1975). Studies on the rumen flagellate *Neocallimastix frontalis*, *J. Gen. Microbiol. 91*: 249-262.
2. Orpin, C. G. (1977). The occurrence of chitin in the cell walls of the rumen organisms *Neocallimastix frontalis*, *Piromonas communis* and *Sphaeromonas communis*, *J. Gen. Microbiol. 99*: 215–218.
3. Bauchop, T. (1983). The gut fungi: colonizers of plant fibre, in *Fibre in Human Nutrition* (G. Wallace and L. Bell, eds.), Royal Society of New Zealand, Wellington, pp. 143–148.
4. Orpin, C. G., and Joblin, K. N. (1988). Anaerobic fungi, in *The Rumen Microbial Ecosystem* (P. N. Hobson, ed.), Elsevier Applied Science, London, pp. 129–150.
5. Teunissen, M. J., Op den Camp, H. J. M., Orpin, C. G., Huis in't Veld, J. H. J., and Vogels, G. D. (1991). Comparison of growth characteristics of anaerobic fungi from ruminant and non-ruminant herbivores during cultivation in a nondefined medium, *J. Gen. Microbiol. 137*: 1401–1408.
6. Orpin, C. G., and Ho, Y. W. (1992). Ecology and function of the anaerobic rumen fungi, in *Recent Advances on the Nutrition of Herbivores* (Y. W. Ho, H. K. Wong, N. Abdullah, and Z. A. Tajuddin, eds.), Malaysian Society of Animal Production, Kuala Lumpur, pp. 163–170.
7. Orpin, C. G. (1993). Anaerobic fungi: taxonomy, biology and distribution in nature, in *Anaerobic Fungi: Biology, Ecology and Function* (D. O. Mountfort and C. G. Orpin, eds.), Marcel Dekker, New York (in press).
8. Heath, I. B., Bauchop, T., and Skipp, R. A. (1983). Assignment of the rumen anaerobe *Neocallimastix frontalis* to the Spizellomycetales (Chytridiomycetes) on the basis of its polyflagellate zoospore ultrastructure, *Can. J. Bot. 61*: 295–230.
9. Munn, E. A., and Orpin, C. G. (1993). Anaeromycetales: a new family to contain the anaerobic fungi, *Mycol. Res.* (in press).
10. Hungate, R. E. (1966). *The Rumen and Its Microbes*, Academic Press, New York.
11. Theodorou, M. K., and Orpin, C. G. (1993). Isolation, culture and survival of anaerobic fungi, in *Anaerobic Fungi: Biology, Function and Ecology* (D. O. Mountfort and C. G. Orpin, eds.), Marcel Dekker, New York (in press).
12. McBee, R. H. (1977). Fermentation in the hindgut, in *Microbial Ecology of the Gut* (R. T. Clarke and T. Bauchop, eds.), Academic Press, New York, pp. 185–222.
13. Clarke, R. T. J. (1977). Methods of studying gut microbes, in *Microbial Ecology of the Gut* (R. T. J. Clarke and T. Bauchop, eds.), Academic Press, New York, pp. 1–33.

14. Scott, R. I., Yarlett, N., Hillman, K., Williams, T. N., Williams, A. G., and Lloyd, D. (1983). The presence of oxygen in rumen liquor and its effect on methanogenesis, *J. Appl. Bact. 55*: 143–149.
15. Yarlett, N., Rowlands, C., Yarlett, N. C., Evans, J. C., and Lloyd, D. (1987). Respiration of the hydrogenosome-containing fungus *Neocallimastix patriciarum, Arch. Microbiol. 148*: 25–28.
16. Lloyd, D., Ellis, E., Hillman, K., and Williams, A. G. (1992). Membrane inlet mass spectroscopy: probing the rumen ecosystem, *J. Appl. Bact.* Symposium supplement 1992, *73*: 155S–163S.
17. Orpin, C. G. (1976). Studies on the rumen flagellate *Sphaeromonas communis, J. Gen. Microbiol. 94*: 270–280.
18. Orpin, C. G. (1977). The rumen flagellate *Piromonas communis*: its life-history and invasion of plant material in the rumen, *J. Gen. Microbiol. 99*: 107–117.
19. Orpin, C. G. (1977). Invasion of plant tissue in the rumen by the flagellate *Neocallimastix frontalis, J. Gen. Microbiol. 98*: 423–430.
20. Patton, R. S., and Chandler, P. T. (1975). In vivo digestibility evaluation of chitinous materials, *J. Dairy Sci. 58*: 473–476.
21. Wubah, D. A., Fuller, M. S., and Akin, D. E. (1991). Resistant body formation in *Neocallimastix* sp., an anaerobic fungus from the rumen of a cow, *Mycologia 83*: 40–47.
22. Milne, A., Theodorou, M. K., Jordan, M. G. C., King-Spooner, C., and Trinci, A. P. J. (1988). Survival of anaerobic fungi in feces, saliva and in pure culture, *Exp. Mycol. 13*: 27–37.
23. Orpin, C. G., and Theodorous, M. K., unpublished observations.
24. Orpin, C. G. (1989). Ecology of rumen anaerobic fungi in relation to the nutrition of the host animal, in *The Roles of Protozoa and Fungi in Ruminant Digestion* (J. V. Nolan, R. A. Leng, and D. Demeyer, eds.), Penambul, Armidale, pp. 29–38.
25. Orpin, C. G., unpublished observations.
26. Lowe, S. E., Theodorou, M. K., and Trinci, A. P. T. (1987). Isolation of anaerobic rumen fungi from the saliva and faeces of sheep, *J. Gen. Microbiol. 133*: 1822–1834.
27. Fonty, G., Gouet, Ph., Jouanny, J. P., and Senaud, J. (1987). Establishment of the microflora and anaerobic fungi in the rumen of lambs, *J. Gen. Microbiol. 133*: 1835–1843.
28. Gordon, G. L. R., and Phillips, M. W. (1989). Comparative fermentation properties of anaerobic fungi from the rumen, in *The Roles of Protozoa and Fungi in Ruminant Digestion* (J. V. Nolan, R. E. Leng, and D. I. Demeyer, eds.), Penambul, Armidale, pp. 127–138.
29. Orpin, C. G. (1986). The effects of haems and related compounds on growth and zoosporogenesis of the rumen phycomycete *Neocallimastix frontalis* H8, *J. Gen. Microbiol. 132*: 2179–2185.
30. Orpin, C. G., and Greenwood, Y. (1986). Nutritional and germination requirements of the rumen phycomycete *Neocallimastix patriciarum, Trans. Br. Mycol. Soc. 86*: 178–181.

31. Orpin, C. G., and Bountiff, L. (1978). Zoospore chemotaxis in the rumen phycomycete *Neocallimastix frontalis, J. Gen. Microbiol. 104*: 113–122.
32. Akin, D. E. (1987). Association of rumen fungi with various forage grasses, *Anim. Food Sci. Technol. 16*: 273–285.
33. Grenet, E., Breton, A., Barry, P., and Fonty, G. (1989). Rumen anaerobic fungi and plant substrate colonization as affected by diet composition, *Anim. Food Sci. Technol. 26*: 55–70.
34. Akin, D. E., and Benner, R. (1988). Degradation of polysaccharides and lignin by ruminal bacteria and fungi, *Appl. Env. Microbiol. 54*: 1117–1125.
35. Lowe, S. E., Theodorou, M. K., Trinci, A. P. J., and Hespell, R. B. (1985). Growth of anaerobic rumen fungi on defined and semi-defined media lacking rumen fluid, *J. Gen. Microbiol. 131*: 2225–2229.
36. Bauchop, T., and Mountfort, D. O. (1981). Cellulose fermentation by a rumen anaerobic fungus in both the presence and absence of rumen methanogens, *Appl. Env. Microbiol. 42*: 1103–1110.
37. Mountfort, D. O., Asher, R. A., and Bauchop, T. (1982). Fermentation of cellulose to methane and carbon dioxide by a rumen anaerobic fungus in a triculture with *Methanobrevibacter* sp. strain RA1 and *Methanosarcina barkeri, Appl. Env. Microbiol. 44*: 128–134.
38. Bauchop, T., and Mountfort, D. O. (1981). Cellulose fermentation by a rumen anaerobic fungus in both the presence and absence of rumen methanogens, *Appl. Env. Microbiol. 42*: 1103–1110.
39. Williams, A. G., Joblin, K. N., and Fonty, G. (1993). Interactions between the rumen chytrid fungi and other microorganisms, in *Anaerobic Fungi: Biology, Ecology and Function* (D. O. Mountfort and C. G. Orpin, eds.), Marcel Dekker, New York (in press).
40. Yarlett, N., Orpin, C. G., Munn, E. A., Yarlett, N. C., and Greenwood, C. A. (1986). Evidence for hydrogenosomes in the rumen fungus *Neocallimastix patriciarum, Biochem. J. 236*: 729–739.
41. Munn, E. A., Orpin, C. G., and Greenwood, E. A. (1988). The ultrastructure and possible relationships of four obligate anaerobic chytridiomycete fungi from the rumen of sheep, *BioSystems 22*: 67–81.
42. Gold, J. J., Heath, I. B., and Bauchop, T. (1988). Ultrastructural description of a new chytrid genus of caecum anaerobe, *Caecomyces equi* gen. nov., *BioSystems 125*: 403–415.
43. Li, J., Heath, I. B., and Bauchop, T. (1990). *Piromyces mae* and *Piromyces dumbonica*, two new species of uniflagellate anaerobic chytridiomycete fungi from the hindgut of the horse and elephant, *Can. J. Bot., 68*: 1021–1033.
44. Muller, M. (1991). Energy metabolism in anaerobic parasitic protists, in *Biochemical Protozoology*, G. H. Coombs and M. J. North, eds.), Academic Press, London, pp. 80–91.
45. Wood, W. A. (1961). In *The Bacteria* (I. G. Gunsalus and R. Y. Stanier, eds.), Academic Press, New York. pp. 72–112.
46. Evans, W. C. (1977). Biochemistry of the bacterial degradation of aromatic compounds, *Nature* (*Lond.*) *270*: 17–22.

47. Orpin, C. G. (1983/84). The roles of ciliate protozoa and fungi in the rumen digestion of plant cell walls, *Anim. Food Sci. Technol. 10*: 121–143.
48. Tchen, T. T., and Bloch, K. (1957). On the mechanism of enzymic cyclisation of squalene, *J. Biol. Chem. 226*: 931–938.
49. Kemp, P., Lander, D., and Orpin, C. G. (1984). The lipids of the anaerobic rumen fungus *Piromonas communis, J. Gen. Microbiol. 130*: 27–37.
50. Lowe, S. E., Theodorou, M. K., and Trinci, A. P. J. (1987). Growth and fermentation of an anaerobic rumen fungus on various carbon sources and effect of temperature on development, *Appl. Env. Microbiol. 53*: 1210–1223.
51. Williams, A. G., and Orpin, C. G. (1987). Glycoside hydrolase enzymes present in the zoospore and vegetative stage of the rumen fungi *Neocallimastix patriciarum, Piromonas communis* and an unidentified isolate grown on a variety of carbohydrates, *Can. J. Bot. 33*: 427–434.
52. Williams, A. G., and Orpin, C. G. (1987). Polysaccharide degrading enzymes formed by three species of anaerobic fungi grown on a range of carbohydrate substrates, *Can. J. Bot. 33*: 418–426.
53. Pearce, P. D., and Bauchop, T. (1985). Glycosidases of the rumen anaerobic fungus *Neocallimastix frontalis* grown on cellulosic substrates, *Appl. Env. Microbiol. 49*: 1265–1269.
54. Lowe, S. E., Theodorou, M. K., and Trinci, A. P. J. (1987). Cellulases and xylanase of an anaerobic rumen fungus grown on wheat straw, wheat straw holocellulose, cellulose and xylan, *Appl. Env. Microbiol. 53*: 1216–1223.
55. Hebraud, M., and Fevre, M. (1988). Characterisation of glycoside and polysaccharide hydrolases secreted by the rumen anaerobic fungi *Neocallimastix frontalis, Sphaeromonas communis* and *Piromonas communis, J. Gen. Microbiol. 134*: 1123–1129.
56. Wood, T. M., Wilson, C. A., McRae, S. I., and Joblin. (1986). A highly active extracellular cellulase from the anaerobic rumen fungus *Neocallimastix frontalis, FEMS Microbiol. Lett. 34*: 37–40.
57. Wood, T. M. (1988). Aerobic and anaerobic fungal cellulases, with special reference to their mode of attack on crystalline cellulose, in *Biochemistry and Genetics of Cellulose Degradation* (J.-P. Aubert, P. Beguin, and J. Millet, eds.), Academic Press, London, pp. 31–52.
58. Barichievich, E. B., and Calza, R. E. (1990). Supernatant protein and cellulase activities of the anaerobic rumen fungus *Neocallimastix frontalis* EB188, *Appl. Env. Microbiol. 56*: 43–48.
59. Li, X, and Calza, R. E. (1991). Fractionation of cellulases from the ruminal fungus *Neocallimastix frontalis, Appl. Env. Microbiol. 57*: 3331–3336.
60. Wood, T. M. (1989). Mechanisms of cellulose degradation by enzymes from aerobic and anaerobic fungi, in *Enzyme Systems for Lignocellulosic Degradation* (M. O. Coughlan, ed.), Elsevier Applied Science, London, pp. 17–35.
61. Li, X., and Calza, R. E. (1991). Cellulases from *Neocallimastix frontalis* EB188 synthesised in the presence of glycosylation inhibitors: measurement of pH and temperature optima, protease and ion sensitivities, *Appl. Microbiol. Biotechnol. 35*: 741–747.

62. Wallace, R. J., and Joblin, K. (1985). Proteolytic activity of a rumen anaerobic fungus, *FEMS Microbiol. Lett. 29*: 19–25.
63. Wallace, R. J., and Munro, C. A. (1986). Influence of the rumen anaerobic fungus *Neocallimastix frontalis* on the proteolytic activity of a defined mixtrure of rumen bacteria grown on a solid substrate, *Lett. Appl. Microbiol. 3*: 23–26.
64. Xue, G. P., Orpin, C. G., Gobius, K. S., Aylward, J. H., and Simpson, G. D. (1992). Cloning and expression of multiple cellulase cDNAs from the anaerobic rumen fungus *Neocallimastix patriciarum* in *E. coli, J. Gen. Microbiol. 138*: 1413–1420.
65. Xue, G. P., Gobius, K. S., and Orpin, C. G. (1993). Isolation of a multifunctional cellulase (celE) from the rumen fungus *Neocallimastix patriciarum. Proceedings of the XVII International Grassland Congress* (in press).
66. Xue, G. P., Gobius, K. S., and Orpin, C. G. (1993). A novel polysaccharide hydrolase CDNA (celD) from *Neocallimastix patriciarum* encoding three multifuntional catalytic domains with high endoglucanase, cellobiohydrolase and xylanase activities, *J. Gen. Microbiol. 138*: 2397–2403.
67. Gilbert, H. J., Hazlewood, G. P., Laurie, J. I., Orpin, C. G., and Xue, G. P. (1992). Homologous catalytic domains in a rumen fungal xylanase: evidence for gene duplication and prokaryotic origin, *Mol. Microbiol. 6*: 2065–2072.
68. Bauchop, T., and Mountfort, D. O. (1985). Production and regulation of cellulose by two strains of rumen anaerobic fungus *Neocallimastix frontalis, Appl. Env. Microbiol. 49*: 1314–1322.
69. Aylward, J. H., Xue, G. P., Simpson, G. D., and Orpin, C. G. (1993). Cellobiohydrolase (CBH) from *Neocallimastix patriciarum*: a membrane associated complex? *Proceedings of the XVII International Grasslands Congress* (in press).
70. Xue, G. P., and Orpin, C. G., unpublished observations.
71. Hall, J., Hazlewood, G. P., Barker, P. J., and Gilbert, H. J. (1988). Conserved, reiterated domains in *Clostridium thermocellum* endoglucanases are not essential for catalytic activity, *Gene 69*: 29–38.
71. Li, X., and Calza, R. E. (1991). Purification and characterization of an extracellular β-glucosidase from the rumen fungus *Neocallimastix frontalis* EB188, *Enz. Microbial Technol. 13*: 1–7.
73. Calza, R. E. (1991). Nascent synthesis and secretion of cellobiase in *Neocallimastix frontalis* EB188, *Curr. Microbiol. 23*: 175–180.
74. Hebraud, M., and Fevre, M. (1990). Purification and characterisation of an extracellular xylosidase from the rumen anaerobic fungus *Neocallimastix frontalis, FEMS Microbiol. Lett. 72*: 11–16.
75. Bornemann, W. S., Hartley, R. D., Morrison, W. H., Akin, D. E., and Ljungdahl, L. G. (1990). Feruloyl and p-coumaroyl esterase from anaerobic fungi in relation to plant cell wall degradation, *Appl. Microbiol. Biotechnol. 33*: 345–351.

12

The Physiology of Drying and Rewetting in Lichens

David H. S. Richardson

Saint Mary's University
Halifax, Nova Scotia, Canada

I. INTRODUCTION

There are about 14,000 lichen species, which account for some 20% of all the fungi so far described. Lichens colonize a remarkable range of habitats with respect to variation in the supply and abundance of water. A small number of lichens live permanently submerged either in fresh water, e.g., *Hydrotheria venosa*, or in the sea below low-tide level, e.g., *Verrucaria serpuloides* (1,2). Others thrive in hot deserts, and a few specialize in living within rocks in the cold deserts of the dry valleys of Antarctica, where the environment is so hostile that not even lichens can tolerate the surface conditions (3). Between such extremes, and more familiar, are the many lichens of the temperate regions, which are adapted to modest temperatures and water regimes. Even in these mesic situations, the lichens are poikilohydrious; i.e., their water content closely reflects that of the surrounding environment. Without the equivalent of a cuticle or stomata to help retain water, they rapidly lose water by evaporation. Their water content decreases to a degree tolerated only by some mosses and a few drought-tolerant Pteridophytes such as *Selaginella lepidophylla*, often termed "resurrection plants" (4,5). The modifications and ecophysiology exhibited by lichens in relation to drought in different climatic regions have been admirably reviewed recently (6,7). It is clear that the ability of particular lichens to take up water and their response to drying and rewetting is a key feature for their survival in the colonized habitat. This chapter will focus on the physiology of this process.

II. WATER UPTAKE

A. Uptake of Liquid Water

Numerous studies have shown that lichens take up water from precipitation very rapidly. Using simulated rain and a wind tunnel, it has been found that the majority of lichens become fully water-saturated within 30 min and some within 5 min (8). The time taken by particular lichen species to reach maximum thallus water content is related to the surface-area-to-weight ratio. In the genus *Umbilicaria*, rhizinae (fungal strands that attach the lower surface of the lichen to the rock) enhance water uptake in species such as *U. vellea* and *U. mammulata*, which absorb water predominantly via the lower surface. In other species, e.g., *U. papulosa* and *U. deusta*, water is taken up equally by both upper and lower surfaces (8) (Fig. 1). The rapid expansion

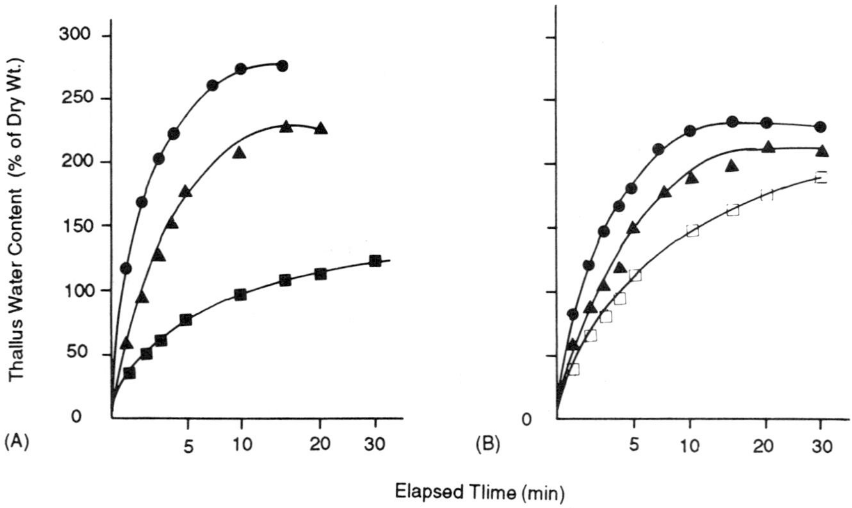

Figure 1 Change in lichen water content versus time for *Umbilicaria mamulata* following removal of the rhizinae and/or blocking of the surface with a silicone-based water repellent. ▲, Samples that had rhizinae removed from lower surface by shaving; ■, samples that were shaved and then had the lower surface blocked by silicone; □, samples that were shaved and then had the upper surface blocked by silicone; ●, untreated control samples. [Redrawn from Larson (8).]

of lower surface (bearing the rhizinae) in *U. vellea* and *U. mammulata* leads to pronounced infolding of thalli at the beginning of a rainstorm. In contrast, *Chrondropsis semiviridis*, a lichen of the arid regions of Australia, is inrolled when dry with a radiation-reflecting lower surface. Following water uptake, presumably via the upper surface, the thallus reflexes and becomes flat (9). Both the structure and morphology (degree of branching) of a lichen can affect the pathway and speed of water uptake. The amount of water held by leafy and shrubby lichens after wetting by liquid water (followed by gently blotting to remove water films covering the surface and between the rhizinae) is generally in the region of 150% of the dry weight. Up to 500% of the dry weight can be held by gelatinous lichens such as *Collema*, which contain cyanobacteria throughout the thallus (10).

B. Uptake of Water Vapor

In addition to water uptake from rain, dew, and runoff, most lichens are able to hydrate beyond their photosynthetic compensation point in humid atmospheres, and this sets them apart from most vascular plants and bryophytes (6). With air at 97% relative humidity and a temperature of 15 °C, the water content of many lichens gradually increases to more than 65% of the dry weight within about 48 hr. Even at relative humidities as low as 75%, lichens have been shown to reach the moisture compensation point of net photosynthesis (11). The uptake of water from humid air is of adaptive significance particularly in arid habitats with a restricted water supply. Various factors can affect such uptake. For example, coastal populations of *Ramalina menzeisii* exhibit more water vapor uptake than inland populations, possibly as a result of thalli being more highly dissected and having surface salt deposits that enhance the water potential of the lichen (12).

As discussed above, most lichens readily absorb water vapor from humid air. Following such water uptake, species containing green algae as photobionts are able to begin net photosynthesis in the light (13). Lichens with cyanobacteria as photobionts seem unable to do this. For this reason, recent studies on lichens in the genus *Pseudocyphellaria* are of particular interest. In some species, a single fungus can associate with either green algae or cyanobacteria, giving a complex thallus in which some lobes are blue-green and others green (14). In such cases (Fig. 2), only the green lobes photosynthesize actively following water vapor uptake. Liquid water is required for the blue-green lobes to show a positive response, and experiments on chlorophyll fluorescence have confirmed this. The fact that respiration takes place in both green and blue-green lobes indicates that this differential response is a feature of the photobiont rather than the fungus (15).

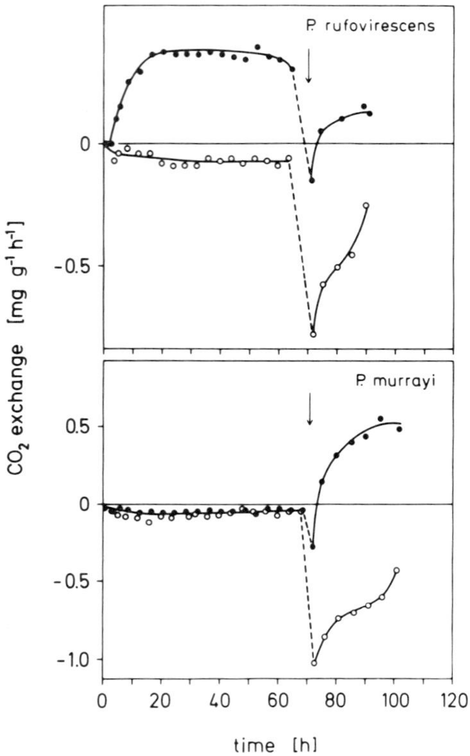

Figure 2 Net photosynthesis and respiration rates of the separated green *Pseudocyphellaria refuscens* and blue-green *Pseudocyphellaria murayi* sectors of a single lichen photosymbiodeme during water vapor uptake (96–97% RH, 15 °C) and subsequent wetting with water (vertical arrow). Closed symbols, net photosynthesis; open symbols, respiration. [From Lange et al. (14).]

C. Mechanism of Water Uptake

When foliose lichens are moistened with liquid water, the water is rapidly taken up by the rhizinae, the medulla, and in the gelatinous matrix of the cortical layers. There is still some uncertainty about the relative amount of water accumulated in these different regions. A number of early studies suggested that large amounts of water are accumulated in water-swollen hyphal cell walls in the medulla (6,16). It has also been suggested that a proportion of the water taken up occupies capillary spaces between the fungal strands of the interior, in addition to that held in the gelatinous matrix of the cortex or cell walls of the medulla (10).

About 10% of lichen species contain cyanobacteria as the photobiont. In those that have been examined by electron microscopy, e.g., *Peltigera*, the fungal hyphae in the central region are covered with a water-repellent layer, but only thin-walled fungal protrusions invade the hydrophilic gelatinous sheaths of the cyanobacterium *Nostoc* (17). However, in lichens containing the green algae *Trebouxia*, the surfaces of both symbionts become covered by a hydrophobic layer. The young algal cells, released from the parent cells, are quickly contacted by a hyphal apex and covered with a thin water-repellent surface layer produced by the fungal partner. *Trebouxia* is the photobiont in some 80% of lichen species. As both symbionts in the algal layer have an outer hydrophobic layer, water should only be able to accumulate in the cell walls and in the cytoplasm of the cells, and this should ensure that the intercellular spaces of the algal layer and medulla are gas-filled, thus optimizing carbon dioxide exchange (17). These ultrastructural studies suggest that water and dissolved nutrients accumulated in the hydrophilic interhyphal gel of the outer layer (cortex) of a lichen are translocated toward the algal cells within the apoplastic space of the fungal cell walls (Fig. 3). It is interesting that some free-living fungi, such as *Schizophyllum commune*, produce at the time of fruit body formation special proteins called hydrophobins (18). These are anchored in the fungal wall of aerial hyphae by disulfide bonds forming complexes that may provide the wall with a hydrophobic surface similar perhaps to that mentioned above on the hyphae of lichen fungi. Other hydrophobins cause the hyphae to adhere to each other during fruit body formation. It would be interesting to investigate lichen thalli for the presence of such compounds.

When lichen thalli are wetted by rain, it is reasonable to assume that the cortical layers first become wet and that water then flows along hyphae either within cells or within the cell walls to the central medulla area of the thallus. Studies on free-living fungi such as *Serpula lacrymans* show that water and solutes like carbohydrates, amino acids, and potassium salts are translocated by a mechanism that involves turgor-driven bulk or mass flow of solutions (19). In lichens, laboratory experiments have shown that photosynthetically fixed radioactive $^{14}CO_2$, in the form of glucose, is translocated a distance of more than 100 μm from the algal layer to the medulla of *Peltigera polydactyla* within 10 min (20). It is possible that photosynthate could be translocated from the cortex to the medulla by bulk flow of solution within the protoplasm of the lichen fungi. In free-living basidiomycetes, transpiration (water loss) from the fruit bodies may enhance translocation of nutrients to these structures (19). To determine the amount and speed of intracellular water flow in lichens, it would be timely to apply the techniques used in the study of free-living fungi.

When some lichens are immersed in liquid water, the medulla appears to remain gas-filled. For example, in the desert lichen *Ramalina maciformis*,

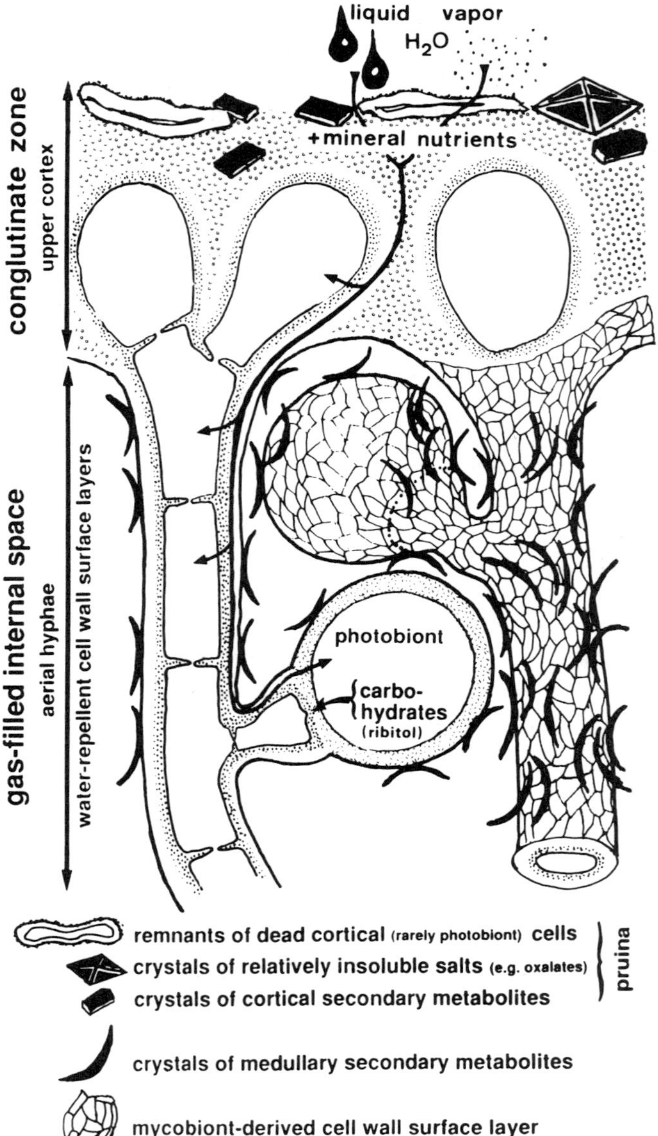

Figure 3 Diagram illustrating the functional aspects of lichens containing the green alga *Trebouxia*. The proposed pathway of water from the outer surface of the lichen to the photobiont and the fungus is shown by the black arrows. The whitish pruina found on the upper surface of some species is shown to be composed of dead cortical cells, crystals of secondary metabolites, and crystals of insoluble salts. [From Honegger (17).]

liquid water is absorbed into the cortex but not into the medulla. This is unwettable owing to the abundance of aromatic lichen compounds and to the presence of hydrophobic layers. However, in the sympatric lichen *Teloschistes lacunosus*, water uptake into the interhyphal spaces of the medulla region appears possible. This species lacks a coherent lower cortex, and following water uptake there is a depression in net photosynthesis at full thallus saturation. In most lichens this depression is caused by the difficulty for carbon dioxide to diffuse through the water-swollen upper and lower cortices (13, 21). The fact that this depression is observed in *Teloschistes* at water saturation, in spite of there being no coherent lower cortex, suggests that the medulla is full of water when such thalli are immersed (3). More studies are clearly required to confirm that water flow to and from the photobiont layer occurs predominantly within the cell walls of the symbionts (17).

Uptake of water vapor from humid air is thought initially to be by the thick gelatinous walls and interhyphal gel of the cortex of lichens (10). This is composed of complex polysaccharides such as lichenan, isolichenan, and galactomannans, which makes lichen hyphae quite unlike those of most free-living fungi (22,23). For example, lichens growing in their natural habitat contain almost twice as much polysaccharide soluble in hot water or alkali as cultured lichen fungi. The cell wall composition of cultured lichen fungi is similar to that of other ascomycete fungi, being a meshwork of chitin microfibrils embedded in insoluble β-1-3-glucan, which makes up about 40% dry weight of the wall (24). The proposal (17) that water and inorganic ions, taken up by interhyphal gel of the cortex, are translocated within the fungal cell wall beneath a hydrophobic layer to the algae requires experimental verification in living lichen thalli.

In lichens containing green algae, sufficient water accumulates in the algal layer following incubation in humid air for net photosynthesis to occur (Fig. 4). This water could result either from flow from the moistened cortex or from direct water vapor uptake by the algal layer. Ultrastructural studies suggest the former. The situation is different in lichens containing cyanobacteria (or cyanobacteria in culture). They can photosynthesize at low water potential but only if liquid water is present. For example, such conditions can be generated by exposing samples to osmotic solutions of low water potential (-4.7 MPa) (e.g., using salt or sorbitol). It is found that cyanobacteria-containing lichens, e.g., *Pseudocyphellaria* sp., show positive net photosynthesis in such solutions but cannot do so when the lichen is maintained in air at the same water potential as the solution (14). Chlorophyll fluorescence studies suggest that the specific osmotic properties of the cyanobacterial cells result in too little water being present for active photosynthesis following water uptake from humid air. *Peltigera rufescens* samples incubated in nearly water-saturated air increase their water content to about 120% dry

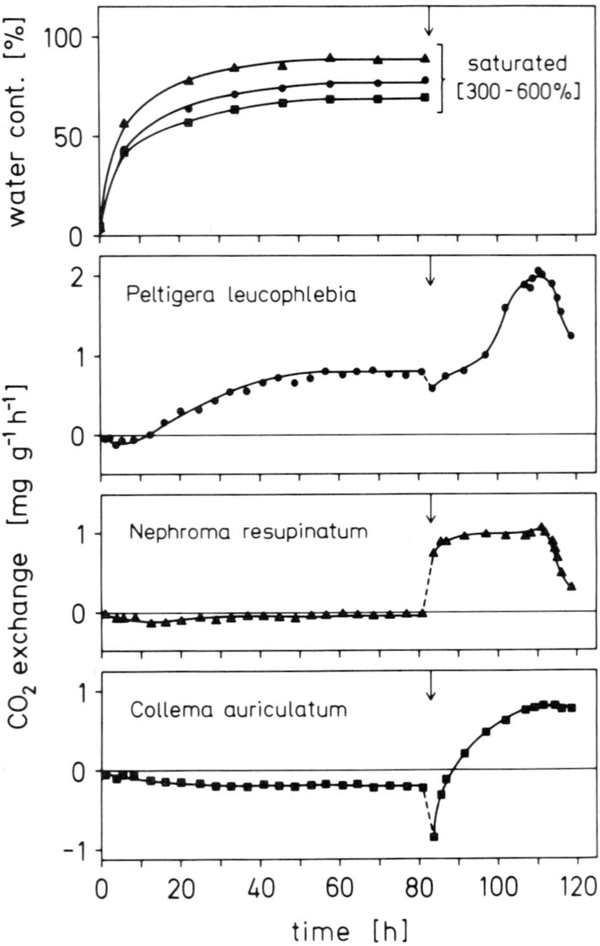

Figure 4 Water vapor uptake from humid air (97% RH, 15 °C) by dry lichens in relation to time (top panel) and carbon dioxide fixation in the light over the same time. The lichens were later sprayed with water (vertical arrow). ●, *Peltigera leucophlebia* contains a green photobiont. The other two species, *Nephroma resupinatum* (▲) and *Collema auriculatum* (■), contain cyanobacterial photobionts. [From Lange et al. (13).]

weight. Although there is a small change in fluorescence which persists over several days, spraying with liquid water is required for complete fluorescence induction on illumination (15). Furthermore, the inability of cyanobacteria in *P. rufescens* to functionally couple phycobilisomes with photosystem 11, when only partly hydrated, could contribute to the failure of lichens with cyanobacterial photobionts to begin net photosynthesis. This contrasts with lichens containing green photobionts, which are able to recouple photosystem 11 with the light-harvesting chlorophyll a/b protein complex upon rehydration following water vapor uptake (25).

The ability of lichens containing green algae to take up water from humid air and become metabolically active is remarkable. Following exposure to humid air, the temperate lichen *Evernia prunastri* was found to exhibit net photosynthesis as low as −18 MPa, *Ramalina menziessii* −22 MPa, and *Dendrographa minor* −38 MPa (26). These lichens were also able to photosynthesize when incubated in concentrations of sorbitol (−5 MPa) or sea salt that would completely inhibit carbon dioxide fixation in leaf slices of xerophytic flowering plants. In the same study, only the blue-green lichen *Pseudocyphellaria anthraspis* was found to be as sensitive as higher plants (with respect to carbon dioxide exchange) to decreases in water potential. Remarkably, *Dendrographa minor*, a lichen that contains the green alga *Trentepohlia* rather than the more common *Trebouxia*, was able to maintain 50% of its maximum gross photosynthesis at −20 MPa (26). Species containing *Trentepohlia* are quite common in Mediterranean and subtropical maritime climates. The product of their photosynthesis, the sugar alcohol erythritol may be the factor conferring an adaptive advantage to *Trentepohlia*-containing lichens in such habitats.

If the hydrophylic cortex becomes saturated with water, following incubation in humid air, can sufficient flow of water occur in the cell walls of the underlying hyphae to enable the green algal cells to begin photosynthesis? Can any proportion of the required water be absorbed directly by the algae in spite of their hydrophobic covering of fungal origin? Is this continuous hydrophobic layer encompassing both fungus and green photobiont required for water flow to occur? In lichens containing cyanobacterial photobionts, is it the lack of this continuity between the water-requiring cyanobacteria and the water-rich cortex which prevents sufficient water flow to the alga or is it the presence of the gelatinous sheath on the cyanobacterial cells which limit the bulk flow of water molecules to them? Experiments designed to answer some of these questions are urgently required.

III. WATER LOSS

From a physiological point of view, lichens face a dilemma. Many colonize habitats like rock and tree bark which are subject to insolation. Without a root system to provide a continued supply of water and the potential for evaporative cooling, the temperature of the lichen may quickly rise to the point where metabolic upset occurs and photosynthesis is curtailed. Fortunately, like other poikilohydrous plants, lichens are resistant to thermally induced damage when dry, and thus it is a distinct advantage if the lichen dries quickly. For example, *Chondriopsis semiviridis* is unaffected by soil temperatures as high as 65 °C when dry but its photosynthesis is halved following exposure to 45 °C for 30 min when moist (27). Lichens possessing strategies that delay drying for a limited period tend to be limited to shaded and semishaded habitats. They are able to photosynthesize for longer periods, which increases the annual net production. Strategies that reduce the rate of drying include a high thallus specific weight, increased thickness of the thallus, the presence of rhizinae, and the formation of clumps or mats. The last strategy is common among ground-dwelling lichens in the boreal forest and sub-Arctic (6). The importance of rhizinae in reducing the rate of water loss is shown in Figure 5 (28). Species of *Umbilicaria* (sensu lato) with rhizinae lose water more slowly than those lacking them. Further differences between species reflect the fact that some species have a rather loose hyphal arrangement (arachnoid) in the medulla while others have a very dense arrangement (prosenplectenchymatous). Species exhibiting the former take half as long to dry to 50% of their water-holding capacity (approximately 30 min under standard conditions) and also take less time to air-dry completely (approximately 100 vs. 150 min) (28). In essence, this leads to two strategies: Those with arachnoid medulla exhibit rapid vapor exchange with the surrounding air but have relatively low saturation capacity. Those with the more dense prosenplectenchymatous medulla, especially if they have rhizinae, retain considerable quantities of water for a comparatively long period. Pustulate species have a series of shallow depressions in the thalli and have a dense medulla except in the regions of the pustules where the algal cells are surrounded by arachnoid hyphae. Such lichens hold almost as much water as species with a prosenplechtenchymatous medulla, but the algae profit by improved gas diffusion conditions (28).

There has been a general assumption that photosynthesis is more sensitive than respiration to low water potentials following loss of water from lichen thalli. However, as chlorophyll fluorescence and infrared gas analysis systems become more refined, it has become possible to show that the algal cells are among the last to become metabolically inactive. From studies on *Umbilicaria antarctica* (29), it is evident that the rate of dark respiration varies with temperature and water content in a complex way (Fig. 6) (29). The following form of the Michaelis-Menten equation was used to interpret the observed pattern:

$$\frac{y}{Y_{max}} = \frac{(X - X_0)^n}{H^n + (X - X_0^n)}$$

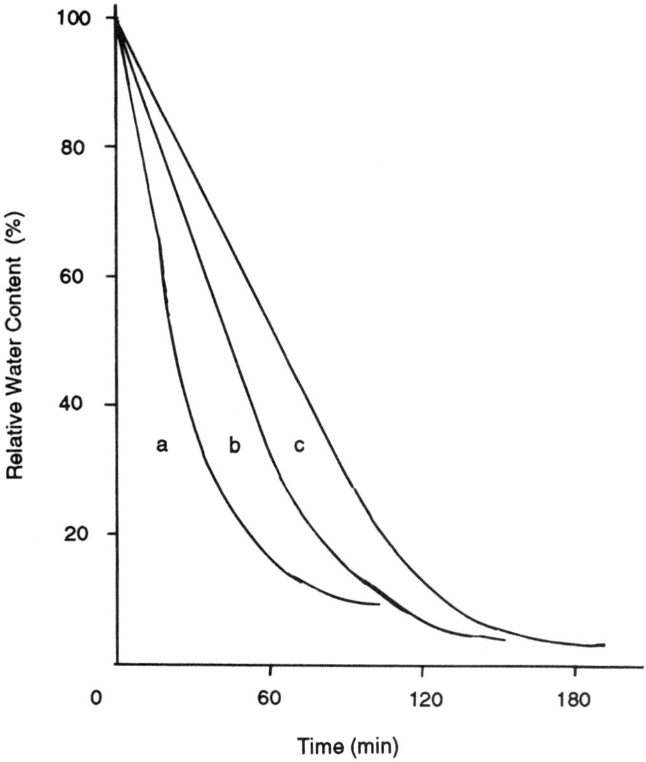

Figure 5 Water loss (at 10°C, 60% RH, and 70 μmol photons $m^{-2} s^{-1}$ PAR) from individual thalli of *Umbilicaria* in relation to their anatomy: (a) *U. grisea*, which lacks rhizinae and has loose hyphae in the medulla; (b) *U. decussata*, which lacks rhizinae but has dense hyphae in the medulla; (c) *U. spodochroa*, which has both rhizinae and dense hyphae in the medulla. [Redrawn from Sancho and Kappen (28).]

where X is the water content, y is the dark respiration rate, Y_{max} the asymptotic dark respiration rate, H the water content yielding half Y_{max}, and X_0 the minimum water content for metabolic activity. The equation reduces to a Michaelis-Menten response curve for $n = 1$. A higher value of n produces a sigmoidal response curve. With the help of the model, it is hypothesized that water loss initially occurs from the upper and lower cortex and from the algal cells. Thus the initial dark respiration rate is directly proportional to the thallus water content demonstrating first-order kinetics and yielding the start slope. Subsequently, water is lost from the free-water space (which might include capillary water from between the fungal hyphae and from water-swollen cell walls) of the medulla. However, the cells of the medulla remain

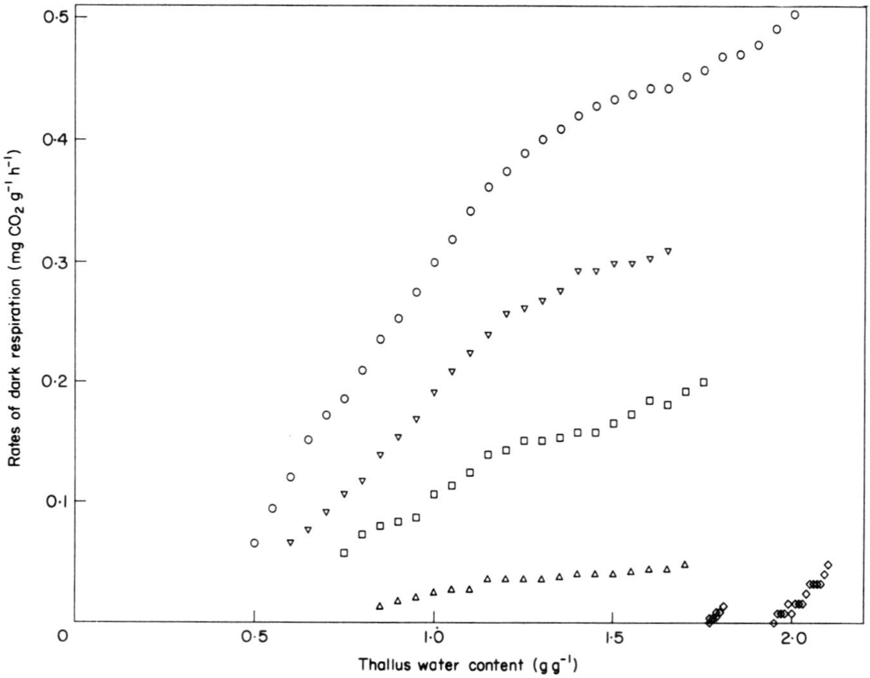

Figure 6 Mean dark respiration rates in *Umbilicaria antarctica* for four temperatures plotted as a function of thallus water content: O, 19 °C; ▽, 13 °C; □, 7 °C; △, 0 °C; ◇, −5 °C. [From Harrisson et al. (29).]

hydrated and active and yield zero-order rate kinetics. Eventually the medulla cells dehydrate and become inactive (this is the initial rate Y_{max}/H demonstrating first-order kinetics). These rate processes, zero and first-order kinetics, together yield a diphasic rectangular hyperbolic curve for dark respiration plotted as a function of water content similar to that actually observed in lichen samples. This interpretation presumes a key role for the medulla region in lichen thalli, and as mentioned above, there has been little recent experimental research to clarify its function in lichen water relations.

The degree of desiccation tolerated varies from lichen to lichen and generally relates to its natural habitat. Thus *Ramalina maciformis* from the Negev desert could regain its full photosynthetic capacity after nearly a year in dry storage, but only 2 weeks of such a regime was sufficient to depress photosynthesis irreversibly in *Roccella fucoides* from the Mediterranean coast (6). Among the most desiccation-sensitive lichens are those from aquatic habitats or those from the tropical and subtropical humid forests. *Pseudocyphellaria dissimilis* from the forest floor of rain forests of New Zealand exhibited a temporary depression of photosynthesis after a drying period of 20 hr at 45–65% relative humidity (30). This lichen appears to be an extreme shade-adapted plant, photosynthesizing positively at very low light levels (e.g., it is light saturated at 20 μmol m^{-2} s^{-1} photosynthetically active radiation compared with 700 μmol m^{-2} s^{-1} for lichens from open deciduous forest). Sustained damage in *P. dissimilis* can be induced by exposure to air at 15–25% relative humidity. This involves cellular damage to both the cyanobacterial photobiont, *Nostoc*, and the fungal partner. Evidence for this comes from the loss of pigments, discoloration of damaged thalli, and increased potassium leakage (30). It is intriguing that *Nostoc* also occurs as a symbiont in lichens such as *Peltigera rufescens* which are found in habitats, such as sand dunes, subject to frequent and significant desiccation. This is analogous to the situation in green alga-containing lichens, where a range of lichens with markedly different pollution tolerance contain the same photobiont, *Trebouxia* (31). It is not known whether it is the strain of photobiont or the environment provided by the surrounding fungus that induces this variability in desiccation or pollution tolerance.

Sugar alcohols (polyols) are synthesized by all the common green algal symbionts of lichens. It has been suggested that the sugar alcohols present in these green algal symbionts may enhance the osmotic potential within the algal cells so that liquid water, albeit in the form of a concentrated sugar solution, is retained at the low water potentials found in dry lichen thalli. Ribitol, the sugar alcohol produced by *Trebouxia*, the most common green algal symbiont of lichens, is found as 2–3% dry weight of the thallus, which could represent 20–30% in the algal layer (14). Cyanobacteria do not produce sugar alcohols and seem to have a different survival strategy. When free-living, the mat-forming cyanobacterium *Nostoc commune* produces only a small amount of the storage carbohydrate glycogen. The major part of photosynthate is channeled into synthesis of additional algal sheath. The sheath may play a role in the recovery process during rehydration, but its primary role is to enhance the capacity of the dry mat to take up and retain liquid water (32). In symbiosis the amount of algal sheath material is much reduced and the cyanobacteria within the lichen lose water to the point where only matric water and proteins remain. Both in cyanobacteria-containing lichens and in epilithic cyanobacteria it is believed that water uptake from humid air is insufficient to establish the liquid pools necessary to restart photosynthesis (J. Garty, personal communication) (14).

IV. CYCLICAL WETTING AND DRYING

A. Respiration and Photosynthesis

Many lichens grow in habitats subjected to periodic rainfall interspersed with dry periods of varying length. Upon rapid wetting, such lichens exhibit a period of enhanced respiration termed resaturation respiration. This has a short-term and long-term component. There is a brief (1–2 min) release of nonmetabolic carbon dioxide. Studies on mosses suggest that this represents carbon dioxide–rich gas displaced from within the cells as the dried shrunken protoplasm takes up water (33,34). This rewetting burst is followed by a period of increased oxygen uptake and carbon dioxide release (Fig. 7) (35). It is presumed that this enhanced period of metabolic activity enables the two lichen symbionts to reestablish the integrity of the cell membranes, to rebuild a functional photosynthetic system, and to develop a normal pattern of carbohydrate flow. There seems to be a seasonal component shown by *Peltigera polydactyla* with regard to the degree of resaturation respiration. Material collected in winter under snow, and then allowed to air-dry, exhibited marked resaturation respiration, whereas that collected in May after a natural dry period failed to show any enhanced respiratory activity on rewetting. Samples collected and dried in June after a 2-week rainy period again showed enhanced respiration. It has been concluded that it is both the speed of dry-

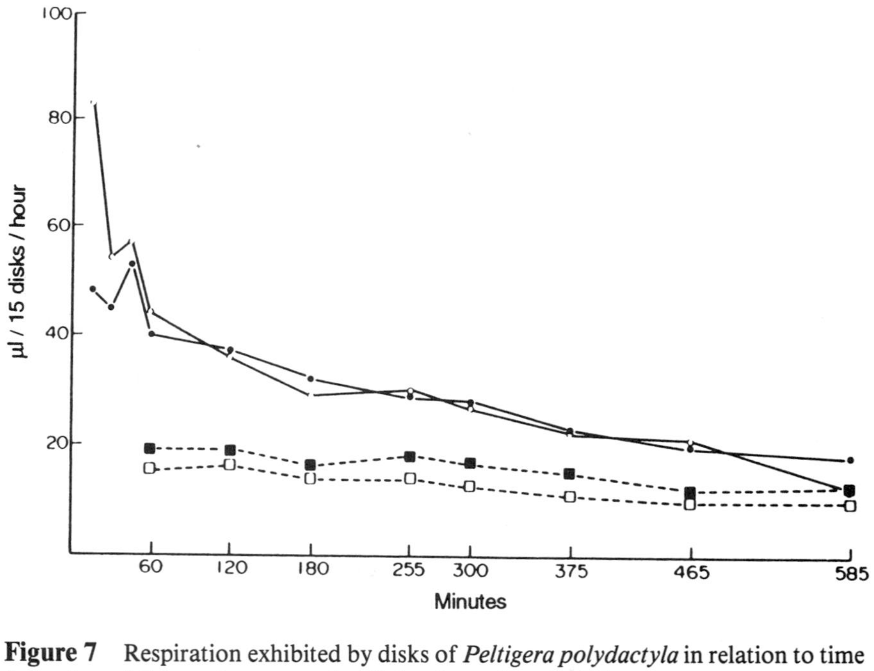

Figure 7 Respiration exhibited by disks of *Peltigera polydactyla* in relation to time following addition of water to dry disks; O, carbon dioxide output; ●, oxygen uptake; □, oxygen uptake by control disks on water; ■, carbon dioxide output by control disks. [From Smith and Molesworth (35).]

ing and the length of the drought period which determine the degree of subsequent resaturation respiration (36). It is interesting that a similar conclusion has been made with respect to resaturation respiration of poikilohydrous Pteridophytes (5).

The effects of a series of rewetting episodes on lichens have been studied using *Parmelia praesignis*. This lichen grows in the mountains of southwest United States, where heavy rains begin in July after the dry months of April, May, and June (37). These rains occur in late afternoon and may occur on several consecutive days. Laboratory simulation of these events revealed that specimens that had been dry for more than a month and then wetted rapidly were still respiring excessively after 5 min and took over an hour to regain full photosynthetic activity. If allowed to dry and then rehydrated 24 hr later, the time taken to regain maximum net photosynthesis was reduced by 50%. On the third day it took only 11 min to reach this level of activity (Fig. 8) (37). This pattern of regaining full photosynthetic activity faster after a series of wetting and drying periods has also been found in other lichens, and the time taken to regain normal photosynthetic activity after the first wetting has been related to the lichen's natural habitat. Thus *Peltigera polydactyla*, a mesic species from woodlands, took 9 hr while *Xanthoria aureola*, a more xeric species from farm roofs, took only 2 hr (35). *Chondriopsis semiviridis* from the more arid parts of Australia took a mere 30 min to regain full photosynthesis (27). Even the most poikilohydrous vascular plants such as *Selaginella lepidophylla* take much longer than these lichens to regain full net photosynthetic activity. In this fern three phases of recovery were observed. In the first phase, lasting about 6 hr, the curled branches unfold, respiration starts, and a low level of photosynthesis begins. In the second phase between 6 and 8 hr, the cells become properly organized, with the central vacuole forming, and thylacoid membranes proliferate in the chloroplasts. The third phase of 12–24 hr is marked by the reorganization of chloroplast membranes into defined grana and continued starch synthesis (4).

The rapid wetting of completely dry lichens by rainwater is probably not a very common event except for lichens such as *Parmelia praesignis*. Usually a period of high humidity or light rain precedes a heavy downpour and lichens become rehydrated more slowly than when immersed in water. Under such conditions, where lichens absorb water vapor from humid air before being saturated with liquid water, resaturation respiration is reduced to the point where it is undetectable (6).

B. Carbohydrate Metabolism

The major soluble carbohydrates in lichens are sugar alcohols (polyols), which generally make up between 1 and 5% dry weight and provide a major respiratory reserve for the partner. Studies on free-living fungi suggest that polyols

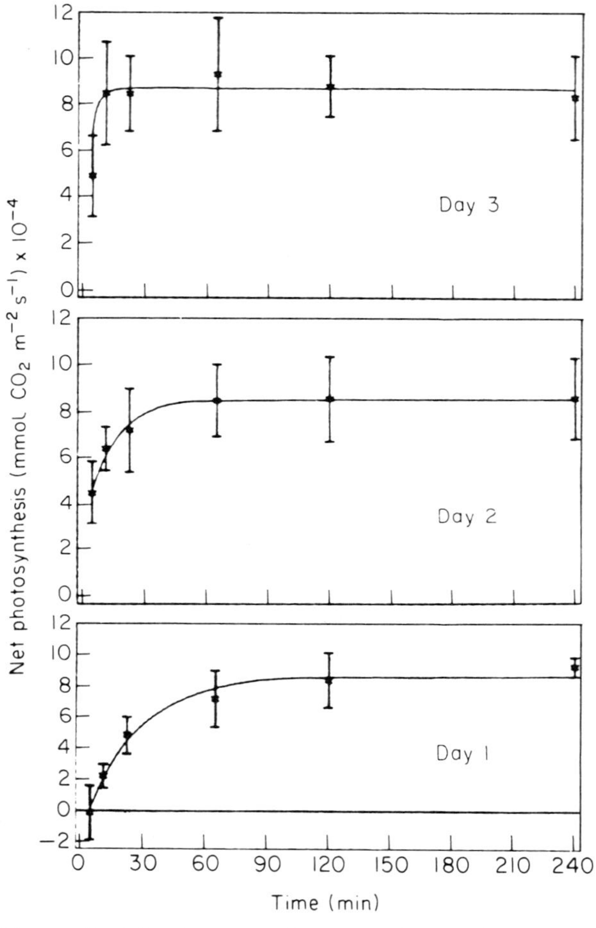

Figure 8 The time taken for *Parmelia praesignis* to regain full photosynthetic activity following one, two, or three wetting cycles. Initially material had been dry for at least 35 days before rehydration (day 1). The same material was then wetted 24 hr later (day 2) and for a third time (day 3). Data are presented as means with 95% confidence intervals (n = 5). [From Link and Nash (37).]

may play several other metabolic roles (38). Mannitol and to a lesser extent arabitol are the main sugar alcohols found within the fungal symbiont of lichens. Ribitol is synthesized and released by *Trebouxia*, the most common green photobiont, and erythritol plays the same role in lichens containing *Trentepohlia*, which is a common symbiont in lichens from tropical and mediterranean areas (39). In lichens that contain cyanobacteria, glucose is believed to be released passively via carrier-mediated diffusion because uncoupling agents and other inhibitors enhance rather than reduce the release. Studies on the internal pools of metabolites, within isolated cyanobacteria from lichens, revealed that over 80% of radioactive label incorporated into the insoluble fraction could be released by the action of amyloglucosidase. This indicates that the glucose release is fed by degradation of a glucose precursor, probably an intracellular glucan (40). Lichen photobionts, whether green algae or cyanobacteria, release a high proportion of their photosynthate (perhaps over 90%) to the fungal partner (20). Studies by Kershaw suggest that carbohydrate flow from alga to fungus occurs at maximum rates when lichens are optimally water-saturated and that much of the photosynthate produced is retained by the algae under drying conditions. Rather than being ecological stressful events to most lichens, it now seems that alternating cycles of wetting and drying may be essential for both partners to gain sufficient carbohydrate for continued survival (36).

As discussed above, sugar alcohols may be required for maintaining liquid water in the dried shrunken protoplasm of the fungal and green algal symbionts of lichens under drought conditions. In addition, large pools of these substances within the cells may also provide the lichen with a respiratory reserve, which can be a physiological buffer against metabolic stress from desiccation (41,42). However, it has been found that in *Hypogymnia physodes* the net loss of polyols on rewetting could not be accounted for by respiration alone (41). A significant leakage of polyols from both symbionts occurs during rehydration as a result of membrane damage caused during desiccation. Such loss of polyols has been documented for air-dried *Cladonia rangiferina* subjected to simulated rainfall (43). The greatest loss took place within the first 15 min, and by 45 min leakage became negligible. In a further study on six different lichens, it was found that the greater the polyol concentration in the species, the greater the rate of leaching during a rain episode (Fig. 9) (43). It was estimated that about 10% of the polyol pool may be lost during each natural wetting event. Although in both comparative and absolute terms the polyol leaching rate is low, the losses may be important. For example, it was estimated that in the Canadian sub-Arctic, lichens add from zero to 0.8% of their biomass per day depending on the rainfall and species. An active period of growth is usually initiated by rainfall, which can result in a loss of 0.01 to 0.7% of biomass due to polyol leaching. Thus a large proportion of

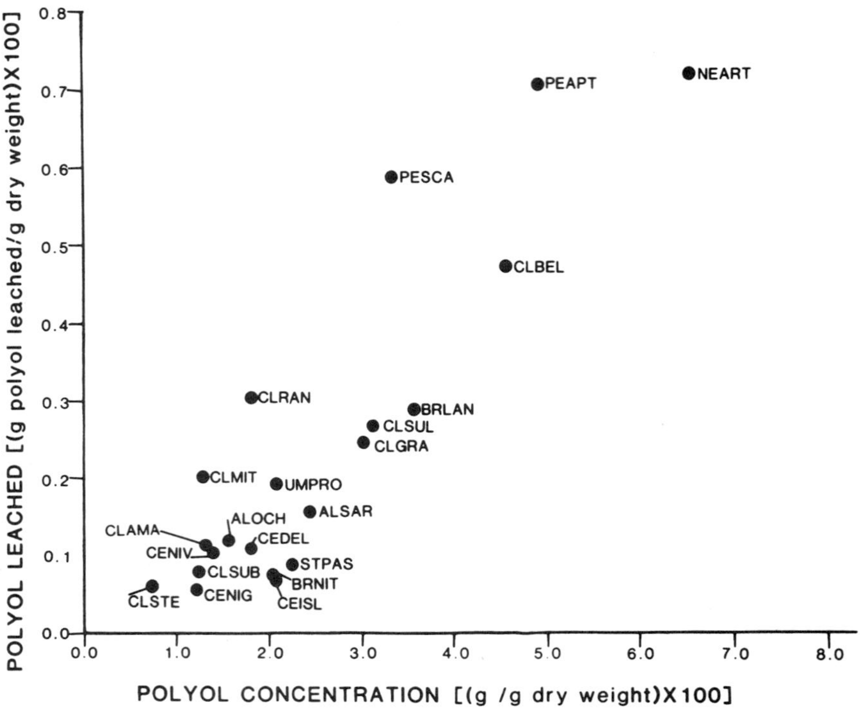

Figure 9 Leaching of polyols in a range of lichen species versus the contained concentration in the particular species: ALOCH, *Alectoria ochroleuca*; ALSAR, *A. sarmentosa*; BRLAN, *Bryoria lanestris*; BRNIT, *Bryoria nitidula*; CEDEL, *Cetraria delisei*; CEISL, *C. islandica*; CENIG, *C. nigricans*; CENIV, *C. nivalis*; CLAMA, *Cladonia amaurocraea*; CLBEL, *C. bellidiflora*; CLGRA, *C. gracilis*; CLMIT, *Cladina mitis*; CLRAN, *C. rangiferina*; CLSTE, *C. stellaris*; CLSUB, *C. subfurcata*; CLSUL, *C. sulphurina*; NEART, *Nephroma arcticum*; PEAPT, *Peltigera aphthosa*; PESCA, *P. scabrosa*; STPAS, *Stereocaulon paschale*; UMPRO, *Umbilicaria proboscidea*. [From Dudley and Lechowicz (43).]

gross daily production must be allocated to replenishing the lost polyol pool since leaching rates are of the same order of magnitude as the biomass increase rates (43). Other stresses besides drought, e.g., air pollution, can lead to increased leaching of inorganic ions and nutrients from lichens (31). In this context, it is interesting that studies on undisturbed and polluted sites at the Bailey Peninsula, Anarctica, have shown that specimens of *Umbilicaria decussata* and *Usnea sphacelata* from localities affected by alkaline dust from a cement mixing facility showed a dramatic reduction, up to 50%, of the expected polyol levels (44).

V. CONCLUDING REMARKS

Over the last decade considerable advances have been made in understanding the water relations of lichens. In particular, studies have revealed the differences between lichens with green and those with cyanobacterial photobionts following water vapor uptake from humid air. However, relatively few genera and species of lichen have been investigated so far, and no doubt further interesting variations remain to be discovered. Physiological and ultrastructural approaches both have a valuable role in developing and testing hypotheses. It is clear that the pathway and flow of water between the symbiont under wetting and drying conditions needs more detailed study and should involve the examination of a range of lichens from mesic and xeric habitats.

There has been much speculation on the role played by sugar alcohols in the stress tolerance of lichens. However, there is little information on the factors that optimize the replenishment of the polyol pools depleted by leaching or respiratory metabolism. The conditions that optimize carbohydrate flow between the symbionts in different lichens also require more research. Recent studies using 13C-NMR have suggested that flow of polyols from alga to fungus is reduced at lower temperatures and thallus moisture levels and terminated by darkness (45). There is a need to determine the degree to which polyols are leached in the natural environment of the lichen and to assess the comparative importance of polyols for resaturation respiration or as compounds for maintaining the osmotic potential of protoplasts during dry periods.

The frequency and severity of resaturation respiration in lichens growing under natural conditions also must be resolved. The importance of climatic and seasonal factors must be determined more fully. In terms of the overall symbiosis, it is interesting that recent evidence on the effects of drought on lichens and their algal components provides no indication that the fungal component of the lichen protects the photobiont from the adverse effects of desiccation once steady-state conditions are achieved (46). Thus evidence that the alga receives any benefit from the association has still to be shown experimentally. Clearly there is much ecophysiological research to be accomplished before we fully understand the physiology of drying and rewetting in lichens. Indeed, a better knowledge of water movement in lichens may be very helpful for understanding such movement in the aerial structures of free-living fungi. Some lichens, such as *Cladonia* and *Usnea ramalina*, are of considerable size. They could provide useful experimental subjects for the radioactive tracer techniques and metabolic analog methods that to date have been employed with rhizomorphs of free-living fungi. The results of such studies could yield results that might be of value in understanding water flow in fungi as a whole as well as extending our understanding of the water relations and ecophysiology of lichens.

ACKNOWLEDGMENTS

I thank the authors and publishers for their kind permission to reproduce the various figures from the original research papers cited. I also thank Professor David Jennings for his most helpful comments on an early draft of this chapter.

REFERENCES

1. Dennis, W. M., Collier, P. A., De Priest, P., and Morgan, E. L. (1981). Habitat notes on the aquatic lichen *Hydrotheria venosa* Russell in Tennessee, *Bryologist 84*: 402–403.
2. Mackenzie-Lamb, I. (1973). Further observations on *Verrucaria serpuloides* M. Lamb, the only known permanently submerged marine lichen, *Occasional papers of the Farlow Herbarium* No. 6, 5 pp.
3. Palmer, R. J., and Friedmann, E. I. (1990). Water relations, thallus structure and photosynthesis in Negev Desert lichens, *New Phytologist 116*: 597–603.
4. Bergstrom, G., Schaller, M., and Eickmeier, W. G. (1982). Ultrastructural and biochemical bases of resurrection in the drought-tolerant vascular plant *Selaginella lepidophylla*, *J. Ultrastruct. Res. 78*: 269–282.
5. Harten, J. B., and Eickmeier, W. G. (1987). Comparative desiccation tolerance of three desert Pteridophytes: response to long-term desiccation, *Am. Midland Naturalist 118*: 337–347.
6. Rundel, P. W. (1988). Water relations, in *CRC Handbook of Lichenology*, vol. II (M. Galun, ed.), CRC Press, Boca Raton, FL, pp. 17–36.
7. Kappen, L. (1988). Ecophysiological relationships in different climatic regions, in *CRC Handbook of Lichenology*, Vol. II (M. Galun, ed.), CRC Press, Boca Raton, FL. pp. 37–100.
8. Larson, D. W. (1981). Differential wetting in some lichens and mosses: the role of morphology, *Bryologist 84*: 1–15.
9. Rodgers, R. W. (1977). Lichens of hot arid and semi-arid lands, in *Lichen Ecology* (M. R. D. Seaward, ed.), Academic Press, London, pp. 210–252.
10. Jahns, H. M. (1984). Morphology, reproduction and water relations—a systems of morphogenetic interactions in *Parmelia saxatilis*, *Nova Hewigia Beiheft 79*: 715–737.
11. Lange, O. L. (1988). Oecophysiology of photosynthesis; performance of poikilohydric lichens and homoiohydric Mediterranean sclerophylls, *J. Ecol. 76*: 915–937.
12. Matthes-Sears, U., Nash, T. H., and Larson, D. W. (1986). The ecology of *Ramalina menzeisii*. IV. In situ photosynthetic patterns and water relations of reciprocal transplants between two sites on a coastal-inland gradient, *Can. J. Bot. 64*: 1183–1187.
13. Lange, O. L., Killian, E., and Ziegler, H. (1986). Water vapour uptake and photosynthesis of lichens: performance differences in species with green and blue-green algae as phycobionts, *Oecologia 71*: 104–110.

14. Lange, O. L., Green, T. G. A., and Ziegler, H. (1988). Water status related to photosynthesis and carbon isotope discrimination in species of the genus *Pseudocyphellaria* with green or blue-green photobionts and in photosymbiodemes, *Oecologia 75*: 494–501.
15. Lange, O. L., Bilger, W., Rimke, S., and Schreiber, U. (1989). Chlorophyll fluorescence of lichens containing green and blue-green algae during hydration by water vapour uptake and by addition of liquid water, *Bot. Acta 102*: 306–313.
16. Stocker, O. (1927). Physiologische und okologische Untersuchungen an Laub- und Strauchflechten, *Flora N. F. 21*: 334–415.
17. Honegger, R. (1991). Functional aspects of the lichen symbiosis, *Annu. Rev. Plant Physiol. Plant Mol. Biol. 42*: 553–578.
18. Wessels, J. G. H. (1990). Fungal growth and development: a molecular perspective, in *Frontiers in Mycology* (D. H. Hawksworth, ed.), C. A. B. International, Wallingford, pp. 27–48.
19. Jennings, D. H. (1987). Translocation of solutes in fungi, *Biol. Rev. 62*: 215–243.
20. Smith, D. C., and Douglas, A. E. (1987). *The Biology of Symbiosis*, Edward Arnold, London.
21. Lange, O. L., and Matthes, U. (1981). Moisture-dependent CO_2 exchange of lichens, *Photosynthetica 15*: 555–574.
22. Gorin, P. A. J., and Iacomini, M. (1985). Structural diversity of D-galacto-D-mannan components isolated from lichens having ascomycetous mycosymbionts, *Carbohydrate Res. 142*: 253–267.
23. Common, R. S. (1991). The distribution and taxonomic significance of lichenan and isolichenan in the Parmeliaceae (lichenized Ascomycotina), as determined by iodine reactions. I. Introduction and methods. II. The genus *Alectoria* and associated taxa, *Mycotaxon 51*: 67–112.
24. Honegger, R., and Bartnicki-Garcia, S. (1991). Cell wall structure and composition of cultured mycobionts from the lichens *Cladonia macrophylla, Cladonia caespiticia*, and *Physcia stellaris* (Lecanorales, Ascomycetes), *Mycol. Res. 95*: 905–924.
25. Bilger, W., Rimke, S., Schreiber, U., and Lange, O. L. (1989). Inhibition of energy-transfer to photosystem II in lichens by dehydration: different properties of reversibility with green and blue-green phycobionts, *J. Plant Physiol. 134*: 261–268.
26. Nash, T. H., Reiner, A., Demmig-Adams, B., Killian, E., Kaiser, W. M., and Lange, O. L. (1990). The effect of atmospheric desiccation and osmotic water stress on photosynthesis and dark respiration of lichens, *New Phytol. 116*: 269–276.
27. Rodgers, R. W. (1971). Distribution of the lichen *Chondriois semiviridis* in relation to its heat and drought resistance, *New Phytol. 70*: 1069–1077.
28. Sancho, L. G., and Kappen, L. (1989). Photosynthesis and water relations and the role of anatomy in Umbilicariaceae (lichenes) from Central Spain, *Oecologia 81*: 473–480.
29. Harrisson, P. M., Walton, D. W. H., and Rothery, P. (1986). The effects of temperature and moisture on dark respiration in the foliose lichen *Umbilicaria antarctica*, *New Phytol. 103*: 443–455.

30. Green, T. G. A., Killian, E., and Lange, O. L. (1991) *Pseudocyphellaria dissimilis*: a desiccation-sensitive, highly shade-adapted lichen from New Zealand, *Oecologia 85*: 498–503.
31. Richardson, D. H. S. (1988). Understanding the pollution sensitivity of lichens, *Bot. J. Linnean Soc. 96*: 31–43.
32. Ernst, A., Chen, T. W., and Boger, P. (1987). Carbohydrate formation in rewetted terrestrial cyanobacteria, *Oecologia 72*: 574–576.
33. Bewley, J. D. (1979). Physiological aspects of desiccation tolerance, *Annu. Rev. Plant Physiol. 30*: 195–238.
34. Richardson, D. H. S. (1981). *The Biology of Mosses*, Blackwell Scientific, Oxford.
35. Smith, D. C., and Molesworth, S. (1973). Lichen physiology. XIII. Effects of rewetting dry lichens, *New Phytol. 72*: 525–533.
36. Kershaw, K. A. (1985). *Physiological Ecology of Lichens*, Cambridge University Press, Cambridge.
37. Link, S. O., and Nash, T. H. (1984). A mathematical description of the effect of resaturation on net photosynthesis in the lichen *Parmelia praesignis* Nyl, *New Phytol. 96*: 257–262.
38. Jennings, D. H. (1984). Polyol metabolism in fungi, *Adv. Microb. Physiol. 25*: 149–193.
39. Richardson, D. H. S. (1985). Carbohydrate transfer between lichen symbionts, in *The Surface Physiology of Lichens* (C. Vicente and D. H. Brown, eds.), Universidad Complutense, Madrid, pp. 25–55.
40. Meindl, D., and Loos E. (1990). Release of glucose by *Nostoc* species isolated from the lichen *Peltigera horizontalis*, *Plant Sci. 72*: 283–288.
41. Farrar, J. F., and Smith, D. C. (1976). Ecological physiology of the lichen *Hypogymnia physodes*. III. The importance of the rewetting phase. *New Phytol. 77*: 115–125.
42. Farrar, J. F. (1988). Physiological buffering, in *CRC Handbook of Lichenology*, Vol. II (M. Galun, ed.), CRC Press, Boca Raton, FL, pp. 101–105.
43. Dudley, S. A., and Lechowicz, M. J. (1987). Losses of polyol through leaching in subarctic lichens, *Plant Physiol. 83*: 813–815.
44. Roser, D. J., Melick, D. R., and Seppelt, R. D. (1992). Loss of polyhydric alcohols and sugars from lichens as an indicator of environmental pollutin at a continental Antarctic locality, *Antarctic Sci. 4*: 185–189.
45. Lines, C. E. M., Ratcliffe, R. G., Rees, T. A. V., and Southon, T. E. (1989). A 13C NMR study of photosynthate transport and metabolism in the lichen *Xanthoria calcicola* Oxner, *New Phytol. 111*: 447–456.
46. Lange, O. L., Pfanz, H., Killian, E., and Meyer, A. (1990b). Effect of low water potential on photosynthesis in intact lichens and their liberated algal components, *Planta 182*: 467–472.

13

Fungal Responses to DNA Damage

Peter Strike and Fekret Osman

The University of Liverpool
Liverpool, England

I. INTRODUCTION

Deoxyribonucleic acid, as the virtually universal repository of genetic information, can be considered as having two essential functions. It must act as the information source for the expression of RNA and proteins involved in all aspects of cell development, maturation, and growth, and in addition, it must ensure the faithful transmission of this genetic information from one generation to the next. Any modification of DNA that alters its coding properties or its normal functions in replication and transcription must therefore be considered potentially deleterious to the organism involved.

It is important to recognize that the primary structure of the DNA molecule is dynamic and subject to constant change. Such changes can lead to alterations in the structure or sequence of nucleotides in the DNA and may have serious effects on DNA function. The biological consequences of these effects may include lethality, mutation, and enhanced levels of recombination. Such damage can occur spontaneously, simply due to normal cellular metabolism and to the chemistry of the bases in the DNA, or it may occur due to lack of accuracy during DNA replication. Alternatively, DNA can be damaged either directly or indirectly by interactions with a wide variety of physical and chemical agents that may be encountered in the environment. Such agents include ultraviolet light (UV), ionizing radiation, and a plethora of chemical mutagens.

It is the role of DNA repair mechanisms to counter this constant attack on the fidelity of the genetic message, to remove lesions that might prevent DNA replication, and to restore the veracity of the genetic information. In recent years it has become increasingly apparent that mechanisms of DNA repair are largely conserved throughout nature. This is perhaps not surprising, as the nature of the damaged molecule is of course the same whether it contains the genetic information for a bacterium or a human cell. It has nevertheless been quite remarkable how mechanisms initially characterized in relatively simple bacterial cells have provided the key to understanding these processes in higher organisms.

In this learning process, the fungi have played their part. The ease with which the yeasts and filamentous fungi can be manipulated has made them ideal organisms in which to explore the genetics and biochemistry of eukaryotic DNA repair processes, and the yeasts in particular have played an essential role in developing the models of repair to their current levels of sophistication. However, it would be quite unfair to regard the fungi simply as model systems for more complex eukaryotes, and it is the intention of this review not only to highlight the similarities between repair processes in all eukaryotes, but also to demonstrate that, as biological systems, repair mechanisms in fungi are worthy of study in their own right.

In any review of this type, it will also be essential to make comparisons with the repair systems in bacterial cells, particularly in *Escherichia coli.* Our understanding of the genetics and biochemistry of repair processes in *E. coli* is considerably greater than that of any other organism, and it is in bacterial cells that, in almost every case, novel repair processes were first identified. Inevitably, therefore, the standard approach to understanding repair in any organism has been first to look for equivalents of those processes found in *E. coli.* That such an approach has been so successful effectively demonstrates the ubiquitous nature of DNA repair mechanisms. An outline of the various types of DNA repair processes, as found in both prokaryotes and eukaryotes is given in Figure 1, and the reader wishing to have more background information on the subject is referred to the seminal text by Friedberg (8).

II. THE NATURE OF DNA DAMAGE

As already indicated, DNA damage can arise either as a consequence of normal cellular metabolism or through reaction with environmental damaging agents. Measurements of hydrolysis rates indicate that apurinic and apyrimidinic (AP) sites arise in DNA, at physiological pH and temperature, at biologically significant rates (1,2), due to breakage of the glycosylic bond between the deoxyribose moiety in the DNA backbone, and the heteroaroma-

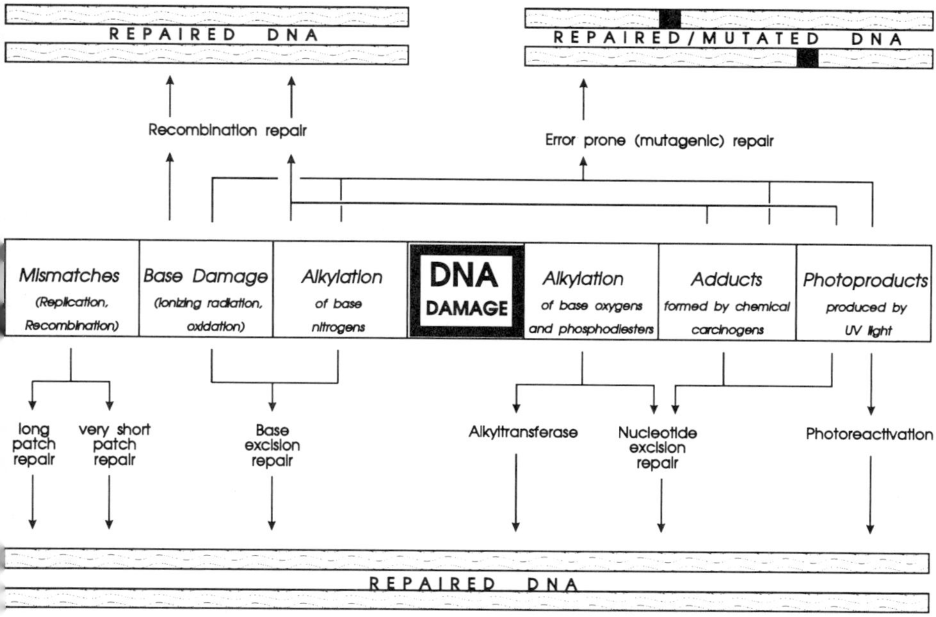

Figure 1 Outline of the variety of DNA repair mechanisms found in both prokaryotes and eukaryotes and their relationship to different types of DNA damage. A number of the repair pathways, although shown as quite separate entities, in fact share a large number of common features. For example, very-short-patch mismatch repair is similar to base excision repair, and long-patch mismatch repair has many of the features of nucleotide excision repair. Equally, rather different repair pathways are sometimes referred to under the same general descriptive title—recombination repair being a case in point. In prokaryotic organisms, this mechanism, also referred to as postreplication repair, involves the recombination-mediated transfer of homologous DNA to repair single-stranded regions created during replication of damaged DNA. Such a mechanism does not appear to operate in eukaryotic organisms—rather, recombination mechanisms are involved primarily in the repair of double-strand breaks. Perhaps the most noticeable feature of the repair scheme is that most types of DNA damage can be repaired in several different ways—each repair system is effectively backed up by a number of others.

tic base. Similarly, hydrolytic deamination of both cytosine and adenine occurs, to give rise to uracil and hypoxanthine, respectively (3,4). Both these types of lesions, if left unrepaired, will cause mispairing during DNA replication. The rates of hydrolysis of glycosylic bonds may be greatly increased if the bases become methylated at certain positions, as may occur by interactions with intracellular *S*-adenosyl methionine, with direct chemical alkylating agents, or with metabolically activated methylating agents

(5, and refs. therein; 6,7). Such methylations may also have a number of other nondesirable effects, depending on the nature of the methylation event. Some methylation products in DNA, such as N^7-methyl guanine, appear largely to be tolerated, but others, such as N^3-methyl adenine, can effectively block DNA replication, while yet others, such as O^6-methyl guanine, cause mispairing and are highly mutagenic (reviewed in 8).

Intracellular DNA damaging processes such as these may be presumed to occur at a fairly constant low rate and are almost certainly responsible for the presence of a number of constitutive repair pathways that effect the repair of this damage. Damage due to environmental agents, however, is likely to be much more variable and will depend heavily on the niche occupied by the organism. Of particular relevance to the fungi may be the large variety of genotoxic compounds found in plant tissue, as fungi are largely responsible for the breakdown of this material under natural conditions. Indeed, it is possible that plants produce such compounds in order to deter predation by other organisms, and any organism that does grow on this material will be obliged to develop DNA repair processes to overcome the inevitable consequences (9). Fungi themselves, and some bacteria, can under suitable conditions produce extremely high concentrations of powerful genotoxic compounds, such as aflatoxin produced by *Aspergillus flavus* and streptozotocin produced by *Streptomyces achromogenes* (10). The producing organisms at least would be expected to have repair mechanisms of such effectiveness as to prevent self-destruction. It is almost certainly in order to deal with these variable environmental hazards that inducible repair pathways have evolved, which are normally kept switched off but which can, under appropriate conditions, be induced to high levels.

In addition to hydrolysis and methylation damage incurred as a consequence of normal cellular metabolism, a wide variety of structural alterations can be caused to DNA by reaction with a large number of well-characterized environmental agents. Of these, exposure to UV light has perhaps been the most extensively studied, and the nature of the damage produced in DNA is well documented. A number of stable photoproducts are produced by short-wavelength UV, including cyclobutane (5-6) dimers, 6-4′ (pyrimidine-2′-one) dimers, pyrimidine hydrates, DNA-protein cross-links, and spore photoproducts. The presence of either cyclobutane (5-6- or 6-4′ (pyrimidine-2′-one) dimers in DNA causes a number of problems, as the damaged bases can no longer pair correctly, and the structure of the DNA duplex is locally distorted. Unless they are somehow removed from the DNA, such lesions either effectively inhibit DNA replication, as has been demonstrated in vitro (11,12), or, in prokaryotic cells at least, they result in large single-strand gaps being left opposite the dimer as the polymerase complex passes (13). A high frequency of single-strand gaps may be lethal, as they can be converted to double-strand breaks by endonucleolytic incision (14), and such breaks may be impossible to repair. Calculations based on the behavior of bacterial cells indicate that one or two unrepaired pyrimidine dimers per haploid genome can be lethal (15).

A number of chemical compounds also produce lesions in DNA that cause local distortions and loss of coding function; among the best characterized is 4-nitroquinoline-1-oxide (4NQO). When activated by cellular metabolism, 4NQO forms adducts by reacting with the purine bases in DNA, and the distortions in the DNA helix so caused are very similar to those produced by UV-induced lesions (16). The effects of both UV (17) and 4NQO (18) have been characterized in yeasts and filamentous fungi, and the biological effects of the two damaging agents are remarkably similar. Both cause increased mutation, mitotic recombination, and lethality. A large number of chemical carcinogens appear to have the same sort of effects as 4NQO, by causing the production of large bulky adducts in cellular DNA.

Alkylating agents represent a quite different group of chemical carcinogens from those that cause bulky lesions. Depending on the exact compound used, alkyl groups may be introduced into a number of different positions, either on the heteroaromatic bases of the DNA or on the phosphate groups in the DNA backbone. Among the most effective of these agents is the monofunctional methylating agent *N*-methyl-*N*′-nitro-*N*-nitrosoguanidine (MNNG), which produces a number of different lesions in the DNA (reviewed in 19). The major product of the reaction under physiological conditions is N^7-methyl guanine, a lesion that appears largely to be tolerated and causes few biological problems. Around 70% of the methylation of DNA by MNNG occurs at the N^7 position of guanine. Two other products, N^3-methyl adenine and O^6-methyl guanine, appear to cause the most severe biological effects. The methyl group in N^3-methyl adenine protrudes into the minor groove of DNA and presents an effective block to DNA replication. Methylation of the O^6 position of guanine causes mispairing during DNA replication, leading to G-C to A-T transition mutations; other minor promutagenic lesions include O^4-methyl thymine. In addition to methylation of the bases in DNA, methylation can also take place at the secondary hydroxyl groups present in the phosphodiester linkage between adjacent nucleosides in DNA, to produce methyl-phosphotriesters. Two stereoisomers of this product can be formed—the R configuration in which the methyl group protrudes into the major groove, and the S configuration in which the methyl group protrudes perpendicular to the helix (discussed in 20). Neither of these lesions presents any biological problems, although, as will be discussed later, the S stereoisomer acts as the key signal to induce DNA repair enzymes. The DNA-damaging effects of MNNG and its genetic effects on a wide range of organisms, including fungi, have been reviewed previously, (21).

A further class of potent environmental DNA-damaging agents includes the ionizing radiations such as X- and gamma rays. Such energetic radiation is produced naturally by radioactive decay in minerals and rocks and is also received constantly from the cosmos. In contrast to UV, ionizing radiation can cause damage to many different cellular components, some of which may then react with DNA. As a consequence, DNA damage following exposure to ionizing radiation may arise either directly or indirectly (22). The energy of ionizing radiation may be directly absorbed by the DNA molecule itself, resulting in strand breakage or base damage, or highly reactive species such as superoxide may arise, which subsequently react with DNA. All four bases are subject to modification by ionizing radiation, including deamination, dehydroxylation, and depurination, but perhaps the most important biological consequence of such radiation is the production of single- and double-strand breaks (23). Depending on the nature of the event that causes the breaks, the broken ends may contain a variety of base damage. The repair of single-strand breaks is relatively simple to achieve, using the genetic information available in the undamaged strand. However, double-strand breaks are much more difficult to repair, and a single such unrepaired break can cause chromosome breakage and in a haploid organism will be lethal. The key role played by recombination mechanisms in the repair of such damage has been elegantly demonstrated in *Saccharomyces cerevisiae* (24), and similar mechanisms are known to operate in both prokaryotes and other eukaryotes.

III. TYPES OF DNA REPAIR MECHANISMS

The number and variety of repair mechanisms are almost as great as the number and variety of lesions that arise in DNA. While some repair enzymes are specific for just one lesion, others will act against a whole spectrum of damage, which might, for example, cause similar distortions in DNA. Some DNA lesions may only be repaired in one specific way, while others may be potentially repairable by several different repair mechanisms. Some repair processes result in the complete removal of the damaged lesion from the DNA, while others are perhaps more accurately described as tolerance mechanisms, in which DNA replication is allowed to proceed despite the presence of DNA damage, and in consequence the amount of damage is simply diluted out through several rounds of replication. The type of repair strategy adopted for any particular lesion will depend on a number of factors including the type of lesion involved; whether one or both strands are damaged; the presence or absence of a homologous undamaged DNA duplex; and the context of the site, i.e., whether it is associated with a particular piece of secondary structure, whether it is located in an actively transcribing gene, or whether it is located close to a replication origin. In eukaryotes, the local status of the chromosome structure and the phase of the cell cycle will also be important factors.

A. Direct Reversal Mechanisms

Perhaps the simplest repair mechanisms to understand are those in which the event that caused the damage is directly reversed. Two such processes are recognized as existing in almost all organisms, from bacteria to the more complex eukaryotes, namely photoreactivation of thymine dimers and repair of certain methyl lesions by methyltransferase enzymes.

1. Photoreactivation

The process of photoreactivation was the first true repair process to be identified, being originally described in the gram-positive bacterium *Streptomyces griseus* (25). However, the demonstration that this was a truly enzymatic process did not occur until cell free extracts from both *E. coli* and yeast were shown to be capable of carrying out the reaction (26,27). The reaction carried out by such extracts is specific for just one type of lesion, the *cis-syn* cyclobutyl pyrimidine dimer, and is mediated by the enzyme DNA photolyase. In the presence of visible light (300–500 nm wavelength—the action spectrum of the enzyme varies somewhat from organism to organism), the enzyme converts the dimerized adjacent bases back to monomers by breaking down the cyclobutane ring, thus directly reversing the event that gave rise to the damage. The photolyase enzyme of *S. cerevisiae* has been extensively characterized (reviewed in 28 and 29), and mutants defective in this enzyme have been described (30). The phenomenon of photoreactivation has been described in *Aspergillus nidulans* (31), and the gene coding for the enzyme has recently been isolated from *Neurospora crassa* (32). The budding yeast *Schizosaccharomyces pombe* is, however, reported not to contain the enzyme (reviewed in 8).

2. Methyltransferase Enzymes

In contrast to photoreactivation, repair of methyl lesions by methyltransferase is a fairly recent discovery. The events leading to its recognition involved the characterization of the so-called "adaptive response" in *E. coli* (33) and the subsequent characterization of the enzymes responsible for this phenomenon (34; reviewed in 35). Adaptation of *E. coli* by exposure to low levels of alkylating agents such as MNNG results in the cells becoming resistant to both the killing and mutagenic effects of much higher doses of the same agent. The gene regulating this response, *ada*, has been cloned and characterized (36,37) and is known to code for the production of a 39-kDa protein species with two active sites involved in the repair of alkylated-DNA lesions. One active site, in the C-terminal part of the molecule, can accept methyl groups from O^6-methyl guanine and O^4-methyl thymine in DNA, while the second site in the N-terminal part of the molecule accepts methyl groups from methyl phosphotriesters (20). In each case, the transfer of the methyl group is to a cysteine residue in the protein, and the covalent attachment of the methyl group to this residue totally inactivates this site. The enzyme therefore plays a stoichiometric role in the reaction and is described as displaying "suicide kinetics," since the inactivated enzyme is no longer available to take part in any further DNA repair (34). The transfer of the methyl group from the O^6MeG and O^4MeT lesions effectively removes the two primary promutagenic lesions created by alkylating agents and thus effectively reduces the level of mutagenesis caused by these compounds.

Transfer of the methyl group from the phosphotriester lesions removes neither a mutagenic nor a potentially killing lesion—instead, this transfer from a relatively innocuous lesion to the Ada protein, represents the key step in the control of the adaptive response (38). Again, the transfer results in the covalent attachment of a methyl group to a cysteine residue in the protein, and this methylation has the effect of converting the Ada protein into a strong transcriptional activator. The activated protein binds to a conserved sequence, the *ada* box, associated with the regulatory regions of two genes, *ada* itself and *alkA*, the gene encoding the inducible 3-methyl adenine DNA glycosylate II enzyme (39). The induction of this enzyme also forms part of the adaptive response, the glycosylate being responsible for the removal of 3-methyl adenine and a number of other minor methylated bases. Removal of these potentially killing lesions is therefore primarily responsible for the increased resistance of adapted cells to the killing effects of alkylating agents, while induction of the methyltransferase reduces the mutation frequency. The role of glycosylases in the repair of alkyl lesions in DNA will be discussed further in a later section.

In addition to the Ada protein, a constitutive methyltransferase protein is also present in *E. coli*, the product of the *ogt* gene (40). This protein has a molecular weight of 19 kDa and repairs both O^6MeG and O^4MeT. This picture of multiple methyltransferase proteins is also seen in another bacterium, the gram-positive *Bacillus subtilis*. In this case also there is a constitutive methyltransferase, a 22-kDa protein encoded by the *dat1* gene (41). In addition, there are two inducible methyltransferases, one of which (AdaB, Mw 22 kDa) has activity against O^6MeG, and AdaA (MW 27 kDa), which has activity against methyl phosphotriesters. As with the *ada* system of *E. coli*, it is methylation of the methylphosphotriester enzyme that triggers induction of the adaptive response.

Methyltransferase activities have also been widely found among eukaryotic organisms and appear to be of largely similar molecular weight—between 20 and 27 kDa (reviewed in 42 and 43). In every case, they are specific for the repair of O^6MeG and O^4MeT lesions and have no activity against methyl phosphotriesters. Induction of these enzymes appears to occur in some organisms and specific tissues, but the level of induction is very small compared to the bacterial systems—of the order of three- to fourfold. Moreover, such induction can be triggered not only by certain alkylating agents, but also by a number of nonalkylating damaging agents such as X-irradiation. It appears therefore that in higher eukaryotes, the control of methyltransferase enzymes is very different from that seen in bacterial cells.

It was thought for a number of years that this situation pertained in all eukaryotes, including the fungi, this interpretation being reinforced by the failure to observe any equivalent of the adaptive response in *S. cerevisiae* (44). Initial attempts to isolate a methyltransferase enzyme from this organism were also unsuccessful, until a low-level constitutive enzyme was identified in exponentially growing cells (45). This activity was specific for the repair of O^6MeG and O^4MeT lesions—as in all other eukaryotes, no enzyme was identified with activity against methyl phosphotriesters. The properties of the yeast methyltransferase account for the initial difficulties in its isolation—the enzyme is extremely heat labile, with a half-life of 4 min at 37°C. The gene for the enzyme, designated *MGT1*, has now been cloned and sequenced and encodes a 21.5-kDa protein. The gene maps to chromosome 4, and the sequence reveals an 88-amino-acid stretch with a high degree of homology to the four bacterial transferases (two from *E. coli*, two from *B. subtilis*) and with the human O^6-methyltransferase (46). Mutants defective in this gene display an increased sensitivity to both the killing and mutagenic effects of alkylating agents, implicating O^6MeG and/or O^4MeT in both these biological processes. The *S. cerevisiae* enzyme is constitutively expressed, and attempts at induction with MNNG result in a small decrease in activity rather than a large induction (45,46).

Recent work in our laboratory, however, has shown that this picture does not apply to all eukaryotes. Pretreatment of mycelium of *A. nidulans* with a low level of alkylating agent results in a considerable increase in resistance to subsequent high challenge doses (47), and a stable alkyltransferase activity can be detected in cell extracts (48). Analysis of the methyltransferase species reveals two proteins in uninduced cells, with molecular weights of 16 and 18.5 kDa, respectively. Induction of the cells by exposure to low levels of alkylating agents results in a dramatic induction of total methyltransferase activity, approximately 100-fold compared to the activity in uninduced cells (Fig. 2). This inducible activity comprises three distinct protein species. Two proteins, one the 18.5-kDa species identified in uninduced cells, the other a 21-kDa protein that can only be detected after induction, have activity against O^6MeG lesions. The third, a protein of 19.5 kDa, has activity against methyl phosphotriesters. To date, *A. nidulans* is the only eukaryote in which a highly inducible methyltransferase system has been demonstrated, and the

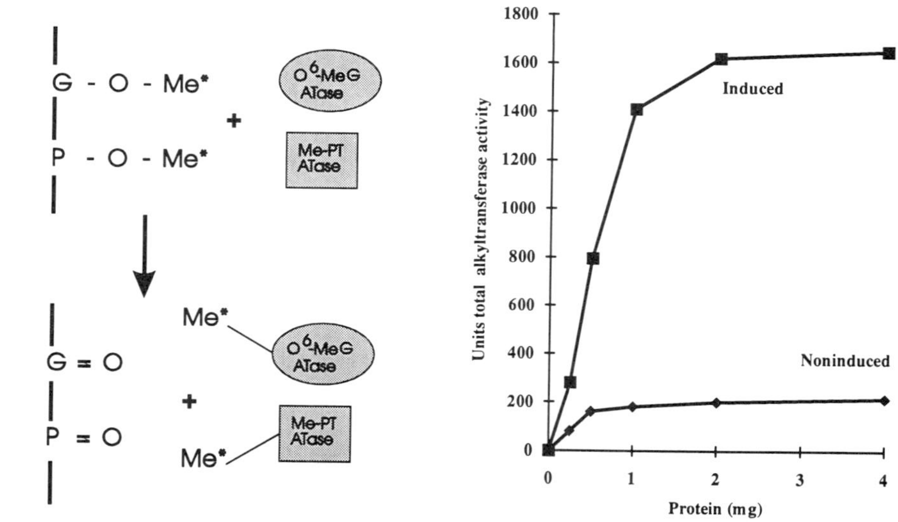

(A) (B)

Figure 2 (A) Mode of action of alkyltransferase proteins in repairing alkylated base oxygens and alkyl phosphotriesters. The scheme shown is most similar to that found in *Bacillus subtilis*, in that the enzyme that repairs O^6 alkylation damage is separate from the enzyme that removes alkyl lesions from alkyl phosphotriesters. A similar situation is known to exist in *Aspergillus nidulans*, in contrast to *Escherichia coli*, where a single protein possesses both activities. In *B. subtilis*, the enzyme alkylated by reaction with alkyl phosphotriesters is then responsible for inducing the high level of expression of the O^6-repairing enzyme. The levels of O^6-repairing enzymes are also highly inducible in *A. nidulans*, although the control mechanism involved is not yet characterized. (B) Levels of alkyltransferase activity measured in extracts of *A. nidulans*, either untreated or pretreated by exposure to a low-level *N*-methyl *N'*-nitro *N*-nitroso-guanidine (0.5 μg/ml) for 3 hr. Enzyme activity is assessed by following the transfer of radiolabeled methyl groups from DNA to protein, as described in Ref. 49.

concurrent presence of a protein with activity against methyl phosphotriesters suggests very strongly that a mechanism similar to that in bacterial cells may exist in some, if not all, eukaryotic species (49).

B. Damage Removal Mechanisms

1. Base Excision Repair

The pioneering work of Setlow and Carrier (50) and Boyce and Howard-Flanders (51) established that pyrimidine dimers induced in DNA by UV light are effectively removed and released as short oligonucleotides. Because whole nucleotides are released from damaged DNA, this process is now referred to as nucleotide excision repair. The mechanism of this process is very well understood in prokaryotic cells (reviewed in 52), but is much less well characterized in eukaryotes. In contrast, it has proved much easier to characterize eukaryotic enzymes involved in a second type of excision repair in which the first step is the release not of oligonucleotides from DNA, but of single damaged bases. Such repair was initially characterized in *E. coli* by Lindahl and his co-workers (53,54) and is now referred to as base excision repair. A number of recent reviews have been published (29,35,55).

The initial step in base excision repair involves the recognition of the damaged base by a specific DNA glycosylase, which subsequently hydrolyzes the sugar-base *N*-glycosylic bond, thus releasing the damaged base and leaving an AP site in the DNA. Glycosylase enzymes exist for a wide variety of base damage, each enzyme being remarkably specific for a single lesion or a small group of closely related lesions. The lesions repaired by base excision repair include not only those created by alkylating agents, but also abnormal bases, such as uracil in DNA and ring-open bases created either by oxygen damage or by ionizing radiation.

The first glycosylase to be identified was the uracil DNA glycosylase in *E. coli* (53). Uracil residues may arise in DNA either as a consequence of a misincorporation event by DNA polymerase or by hydrolytic deamination of cytosine residues. The importance of this enzyme in maintaining the fidelity of the genetic message can be inferred from the high spontaneous mutation rates in *ung* mutants, which lack this enzyme. In wild-type cells, in the presence of *ung* glycosylase, the deaminated cytosines will be removed before DNA replication can fix the G-C → A-T transition mutation (56).

Specific glycosylases have now been identified in *E. coli* for a number of different lesions. A hypoxanthine-specific glycosylase has been identified (4,57), which presumably has a role similar to the uracil glycosylase, since hypoxanthine will arise in DNA as a consequence of deamination of adenine. *E. coli* possesses two glycosylases capable of removing 3-methyl adenine from DNA. The *tag* gene product is expressed constitutively and is specific for the removal of 3-methyl adenine (58). DNA glycosylase II, the product of the *alkA* gene, has already been referred to and is part of the inducible adaptive response (59). This enzyme has a much broader specificity than the *tag* enzyme and can remove not only 3-methyl adenine, but also 7-methyl adenine, 7-methyl guanine, and O^2-methyl pyrimidines (60,61). It seems likely that this enzyme recognizes particular distortions in the DNA helix, rather than one specific damaged base.

A second *E. coli* glycosylate with a relatively wide specificity is encoded by the *nth* gene and was originally characterized as a urea DNA-glycosylase (62). In fact, the enzyme has activity against a variety of thymine hydrates and their breakdown products, which include urea- and *N*-substituted urea derivatives. Such products can be created either by powerful oxidizing agents or by UV or ionizing radiation (63,64). In addition to glycosylase activity, this enzyme also possesses Type 1 AP endonuclease (or, more correctly, AP lyase) activity, breaking the phosphodiester bond on the 3′ side of the AP site produced by the glycosylase action. This strand break can then be processed by a type II AP endonuclease to generate the correct ends for DNA repair synthesis (see below).

Ring-open base derivatives may also arise as a consequence of methylation of guanine at the N^7 position. The product of this reaction, 2,6-diamino-4-hydroxy-5-*N*-formamidopyrimidine (FaPy) is removed from DNA by another specific glycosylase, the product of the *fpg* gene (65,66). This same enzyme is also able to remove ring-open purines produced by the addition of aflatoxin B1, at the N^7 position (67).

The FaPy glycosylase and the *nth* glycosylase differ from the uracil glycosylase and the two 3-methyl adenine glycosylases in that the former both have associated AP endonuclease activity. This enables these enzymes with a relatively broad substrate specificity to carry out in a coordinated way both of the first two steps of base excision repair, while the action of the uracil and 3-methyl adenine glycosylases must be followed by a separate AP endonucleolytic step. Two other glycosylases have been identified that also carry associated AP endonuclease activity, namely the two thymine-dimer specific enzymes found in *Micrococcus luteus* (68) and bacteriophage T4, respectively. The T4 enzyme has been extensively characterized (reviewed by Friedberg, 8) and correlations between structure and function analyzed (69).

The AP sites created by the glycosylases that do not have an associated AP endonuclease, and the strand breaks produced by type I AP endonucleases, must subsequently be processed by other enzymes to yield the correct ends for repair synthesis. In *E. coli*, the major AP endonuclease/3′ diesterase is endonuclease II, an activity associated with exonuclease III and encoded by the *xth* gene. It is also sometimes referred to in the literature as endonuclease VI. AP endonucleases have been reviewed by Lindahl (5) and by Sancar and Sancar (29). The action of the *xth* gene product on AP sites is to produce a 5′ incision adjacent to the AP site (a type II mechanism). The new 5′ terminal deoxyribose-phosphate can then be excised by DNA deoxyribophosphodiesterase (70), leaving 3′-OH and 5′-P ends suitable to prime DNA repair synthesis and ligation. A second AP endonuclease, designated endonuclease IV and encoded by the *nfo* gene, plays only a minor role in the repair of most AP lesions in *E. coli*, as it normally represents $<5\%$ of the total AP endonuclease activity in the cell. However, expression of *nfo* is induced by oxidizing agents, and the enzyme plays a key role in the repair of lesions caused by such agents.

Endonuclease IV is also a type II endonuclease and generates ends suitable for subsequent repair. However, if the AP site is cleaved by a type I AP endonuclease (AP lyase) such as endonuclease III, the ends must then be further processed by a type II enzyme to yield suitable substrates for completion of repair. Both *xth* and *nfo* gene products play an important role in processing the 3′-ends of radiation-damaged DNA to remove a variety of base and sugar fragments, leaving a 3′-OH terminus suitable for repair synthesis, this 3′-diesterase activity being a characteristic property of both enzymes.

This process of base excision repair has now been demonstrated to occur in a wide variety of organisms, including the yeasts and filamentous fungi. In *S. cerevisiae*, the presence of a uracil DNA glycosylase has been reported (71), and the gene has recently been cloned and sequenced (72). The behavior of mutants of *S. cerevisiae* that lack this enzyme (73) indicates that its main in vivo role is in the repair of cytosine deamination damage.

The presence of a 3-methyl adenine DNA glycosylase in *S. cerevisiae* has been demonstrated in two laboratories (74–77), and by using functional complementation in an *E. coli tag ada* mutant strain, it has been possible to isolate the yeast gene, now designated *MAG*. The MAG DNA glycosylase specifically protects yeast cells against the killing effects of alkylating agents, but does not protect cells against mutation. In fact, rather unexpectedly, *mag* mutants displayed a reduced mutagenic response to alkylating agents, implying that the enzyme might actually generate promutagenic lesions during the repair process (76). The sequence of the *MAG* gene revealed striking similarities between the MAG protein and the *E. coli* AlkA protein, but no homology to the *tag* gene product. Like the *E. coli* enzyme, the MAG protein can remove both 7-methyl guanine and 3-methyl adenine from alkylated DNA (74). Also like the *alkA* gene, expression of *MAG* can be induced by exposure of the cells to low levels of MNNG. Measurement of transcript levels indicates that an increase in expression of approximately 15-fold can be achieved within 1–2 hr of exposure. The mechanism underlying the induction process has yet to be characterized, although it has been shown that the MGT1 transferase plays no part in the process (46). The product of the *MAG* gene is a 34-kDa protein, as predicted from the gene sequence and confirmed by SDS page. The gene maps to chromosome 5 and appears to correlate to the gene previously identified in the *mms-5* MMS-sensitive mutant. The defective functions in a number of other yeast MMS-sensitive mutants have yet to be identified (78,79).

An enzyme capable of carrying out the second stage of base excision repair has also been isolated from *S. cerevisiae*, namely an AP endonuclease designated Apn1 (80,81). This enzyme is the major AP endonuclease in yeast and is very similar in properties to endonuclease IV, the minor AP endonuclease in *E. coli*. The gene for the enzyme, *APN1*, has been isolated and sequenced and displays striking homology to the *E. coli nfo* gene, but not to *xth* (82). Like endonuclease II (Xth) and endonuclease IV (Nfo) of *E. coli*, the APN1 enzyme is a type II AP endonuclease and has a 3′-diesterase activity capable of removing fragmented sugar residues from the ends of damaged DNA. Unlike endonuclease II (Xth), it has no associated 3′-5′ exonuclease activity on undamaged DNA. Cross-species complementation studies demonstrate that the APN1 protein can effectively substitute for endonuclease IV function in *E. coli*, but cannot substitute for endonuclease II (Xth) (83).

The characterization of glycosylase and AP endonuclease enzymes in other fungi has not been extensive. A 3-methyl adenine DNA glycosylase has been detected in extracts prepared from *A. nidulans* (48), and unlike the *S. cerevisiae* enzyme, expression of this protein is not induced by alkylating agents (S. Farrington, S. Shawcross, and P. Strike, unpublished). These enzymes are found widely distributed throughout all eukaryotes, however, and it would be a surprise if they were not present in most fungal species.

2. Nucleotide Excision Repair

As indicated in the previous section, nucleotide excision repair was among the first repair processes to be characterized in bacterial cells. However, perhaps because of the complexity of the steps involved, progress in characterizing similar processes in eukaryotic cells has been slow. In *E. coli* both the genetics and biochemistry of the nucleotide excision repair pathway have been extensively characterized, and the subject has recently been comprehensively reviewed (52). The early steps in the process are catalyzed by a complex of three proteins, UvrA, UvrB, and UvrC. These are large proteins of 103,874, 76,118, and 66,038 Da, respectively, and both UvrA and UvrB possess ATPase activity. In the presence of ATP, a complex comprising two molecules of UvrA and one molecule of UvrB displays DNA helicase activity. This activity can scan potentially damaged DNA, and guided by the affinity of the UvrA protein for DNA damage, a complex is assembled at the damage site. The initial stage in complex assembly requires the loading of the UvrB protein onto the DNA, and perhaps the simultaneous release of the UvrA subunits (84–86). Once assembled, this damaged DNA/UvrB site serves as a target for the third component of the complex, the UvrC protein, the introduction of which leads to the production of single-strand breaks in specific positions relative to the damage. The eighth phosphodiester bond 5′ to the damage and the fourth or fifth bond 3′ to the damage are both cleaved, and the oligonucleotide containing the damage is then released by the combined action of the UvrD helicase and DNA polymerase I. The coordinated repair synthesis performed by these proteins at once correctly restores the DNA sequence of the damaged strand, displaces the damage-containing oligonucleotide, and promotes the turnover of the Uvr proteins.

In eukaryotic cells, an equivalent detailed biochemical analysis of excision repair does not exist, although a number of important steps have been taken. It is quite clear from genetic studies that excision repair mechanisms operate in eukaryotic cells in much the same way as in prokaryotes. Mutants of *S. cerevisiae* defective in DNA repair have played a key role in establishing the presence of the repair pathway and have also been particularly important in identifying human genes with a similar role. This latter topic will be discussed in more detail in a later section.

Over 40 distinct genetic loci are identified in *S. cerevisiae* as being involved in various aspects of DNA repair. In addition, a number of similar loci (twenty-three) have been identified in *Sch. pombe*, and a smaller number have been identified in other fungi. There are excellent reviews of DNA repair in *S. cerevisiae* and *Sch. pombe*, which cover the topic in much greater detail than would be appropriate here (87–92). The *S. cerevisiae* genes are commonly allocated to one of three functional or epistatic groupings, the archetypal members of each being *RAD3*, *RAD6*, and *RAD52*, respectively. Members of the *RAD3* group have been shown to be involved primarily in excision repair processes, the *RAD6* group are involved in mutagenesis and postreplication repair (discussed later), and the *RAD52* group defines genes involved in recombination.

The *RAD3* group of genes, involved in nucleotide excision repair, contains at least 17 different complementation groups, each of which contains mutants sensitive to UV light and bulky-lesion-producing chemical carcinogens. Five of these groups, represented by the *rad1, 2, 3, 4,* and *10* mutants, display a total defect in the incision step, which initiates excision repair, and one further gene likely to be involved in incision is the yeast homolog of the human *ERCC3* gene, *ERCC3*^{sc}. Loss of this function in humans results in a block in incision, and it seems probable that the yeast homolog will perform a similar role. Six other groups, represented by the *rad7, 14, 16, 23, 24,* and *mms19* mutants, display only a partial defect in incision, which is not due to the leakiness of the alleles, but rather is a true representation of the phenotype of these mutants. One possible explanation may be that the products of these genes are involved in excision repair only under certain circumstances—perhaps, for example, in the preferential repair of actively transcribed DNA (see later), but not in the repair of nontranscribed DNA. There is also an indication that some repair proteins, for example RAD3 and RAD6, are needed only for the repair of nuclear chromosomal DNA and are not required for the repair of transforming plasmids (93,94). Other genes that are almost certainly involved in the excision repair process but are not strictly members of this epistasis group are *CDC9*, which is known to code for DNA ligase, and the *POLI* gene, which encodes the 180-kDa yeast DNA polymerase I. Although many of these genes have now been cloned and sequenced, their biochemical properties have really only just begun to be characterized. A number of the *RAD3* group of gene products contain acidic regions particularly in the C-termini of the proteins, and it seems highly likely that these regions will confer the ability to bind to the very basic histones present in chromatin. Both the RAD3 protein and the $ERCC3^{sc}$ protein contain a sequence matching the consensus nucleotide-binding motif found in many ATPases, and the RAD3, RAD4, RAD10, and $ERCC3^{sc}$ proteins all contain regions matching the helix-turn-helix motif associated with many DNA-binding proteins. Both RAD3 and $ERCC3^{SC}$ contain sequences that are conserved in both DNA and RNA helicases, and in the case of RAD3, an ATP-dependent DNA helicase activity has been demonstrated (95,96). RAD1 and RAD7 proteins have been shown to share tandemly repeated leucine-rich motifs (LRMs), which are usually involved in mediating specific protein/protein interactions (97). It is also clear that many of the *RAD* gene products will have multiple roles in repair processes. The *RAD1* and *RAD10* gene products, for example, have been shown to be involved not only in excision repair, but also in the processing of mitotic recombination intermediates (98). Moreover, there is likely to be competition between different repair pathways for the same DNA lesion, perhaps making phenotypic characterization of double mutants difficult to interpret (99). Unfortunately, there has been no success to date in devising an in vitro excision repair assay for yeast, such as that which initially allowed an analysis of the *E. coli uvr* genes (100), or that which has recently been developed for mammalian cells (101). Such an assay would appear to be an essential prerequisite for the further characterization of the yeast excision repair system.

Among the loci identified in *Sch. pombe* as being involved in excision repair, it is now clear that a number can be directly related to their *S. cerevisiae* analogs. Eight *Sch. pombe rad* mutants have been assigned to the *rad5* epistasis group involved in excision repair, and include *rad2, 5, 10, 11, 13, 15, 16,* and *17*. Some of these mutants may be allelic—for example, *rad10* and *rad16* (102)—although others represent distinct genetic loci, and homology and functional complementation tests have allowed the *S. cerevisiae* analogs to be identified as shown below.

S. cerevisiae	*Sch. pombe*
RAD1	*rad10*
RAD2	*rad13*
RAD3	*rad15*
RAD4	Unnamed
(*RAD6*	*rhp6*)
(*CDC9*	*cdc17*)

C. Other Repair Mechanisms Revealed by the Characterization of Mutants

1. Analysis of Mutants in *S. cerevisiae*

Most of the *S. cerevisiae* genes that cannot be assigned to the *RAD3* functional group can be assigned to one of two other groups, typified by *RAD6* and *RAD52*, respectively. The *RAD6* group appears to contain genes involved primarily in error-prone pathways of repair, while the genes belonging to the *RAD52* group are involved in recombination.

The loci so far assigned to the *RAD6* group include *RAD5*, *6*, *8*, *9*, *15*, and *18*; *MMS3*, isolated as a mutant sensitive to methyl methane sulfonate; *PSO1*, isolated as a mutant sensitive to photoactivated psoralen; *CDC7*, *8*, and *40*, isolated as conditionally defective in mitotic cell division; *REV1*, *3*, *5*, *6*, *7*, and *UMR1*, *2*, and *3*, which were isolated as mutants showing altered levels of spontaneous or DNA-damage-induced mutation; and finally *RADH*, which was isolated as a suppressor of the UV sensitivity of *RAD18* (103). The mutants in this group are typically sensitive to UV, X-rays, and alkylating agents and show a deficiency in damage-induced mutagenesis. Most do not affect sporulation, and none appear to have a major effect on meiotic recombination. However, many do display increased levels of mitotic recombination. Four divisions of the *RAD6* group have been suggested on the basis of phenotypic properties (104), with the first group containing only *RAD6* itself. Mutations in *RAD6* result in a complex phenotype, including defective meiosis and sporulation, inefficient dimer excision and repair of single-strand breaks, as well as the defect in induced mutation. The second subgroup contains *RAD18*, which may be involved with *RAD6* in an error-free repair pathway; the third group contains *RAD9* and *RAD15*, which are thought to be able to function in either an error-free or an error-prone repair pathway, depending on the type of lesion encountered; and finally the *REV* loci, which were initially thought to be concerned only with error-prone repair, but which now appear also to be involved in an error-free component of repair.

Of the genes in this group, the *RAD6*, *RADH*, *RAD18*, *REV1*, and *REV3* genes are the only ones to be so far cloned and sequenced. The *RAD6* protein is the best characterized biochemically, and it has been shown to function as a ubiquitin-conjugating enzyme, which efficiently polyubiquitinates histones H2A and H2B (105). It is speculated that such histone modification plays a key role in allowing access of DNA repair enzymes to damage sites, and also in coordinating such processes as sporulation. The C-terminus of RAD6 contains a region in which 20 of 23 amino acids are acidic, including an uninterrupted stretch of 13 aspartates. This acidic region directs ubiquitin-conjugating activity to the highly basic histones and is essential for sporulation, as judged by the behavior of deletion mutants. However, the region is not essential for DNA repair or damage-induced mutagenesis and indeed is missing altogether from the equivalent protein found in more complex eukaryotes. Its role in the control of sporulation may be linked to protein degradation mediated by the N-end-recognizing E3 enzyme (106). The *RAD6* gene is conserved in *Sch. pombe*, where the gene is designated *rhp6*. No other member of this epistasis group has yet been identified in *Sch. pombe*. Sequencing of the *RADH* gene has revealed motifs characteristic of ATPase and DNA helicase activity, while the sequence of *RAD18* reveals that the protein contains several zinc finger domains, typical of DNA-binding proteins. This may indicate a role for *RAD18* as a direct component of a DNA repair complex—similar to the zinc finger domains found in the *E. coli* UVRA protein—or may indicate that the protein has a regulatory role. Many eukaryotic transcriptional regulators are known to interact with specific DNA sequences via zinc finger domains. The *REV1* gene encodes a strongly basic protein of 112,239 Da, one stretch of 150 amino acids sharing homology with the *E. coli umu*C protein, which is essential for UV-induced mutagenesis in that organism (107). The *REV3* gene in contrast encodes a 172,956-Da protein with a high degree of homology to several eukaryotic DNA polymerases. Unlike *POL1*, *REV3* is not essential for cell viability, and it may be that this polymerase is responsible for the low-fidelity base incorporation required during translesion synthesis (108).

Thus it is likely that the *RAD6* group contains genes that encode a ubiquitin-conjugating enzyme, a DNA helicase, a DNA-binding protein, and a DNA polymerase. Just such a collection of proteins would be required for mutagenic processing of damaged DNA, although the details of this process remains to be determined.

The *RAD52* epistasis group represents the third major subdivision of the radiation-sensitive mutants of *S. cerevisiae* and contains the loci *RAD50*, *51*, *52*, *53*, *54*, *55*, *56*, *57*, and *PSO1* and *3*. The mutants in this group are particularly sensitive to X-rays and monofunctional alkylating agents, with moderate sensitivity to UV. Many of these genes are required for recombination and for the repair of strand breaks. All of the mutants except *rad50* are deficient in spontaneous mitotic recombination, and *rad50* shows highly elevated levels (i.e., is hyperrec). All are deficient in damage-induced mitotic recombination, except *rad55*, which is hyperrec. All are defective in meiotic sporulation, except *rad56* (109), and all of those tested are defective in meiotic recombination. A new series of mutants with defects specifically in the early steps of homologous recombination has recently been isolated (110) and offers the prospect of a more comprehensive analysis of recombination processes in yeast. Also, a mutant with pleiotropic effects on repair, sporulation, mitosis, and meiosis has been isolated and defines the gene *HRR25*. A number of the genes have been cloned and sequenced, but to date little biochemical characterization of the proteins has taken place. The *HRR25* gene product, however, has been characterized as a protein kinase, a member of the superfamily of such proteins that play a key role in coordinating cell division and growth processes (111).

The *RAD50* gene codes for a 153-kDa protein that contains a purine-nucleotide binding site, as well as sequences typically asociated with the formation of helical coils. The presence of these sequences suggests that this protein might play a part in a contractile system that brings together homologous chromosomes prior to pairing (112,113) and could thus play a role in both meiotic recombination and the repair of double-strand breaks.

The *RAD52* gene encodes a 60-kDa protein that can complement the recombination defect in bacteriophage T4 gene 46 and gene 47 mutants. As these mutants are defective in an exonuclease activity, this appears to be a likely function for the RAD52 protein. The *RAD54* gene codes for a large protein of over 100,000 Da, which contains a nucleotide-binding domain, a possible leucine-zipper domain, and nuclear localization signals. The putative RAD57 protein also contains a nucleotide-binding domain (114), as does the RAD51 protein. As ATPase activity is often associated with helicase activity, it seems possible that a number of such enzymes will be involved in recombination-mediated DNA repair in *S. cerevisiae*.

Although a significant number of genes have been identified as being involved in recombination repair on *S. cerevisiae*, none of them has so far been correlated with an analog in *Sch. pombe*. Given the extensive conservation of repair functions involved in excision repair, this might be considered rather surprising. However, there are major differences in the cell cycles of *S. cerevisiae* and *Sch. pombe*, which must have major consequences for the possibility of recombinational repair. In contrast to *S. cerevisiae, Sch. pombe* spends about 75% of its life cycle in G2, when there are two copies of each chromosome present as sister chromatids, and under such circumstances recombination processes might be expected to be favored. In *S. cerevisiae*, the time spent in G2 is between 20 and 25%, while in the filamentous fungi such as *A. nidulans*, G2 occupies about 40% of the life cycle. Clearly, these differences in the time spent in G2 will have an effect on the strategies adopted by the cells to repair their DNA and might lead to significant divergence in the gene products involved in these processes in these different organisms. To date, as correlation with the *S. cerevisiae* genes has not proved feasible, the *Sch. pombe* genes that are not included in the *rad5* excision repair epistasis group have been divided into two functional groups, the *rad1* group, involved in G2 arrest and recombinational repair, and the *rad21* group, involved in repair of radiation damage. The *rad21* group contains just the *rad21* and *rad22* genes, while the *rad1* group contains *rad1, 3, 4, 6, 8, 9, 17, 18, 19,* and *20*, as well as *rhp6*, the equivalent of the *S. cerevisiae RAD6* gene. A number of *Sch. pombe rad* genes, including *rad7, 12, 14,* and *23*, do not fit easily into either of the above functional groups, and so remain unclassified. Most of the *rad* mutants have been complemented with *Sch. pombe* libraries, and a number of the genes have been subcloned and sequenced (reviewed by Subramani, 92) or assigned to chromosomal *Not1* restriction fragments (115). However, no definitive biochemical role has yet been assigned to any of the gene products. A collection of *Sch. pombe* mutants isolated specifically as defective in mitotic recombination has recently been reported (116) and it is likely that this will become an increasingly active area of study. A similar set of meiotic recombination mutants has also been isolated (117).

2. DNA-Repair-Related Genes in *A. nidulans*

Compared to *S. cerevisiae*, DNA repair studies in *A. nidulans* are perhaps remarkable for their scarcity, given the very well-characterized genetic systems of this organism. A number of mutants have been isolated in different laboratories as showing increased sensitivity to UV radiation (see, for example, 118–125), and most have been characterized for their effects on recombination, fertility, and mutagenesis. Complementation tests with these mutants have defined nine distinct genetic loci, *uvs-A–J* (126), and all except *uvsG* were shown to be sensitive to MMS as well as UV. In this same study, 11 new DNA repair loci, the *mus* genes, were identified by the isolation of mutants sensitive to methyl methane sulfonate. The further characterization of these mutants has not yet been reported. Five further mutants, specifically sensitive to alkylating agents, were isolated by Swirski and his colleagues (48) and shown to be nonallelic to the previously isolated mutants. Enzyme assays for glycosylase and O^6-methyl guanine DNA methyl transferase have been conducted on two of these mutants, *sagA1* and *sagC3*, but no defects have been identified. The *sagA1* coding sequence has, however, been recently isolated (S. Farrington, P. Hooley, and P. Strike, unpublished), and the determination of its sequence is currently in progress. It has been demonstrated that expression of the *sagA1* gene is inducible by alkylating agents, and two of the *sag* mutants, *sagB* and *sagC*, have been shown to display, in homozygous diploids, a hyperrec phenotype for spontaneous, UV- and MNNG-induced mitotic crossing over.

Recombination defective mutants have been selected directly by screening for mutants defective in intragenic recombination between *adE20* on chromosome, 1, and an *adE8* allele on a segment of chromosome 1 translocated onto chromosome 2 (127). The isolation of six recombination defective mutants was reported, although it was demonstrated that none of these mutants displayed increased sensitivity to UV (128). One mutant with a hyperrecombination phenotype was also isolated, which was subsequently shown to be an allele of *uvsB*.

By studying the sensitivities of double-mutants strains, Kafer and Mayor (126) established that the *uvs* mutants could be assigned to three epistasis groups containing *uvsF* and *uvsH; uvsB* and *uvsD*; and *uvsC* and *uvsE*, respectively. Unfortunately, it was not possible to make this study exhaustive, as many of the double-mutant combinations appeared to cause inviability. Nevertheless, the phenotypes of the mutants suggested involvement in particular repair pathways. *uvsF* and *uvsH* mutants were sensitive to UV, both as resting conidia and at the 7-hr germling stage. They were both hypersensitive to monofunctional alkylating agents, but not to X-rays, and both exhibited increased levels of damage-induced mutation. Neither mutation affected meiotic recombination, although both caused an increase in spontaneous mitotic inter- and intragenic recombination. Based on these observations, a role for the *uvsF* and *uvsH* gene products in excision repair has been suggested. A DNA sequence containing *uvsF* has recently been cloned by complementation of the mutant with a genomic library (129), but no further characterization has been reported.

The *uvsB* and *uvsD* mutants shared many of the features of *uvsF* and *H*, including the cross-sensitivity to UV and alkylating agents. However, the sensitivity to alkylating agents was much more marked than in *uvsF* and *H* and included sensitivity to EMS (130); the mutants were virtually sterile in genetic crosses, and such progeny as did arise from meiosis appeared as abnormal segregants. Their most characteristic phenotype was the production of abnormally growing unstable mitotic segregants from homozygous diploids, and it has been suggested (126) that these genes could be essential for the maintenance of integrity of chromosome structure. A recently isolated semidominant allele of *uvsB* indicates a role for this gene product in preventing spontaneous mutation (131).

The third mutant pair, *uvsC* and *uvsE*, displayed a phenotype very similar to the *RAD52* group mutants of *S. cerevisiae*, with increased sensitivity to ionizing radiation and monofunctional alkylating agents, and slight sensitivity to UV at the 7-hr germling stage. Both mutants were recombination defective for induced mitotic recombination, sterile in meiotic crosses, and caused an increase in chromosomal missegregants in diploids.

Because of the paucity of repair mutants available in *A. nidulans*, we have recently isolated a large number (>200) of mutants that display increased sensitivity to both alkylating agents (MNNG) and a UV-mimetic chemical mutagen (4NQO) (132–134). Complementation mapping and segregation analysis of the mutants has so far revealed 29 distinct genetic loci, and the effect of mutations in these genes on intragenic genetic recombination has been tested using a newly developed and relatively simple assay (Fig. 3). One of the mutants, *nuv11*, has been shown to be recombination deficient for meiotic inter- and intragenic recombination and to be hyperrec for spontaneous mitotic recombination. Seven mutants, *nuv2, 4, 8, 110, 112, 114,*

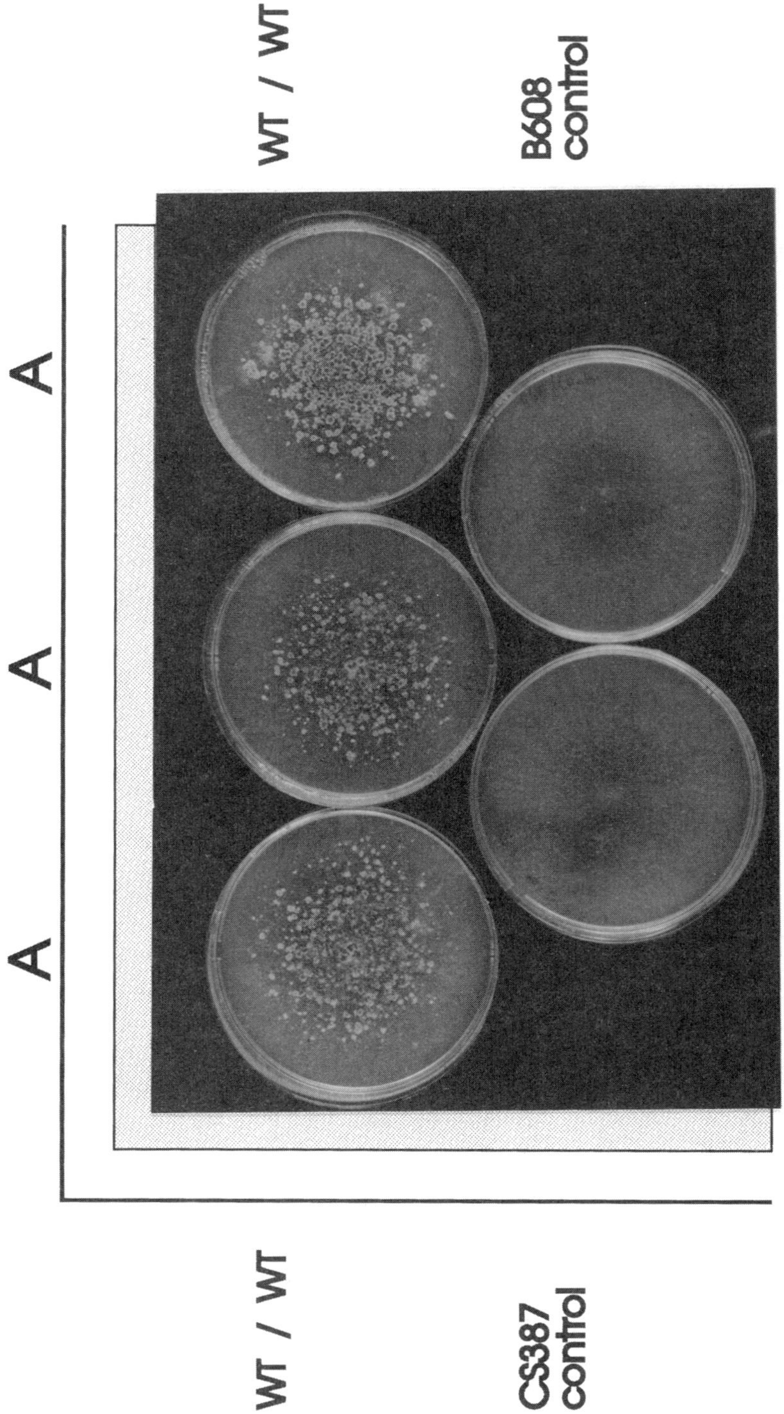
WT / WT
B608
control
A
A
A
C
B
WT / WT
CS387
control

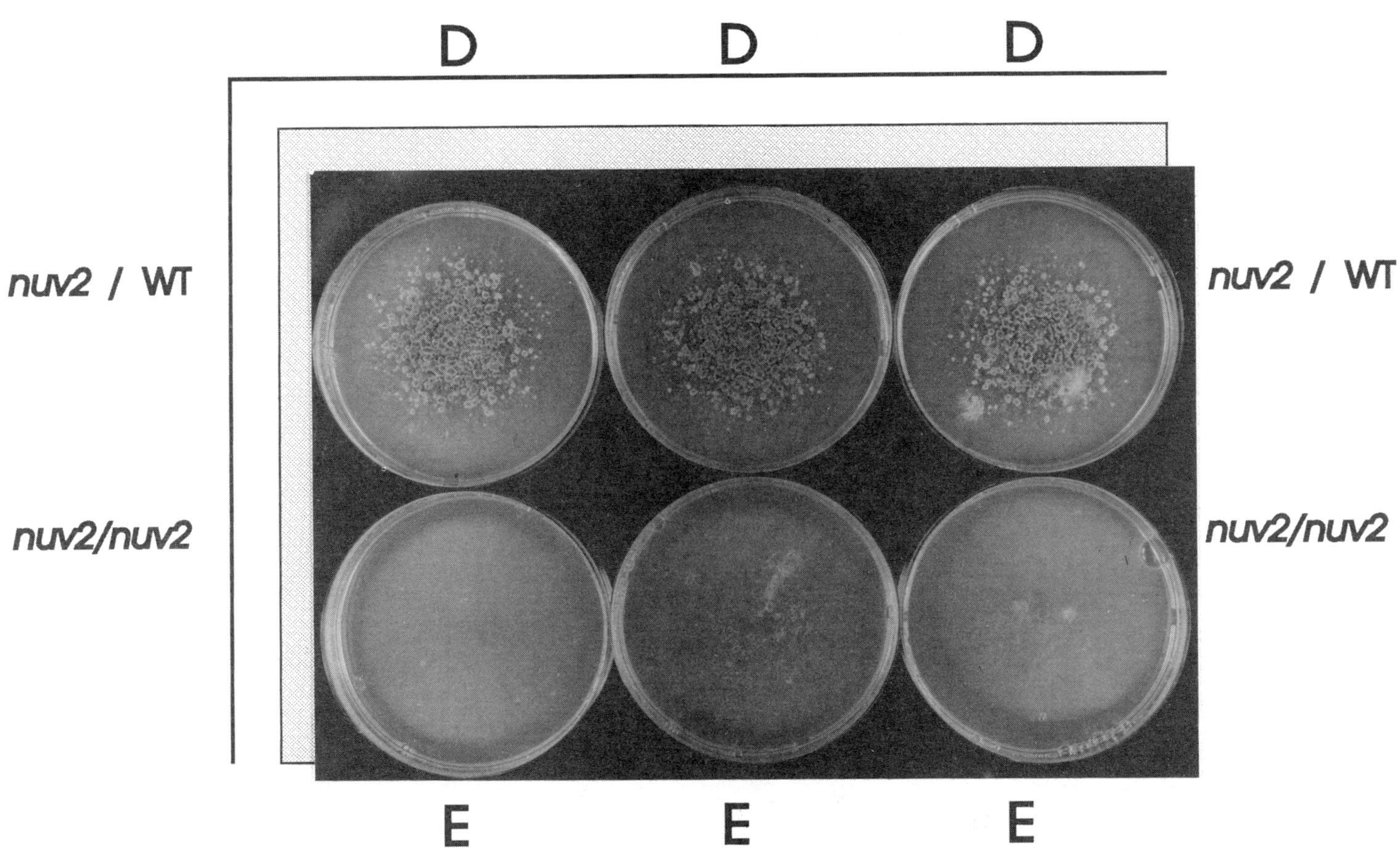

Figure 3 (Legend on p. 320.)

and *117*, were defective for spontaneous mitotic recombination, while three others, like *nuv11*, were hyperrec for this process. Several of the mutations affected the stability of diploid strains, giving rise to haploid and/or unbalanced segregants. Although analysis of these mutants is at an early stage, one gene, *nuv11*, has been cloned and sequencing is in progress. It is already apparent that these mutants display both similarities to and significant differences from the *S. cerevisiae* recombination mutants, and that the life cycle of *A. nidulans*, allowing as it does the creation of stable diploids and analysis by both sexual and parasexual crosses, makes it an ideally suited organism for these studies.

3. DNA-Repair-Related Genes in *N. crassa*

Repair processes in *N. crassa* have been studied somewhat more extensively than those in *A. nidulans*, but not to anything like the extent of those in the two yeast species. Mutants have been isolated as sensitive to UV (the *uvs* and *upr* mutants), and as sensitive to gamma rays (*gs* mutants) (reviewed by Schroeder, 135). Many mutants have also been isolated as sensitive to MMS (*mus* mutants) (for example, 136–140). Some mutants initially isolated as defective in nuclease secretion, the *nuc* and *nuh* mutants, (141,142), and as defective in meiosis, the *mei* mutants (143,144), were subsequently shown to be abnormally sensitive to DNA damage.

Figure 3 Simple assay for mitotic recombination in *Aspergillus nidulans*. The assay is based on a diploid constructed between two haploid test strains, B608 (*wA4, puA2, niiAniaD608*) and CS387 (*yA2, luA1, niaD26*), each carrying a nonoverlapping deletion within the *niaD* gene on chromosome 8. The two strains are deficient in nitrate reductase, the *niaD* gene product, and the B608/CS387 diploid cannot therefore utilize nitrate as sole nitrogen source. However, commercial agar preparations contain traces of reduced nitrogen, which allow nitrate-nonutilizing strains to grow with a characteristic residual growth phenotype, i.e., sparse, nonsporulating, mycelial growth. Sporulating, nitrate-utilizing subcolonies arise spontaneously within the characteristically sparse mycelium of centrally inoculated B608/CS387 diploid colonies growing on nitrate-containing medium (A). These subcolonies have been shown to be diploid $niaD^+$ recombinants, and to have arisen either by crossing over or gene conversion. As controls, heteroallelic diploids were constructed between each test strain and a strain carrying an overlapping *niaD* deletion. In plating tests with these diploids, B608/B114 (B) and CS387/B317 (C), no nitrate-utilizing subcolonies arose. Recombination-defective *nuv* mutants were identified by comparing the frequency of nitrate-utilizing subcolonies arising from B608/CS387 diploids either heterozygous or homozygous for a *nuv* mutation. Plates demonstrating the rec^- phenotype of *nuv2* are shown: (D): B608 *nuv2*/CS387; (E): B608 *nuv2*/CS387 *nuv2*. The assay proved equally well suited to identifying mutants displaying increased levels of mitotic recombination (hyperrec mutants—data not shown).

Although far from complete, complementation analysis has identified at least 20 distinct genetic loci controlling DNA damage sensitivity in *N. crassa*. Many of the strains have been characterized for cross-sensitivity to other DNA damaging agents, for effects on spontaneous and induced mutagenesis, and on meiotic fertility and recombination. Diploid and disomic strains are unstable in *N. crassa*, so that effects on homologous mitotic recombination cannot easily be determined. However, mitotic events could be analyzed by testing for effects on the stability of duplication strains (145,146). Only hyper-rec types have been found for mitotic recombination using this approach. Several mutants cause reduced fertility or sterility in sexual crosses, but no mutant defective in meiotic recombination has been found. However, a number of the *mms* mutants display cross-sensitivity to agents such as bleomycin and gamma rays and must therefore be considered to be potentially defective in recombination repair (139).

As with *A. nidulans*, assignment of mutations to repair pathways has proved difficult because many of the double mutants are inviable. However, it has been possible to describe three different epistasis groups using this approach (147–151). The phenotypes of many of the mutants have been reviewed by Schroeder (146).

Mutants in the first epistasis group, *uvs2, uvs4, upr1,* and *mus8*, were all initially characterized as involved in excision repair. All were sensitive to UV and alkylating agents but, unlike excision repair mutants in *E. coli* and *S. cerevisiae*, were also significantly sensitive to X-rays. All had normal levels of spontaneous mutation, and *uvs2* and *upr1* had elevated levels of damage-induced mutation, as is typical for mutants defective in error-free excision repair. However, *uvs4* and *mus8* were defective in damage-induced mutation. None affected stability in duplication strains, and all except *mus8* showed normal fertility and meiotic recombination. *uvs2* and *upr1* were initially reported to be defective for in vivo excision of pyrimidine dimers (152). However, later assays (153,154) showed clearly that all these mutants are proficient in dimer removal, and it is now less clear where the defects may lie. *uvs2* has been shown to have an effect on spontaneous reversion at the *ad-3* locus, and it has been proposed that it may be involved in a process that normally removes the types of damage leading to spontaneous mutation (155). It is, however, normal for the process of postreplication repair (156), which appears in *N. crassa* to be mechanistically similar to *S. cerevisiae* (157).

Mutants in the second epistasis group, *uvs3, mus9*, and *mus11*, are thought to be defective in error-prone bypass synthesis. They are sensitive to UV, X-rays, and alkylating agents, exhibit increased levels of spontaneous mutation, but are defective in damage-induced mutation. All cause sterility in sexual crosses and display enhanced instability in duplication strains (158). *N. crassa* displays a phenomenon in which MNNG-induced mutation rates are increased as a consequence of heat shock treatment. However, none of the known DNA repair mutants appear to affect this process (159).

The third epistasis group has been postulated to contain mutants involved in recombinational repair, although many of the phenotypes are difficult to categorize. The group includes mutants *uvs6, mus7*, and *mus10, mus7* shows no abnormal sensitivity to UV, *mus10* is slightly sensitive, and *uvs6* shows significant sensitivity. Only *uvs6* is highly sensitive to X-rays, and *mus10* was atypical in displaying only slight sensitivity to alkylating agents. None of the mutants affected spontaneous or damage-induced mutagenesis, *uvs6* and *mus7* were sterile, and *mus10* showed reduced fertility.

To date, none of these genes have been cloned, nor has any particular biochemical function been assigned to any of the genes. However, changes have been observed in polypeptides produced following UV irradiation, two proteins showing altered behavior in a *mei-3* mutant, and one protein appearing changed in the excision repair mutant *upr1*. Two changes were detected in the allelic mutants *uvs3* and *nuh4*, with one polypeptide missing and another showing an alteration in pI, becoming more basic (160,161). Whether expression of these proteins is controlled by a process perhaps equivalent to the bacterial SOS system has yet to be determined, although, as will be discussed below, there is plenty of evidence that expression of many fungal DNA repair genes is inducible by DNA damage.

4. DNA Repair Genes in Other Fungi

DNA repair studies of any depth have really only been carried out in the two yeast species and the two filamentous fungi described so far. Recombination studies have been rather more extensive, as fungi such as *Sordaria* and *Coprinus* are particularly well suited to studying these processes—however, molecular studies in these species have not been extensive. One system that has been studied in detail is the *rec1* system of *Ustilago maydis*, which plays a key role in both genetic recombination and the repair of double-strand breaks. Indeed, the *rec1* mutants were the first mutants in any eukaryote to be specifically isolated as sensitive to a broad range of DNA-damaging agents (162, 163) and then to be characterized for their phenotypic effects (164,165). The gene encoding *rec1* has been cloned and sequenced (166,167), and shown to code for a 56.8-kDa protein, which contains a potential zinc finger sequence and two nuclear localization sequences. It does not encode the 70-kDa recombination protein, which has been characterized in this organism (168, 169), although levels of this activity are reduced in the *rec1* mutant. A second recombination gene from *U. maydis, rec2*, has also recently been cloned (170) and is clearly different from *rec1*. The *rec1* mutant has been shown to be defective in the repair of double-strand breaks, a process that has been well characterized in *S. cerevisiae* and a number of other fungi, and also in the slime mold *Dictyostelium discoideum* (171).

IV. FACTORS AFFECTING DNA REPAIR

A number of factors affect the efficiency of repair processes, in fungi as in other organisms. They include preferential repair of actively transcribing DNA, the coupling of repair gene expression to the cell cycle, and the inducibility of repair genes directly by DNA-damaging agents. It is not possible in the context of this review to provide a comprehensive coverage of these topics, but it is important to outline them here to give an indication of the cellular context in which repair processes must work.

A. Preferential Repair

Preferential repair of transcribed DNA was first described in mammalian cells, where it was observed that in general, post-UV irradiation RNA transcription resumed long before thymine dimers were completely removed from the DNA (reviewed in 172,173). One possible explanation for this observation could be that actively transcribing genes are repaired preferentially compared to nontranscribing genes, and indeed good evidence for such preferential repair now exists not only in higher eukaryotes, but also in lower eukaryotes and in bacterial cells. The preferential repair has been shown, at least in the case of the mammalian DHFR gene, to be associated only with the transcribed, or template, strand (174). Preferential repair of actively transcribing genes in a lower eukaryote was first described for the *MATα* mating-type locus in *S. cerevisiae* (175), and it has been shown that this preferential repair occurs only in the G1 and S phases of the cell cycle. In G2, both active and inactive forms of the mating-type locus (*MATα* and *HMLα*) are repaired with equal efficiency. It is assumed that in G2, a chromatin structure exists that does not allow the repair complex to discriminate between transcriptionally active and inactive DNA; alternatively, although perhaps less likely, there may be a specific repair system that is only functional during G2 (176). Mutations in *RAD9, RAD16*, and *RAD24* appear to impair the repair of damage in the nontranscribed *HMLα* gene, but do not affect the repair of the active *MATα* gene (177). *RAD16* and *RAD24* have been assigned to the *RAD3* excision repair group, while *RAD9* is now known to control the cell cycle response to DNA damage. If *S. cerevisiae* is irradiated during G2, the cell cycle is arrested at a point immediately prior to mitosis to permit completion of repair. The *RAD9* gene is essential for damage-induced cell cycle arrest to take place (178). Detailed repair of particular UV-induced lesions has been elegantly studied in a small yeast minichromosome system (179) where it has been possible to confirm preferential repair of the active *URA3* gene, compared to the nontranscribed flanking regions. Moreover, in this case it was also possible to show that repair was targeted primarily at the transcribed template strand.

B. Cell Cycle Effects

As discussed previously, it is to be expected that the position in the cell cycle will have an effect, particularly on recombination repair, which requires the presence of two suitable homologs. Some repair-related enzymes are coordinately expressed with the cell cycle—for example, the *CDC9* product in *S. cerevisiae*, DNA ligase, is increased in late G1 during normal DNA replication (180). The *RAD2* excision repair gene of *S. cerevisiae* is not cell cycle regulated during mitosis, but does display an increased steady-state level during meiosis (181), as do the other excision repair genes *RAD7* and *RAD23* and the ubiquitin-conjugation gene *RAD6* (182–185). A similar increase during meiosis is displayed by the recombination repair genes *RAD52* and *RAD54* (186) and by the *RAD18* gene, which belongs to the *RAD6* epistasis group (187).

There is a strong correlation between those genes which display increased levels of expression during meiosis and those whose expression is inducible by DNA damage (see below), and it is clear that in a number of cases, a common mechanism of control is involved.

C. Induction of Repair Enzymes by DNA Damage

In previous sections, the induction of the *S. cerevisiae* 3-methyl adenine DNA glycosylase and the *A. nidulans* O^6MeG DNA methyltransferase has been discussed. In addition to these inducible genes, it is apparent that a large number of others are also induced by DNA damage. At present, the mechanisms underlying the control of this induction are not understood, although it is tempting to think that perhaps a global control network equivalent to the *E. coli* SOS system (reviewed by Walker, 188) will exist in fungi. UV-inducible proteins have been detected in extracts of cells of *S. cerevisiae*, one of which appears to have antigenic cross-reactivity to the *E. coli* RecA protein (189). With the cloning of the various *RAD* genes, measurements of their induction have been much simplified. The *RAD2, RAD6, RAD7, RAD23*, and *RAD54* genes have all been shown to be inducible by DNA damage (182–184,190–192), and a number of other poorly characterized *DIN*- and *DDR*-inducible genes have been detected by direct screening (193,194). Both the DNA ligase (*CDC9*) and DNA polymerase I (*CDC17*) genes are induced by DNA damage (180,195), as are the *UB14* and *RNR2* genes (196,197). However, not all functionally related genes are coordinately induced—expression of *RAD10* is not induced by DNA damage despite the fact that it belongs to the same epistasis group as *RAD2* and appears to share many of the same promoter-associated control sequences (198), nor is expression of *RAD3* inducible (198). Little work has been done on repair genes cloned from any other fungi, although, as described earlier, a number of damage-inducible proteins have been described in *N. crassa*.

V. OTHER EUKARYOTIC ANALOGS

The idea that cell components could be highly functionally conserved became a reality with the demonstration by Lee and Nurse (200) that the *cdc*2 defect in *Sch. pombe* could be complemented by a human cDNA clone. Since that time, it has become apparent that DNA repair genes are among those functions most highly conserved, and homologies have been recognized even between bacterial and human genes. The fungi form a particularly important system for recognizing these homologies, not only because their defects can often be directly complemented by genes of more complex eukaryotes, or vice versa (see 201 for example), but also because they can often be complemented by prokaryotic equivalents (202,203). Because of the evolutionary distance that exists, for example, between *S. cerevisiae* and *Sch. pombe*, or between *S. cerevisiae* and the filamentous fungi, once genes are isolated that are functionally conserved between them, it is often possible to identify essential regions by sequence conservation. This information may then facilitate the isolation of similar coding sequences from higher eukaryotes. The extent of these homologies can perhaps best be demonstrated by comparing human, yeast, and bacterial O^6MeG methyl transferases, where a region of five amino acids containing the cysteine active site is completely conserved. The sequence **(I/V)PCHRV(I/V)** can be used to create heterologous probes, which can then be used to identify methyltransferase genes in almost any organism.

The functional homologies between the *S. cerevisiae RAD* genes, the *Sch. pombe rad* genes, and their human analogs have recently been highlighted in reviews by Hoeijmakers and Bootsma (204), by Friedberg and co-workers (89–91), and by Subramani (92). However, it is apparent that the power of this approach is only beginning to be exploited. It is likely that a large number of higher eukaryotic repair and recombination functions may be isolated, either by functional complementation in lower eukaryotes or by sequence homologies. We have recently demonstrated that the defect in the *A. nidulans nuv11* gene can be complemented by a clone from a *Drosophila melanogaster* cDNA library (R. Jones and P. Strike, unpublished), which may prove to be a convenient route to the isolation of the human genes involved in recombination.

VI. CONCLUDING REMARKS

We have attempted in this review to give a flavor of the very rapidly developing field that is now DNA repair in fungi. From the early isolation of mutants in a number of different species, progress has undoubtedly been most comprehensive in *S. cerevisiae*. The recent demonstration that this organism contains a pyrimidine dimer-specific DNA glycosylase/lyase gives the clearest possible indication that it will continue to be a major source of interesting and surprising results (205). However, as we have shown, a large number of interesting mutants are available in other species, and the realization that repair functions are often highly conserved makes study of some of the less well-characterized species even more attractive. The life cycles of some species are particularly well suited to the study of certain aspects of DNA repair —the study of mitotic recombination in *A. nidulans* for example. Moreover, there are also major differences between the various fungal species—the lack of an inducible O^6MeG alkyltransferase in *S. cerevisiae* contrasts with the presence of a number of such enzymes in *A. nidulans*, for example. It is clear that if we are to develop a thorough understanding of repair processes in the fungi, we must consider a variety of different species. The developments in molecular genetic techniques mean that it is now feasible to isolate genes by complementation in a large number of species, and even in those species where the genetic systems are not well developed, the availability of PCR means that homologous sequences can be identified and isolated. In any review of this type, it is inevitable that a number of areas will receive less coverage than they deserve, or even be omitted altogether. In the latter category, in this instance, lies coverage of mismatch repair, which is a key process in phenomena such as gene conversion. Over recent years, this subject has been elegantly investigated in bacterial cells, and work in *S. cerevisiae* is just beginning to make ground (206). Together with an understanding at the biochemical level of the roles played by the proteins identified genetically, this would seem to be an inevitable area for progress in the near future.

ACKNOWLEDGMENTS

The work in our laboratory on *A. nidulans* has been generously supported by the Wellcome Trust and by the North West Cancer Research Fund. The excellent contributions of Richard Swirski, Susan Shawcross, Paul Hooley, Susan Farrington, Sean Baker, Cath Cotton, and Robert Jones, who have all worked on this project over the years, are gratefully acknowledged.

REFERENCES

1. Lindahl, T., and Nyberg, B. (1972). Rate of depurination of native deoxyribonucleic acid, *Biochemistry 11*: 3610.
2. Lindahl, T., and Karlstrom, O. (1973). Heat induced depyrimidination of deoxyribonucleic acid in neutral solution, *Biochemistry 12*: 5151.
3. Lindahl, T., and Nyberg, B. (1974). Heat induced deamination of cytosine residues in deoxyribonucleic acid, *Biochemistry 13*: 3405.
4. Karran, P., and Lindahl, T. (1980). Hypoxanthine in deoxyribonucleic acid: generation by heat-induced hydrolysis of adenine residues and release in free form by a deoxyribonucleic acid glycosylase from calf thymus, *Biochemistry 19*: 6005.

5. Lindahl, T. (1979). DNA glycosylases, endonucleases for apurinic/apyrimidinic sites and base excision repair, *Prog. Nucleic Acids Res. Mol. Biol. 22*: 135.
6. Ryberg, B., and Lindahl, T. (1982). Non-enzymatic methylation of DNA by *S*-adenosyl-L-methionine is a potentially mutagenic reaction, *EMBO J 1*: 211.
7. Barrows, L. R., and Magee, P. N. (1982). Non-enzymatic methylation of DNA by *S*-adenosylmethionine in vitro, *Carcinogenesis 3*: 349.
8. Friedberg, E. C. (1985). *DNA Repair*, W. H. Freeman, San Francisco.
9. Ames, B. N. (1983). Dietary carcinogens and anticarcinogens, *Science 221*: 1256.
10. Herr, R. R., Jahnke, H. K., and Argoudelis, A. D. (1967). The structure of streptozotocin, *J. Am. Chem. Soc. 89*: 4808.
11. Chan, G. L., Doetsch, P. W., and Haseltine, W. A. (1985). Cyclobutane pyrimidine dimers and (6-4) photoproducts block polymerization by DNA polymerase I, *Biochemistry 24*: 5723.
12. Livneh, Z. (1986). Mechanisms of replication of ultraviolet-irradiated single-stranded DNA by DNA polymerase III holoenzyme of *Escherichia coli, J. Biol. Chem. 261*: 9526.
13. Rupp, W. D., and Howard-Flanders, P. (1968). Discontinuities in the DNA synthesized in an excision defective strain of *Escherichia coli* following ultraviolet irradiation, *J. Mol. Biol. 31*: 291.
14. Wang, T. V., and Smith, K. C. (1986). Postreplicational formation and repair of DNA double-strand breaks in UV-irradiated *Escherichia coli uvrB* cells, *Mutation Res. 165*: 39.
15. Cox, B. S., and Game, J. C. (1974). Repair systems in *Saccharomyces cerevisiae, Mutation Res. 20*: 257.
16. Bailleul, B., Galiegue, S., and Louchex-Lefebvre. M. H. (1981). Adducts from the reaction of 0,0′-diacetyl or 0-acetyl derivatives of the carcinogen 4-hydroxyquinoline 1-oxide with purine residues, *Cancer Res. 41*: 4559.
17. Bainbridge, B. W. (1981). The effects of ultraviolet radiation on the nuclei of yeasts and filamentous fungi, in *The Fungal Nucleus* (K. Gull and S. G. Oliver, eds.), Cambridge University Press, Cambridge, p. 258.
18. Morpugo, G., Bellincampi, D., Gaulandi, G., Baldinelli, L., and Crescenzi, O. S. (1979). Analysis of mitotic non-disjunction with *Aspergillus nidulans, Environ. Health Perspect. 31*: 81.
19. Saffhill, R., Margison, G. P., and O'Connor, P. J. (1985). Mechanisms of carcinogenesis induced by alkylating agents. *Biochim. Biophys. Acta 823*: 111.
20. McCarthy, T. V., and Lindahl, T. (1985). Methylphosphotriesters in alkylated DNA are repaired by the Ada regulatory protein of *Escherichia coli, Nucleic Acids Res. 13*: 2683.
21. Gichner, T., and Veleminsky, J. (1982). Genetic effects of *N*-methyl-*N*′-nitro-nitrosoguanidine and its homologs, *Mutation Res. 99*: 129.
22. Ward, J. F. (1975). Molecular mechanisms of radiation induced damage to nucleic acids, *Adv. Radiat. Biol. 5*: 181.
23. Sonntag, C., von Hagen, U., Schon-Bopp, A., and Schulte-Frohlinde, D. (1981). Radiation induced strand breaks in DNA, in *Advances in Radiation Biology*, vol. 9 (T. J. Ratt, and H. Adler, eds.), Academic Press, New York, p. 109.

24. Szostak, J. W., Orr-Weaver, T. L., Rothstein, R. J., and Stahl, F. W. (1983). The double strand break model for recombination, *Cell 33*: 25.
25. Kelner, A. (1949). Effect of visible light on the recovery of *Streptomyces griseus* conidia from ultraviolet irradiation injury, *Proc. Natl. Acad. Sci. USA 35*: 73.
26. Rupert, C. S., Goodgal, S. H., and Herriot, R. M. (1958). Photoreactivation in vitro of ultraviolet inactivated *Haemophilus influenzae* transforming factor, *J. Gen. Physiol. 41*: 451.
27. Rupert, C. S. (1960). Photoreactivation of transforming DNA by an enzyme from bakers yeast, *J. Gen. Physiol. 43*: 573.
28. Sutherland, B. M. (1981). Photoreactivating enzymes, in *The Enzymes*, vol. XIV (P. D. Boyer, ed.), Academic Press, New York, p. 481.
29. Sancar, A., and Sancar, G. B. (1988). DNA repair enzymes, *Annu. Rev. Biochem. 57*: 29.
30. Sancar, G. B. (1985). Expression of a *Saccharomyces cerevisiae* photolyase gene in *Escherichia coli, J. Bacteriol. 161*: 769.
31. Fortuin, J. J. H. (1971). Another two genes controlling mitotic recombination and recovery from UV damage in *Aspergillus nidulans*. III. Photoreactivation of UV damage in *uvsD* and *uvsE* mutants, *Mutation Res. 13*: 131.
32. Yajima, H., Inou, H., Oikawa, A., and Yasui, A. (1991). Cloning and functional characterization of a eukaryotic DNA photolyase gene from *Neurospora crassa, Nucleic Acids Res. 19*: 5339.
33. Samson, L., and Cairns, J. (1977). A new pathway for DNA repair in *Escherichia coli, Nature (Lond.)267*: 281.
34. Lindahl, T., Demple, B., and Robins, P. (1982). Suicide inactivation of the *E. coli* O^6-methylguanine-DNA-methyltransferase, *EMBO J. 1*: 1359.
35. Lindahl, T. (1982). DNA repair enzymes, *Annu. Rev. Biochem. 51*: 61.
36. Margison, G. P., Cooper, D. P., and Brennand, J. (1985). Cloning of the *E. coli* O^6methylguanine and methylphosphotriester methyltransferase gene using a functional DNA repair assay, *Nucleic Acids Res. 13*: 1939.
37. Demple, B., Sedgwick, B., Robins, P., Totty, N., Wakefield, M. D., and Lindahl, T. (1985). Active site and complete sequence of the suicide methyltransferase that counters alkylation mutagenesis, *Proc. Natl. Acad. Sci. USA 82*: 2685.
38. Teo, I., Sedgwick, B., Kilpatrick, M. W., McCarthy, T. V., and Lindahl, T. (1986). The intracellular signal for induction of resistance to alkylating agents in *Escherichia coli, Cell 45*: 315.
39. Nakamura, T., Tokumoto, Y., Sakumi, K., Koike, G., Nakabeppu, Y., and Sekiguchi, M. (1988). Expression of the *ada* gene of *Escherichia coli* in response to alkylating agents. Identification of transcriptional regulatory elements, *J. Mol. Biol 202*: 483.
40. Potter, P. M., Wilkinson, M. C., Fitton, J., Carr, F. J., Brennand, J., Cooper, D. P., and Margison, G. P. (1987). Characterization and nucleotide sequence of *ogt*, and the O^6-alkyguanine-DNA-alkytransferase gene of *E. coli, Nucleic Acids. Res. 15*: 9177.
41. Morohoshi, F., Hayashi, K., and Munakata, N. (1989). *Bacillus subtilis* gene coding for constitutive O^6-methylguanine-DNA-methyltransferase, *Nucleic Acids Res. 17*: 6531.

42. Frosina, G., and Abbondandolo, A. (1985). The current evidence for an adaptive response to alkylating agents in mammalian cells with special reference to experiments with in vitro cell cultures, *Mutation Res. 154*: 85.
43. Pegg, A. E. (1990). Mammalian O^6-alkylguanine DNA alkytransferase: regulation and importance in response to alkylating carcinogenic and therapeutic agents, *Cancer Res. 50*: 6119.
44. Maga, J. A., and McEntee, K. (1985). Response of *Saccharomyces cerevisiae* to *N*-methyl-*N'*-nitro-nitrosoguanidine: mutagenesis, survival and *DDR* gene expression, *Mol. Gen. Genet. 200*: 313.
45. Sassanfar, M., and Samson, L. (1990). Identification and preliminary characterization of an O^6-methylguanine DNA methyltransferase in the yeast *Saccharomyces cerevisiae, J. Biol. Chem. 265*: 20.
46. Xiao, W., Derfler, B., Chen, J., and Samson, L. (1991). Primary sequence and biological functions of a *Saccharomyces cerevisiae* O^6-methylguanine/O^4-methylthymine DNA repair methyltransferase gene, *EMBO J. 10*: 2179.
47. Hooley, P., Shawcross, S. G., and Strike, P. (1988). An adaptive response to alkylating agents in *Aspergillus nidulans, Current Genet. 14*: 445.
48. Swirsky, R. A., Shawcross, S. G., Faulkner, B. M., and Strike, P. (1988). Repair of alkylation damage in the fungus *Aspergillus nidulans, Mutation Res. 193*: 255.
49. Baker, S. M., Margison, G. P., and Strike, P. (1992). Inducible alkyltransferase DNA repair proteins in the filamentous fungus *Aspergillus nidulans, Nucleic Acids Res. 20*: 645.
50. Setlow, R. B. and Carrier, W. L. (1964). The disappearance of thymine dimers from DNA. An error correcting mechanism, *Proc. Natl. Acad. Sci. USA 51*: 226.
51. Boyce, R. P., and Howard-Flanders, P. (1964). The release of UV-induced thymine dimers from DNA in *E. coli* K12, *Proc. Natl. Acad. Sci. USA 51*: 293.
52. van Houten, B. (1990). Nucleotide excision repair in *Escherichia coli, Microbiol. Rev. 54*: 18.
53. Lindahl, T. (1974). An *N*-glycosylase from *Escherichia coli* that releases free uracil from DNA containing deaminated cytosine residues, *Proc. Natl. Acad. Sci. USA 71*: 3649.
54. Lindahl, T., Ljungquist, S., Siegert, W., Nyberg, B., and Sperens, B. (1977). DNA-glycosidases. Properties of uracil-DNA glycosidase from *Escherichia coli, J. Biol. Chem. 252*: 3286.
55. Sakumi, K., and Sekiguchi, M. (1990). Structures and functions of DNA glycosylases, *Mutation Res. 236*: 161.
56. Duncan, B. K., and Weiss, B. (1982). Specific mutator effects of *ung* (uracil-DNA glycosylase) mutations in *Escherichia coli, J. Bacteriol. 151*: 750.
57. Karran, P., and Lindahl, T. (1978). Enzymatic excision of free hypoxanthine from polydeoxynucleotides and DNA containing deoxyinosine monophosphate residues, *J. Biol. Chem. 253*: 5877.
58. Steinum, A-L., and Seeberg, E. (1986). Nucleotide sequence of the *tag* gene from *Escherichia coli, Nucleic Acids Res. 14*: 3763.
59. Evensen, G., and Seeberg, E. (1982). Adaptation to alkylation resistance involves the induction of a DNA glycosylase, *Nature* (*Lond.*) *296*: 773.

60. Thomas, L., Yang, C-H., and Goldthwait, D. A. (1982). Two DNA glycosylases in *Escherichia coli* which release primarily 3-methyladenine, *Biochemistry 21*: 1162.
61. McCarthy, T. V., Karran, P., and Lindahl, T. (1984). Inducible repair of *O*-alkylated DNA pyrimidines in *Escherichia coli, EMBO J. 3*: 545.
62. Breimer, L. H., and Lindahl, T. (1984). DNA glycosylase activity for thymine residues damaged by ring saturation, fragmentation or ring contraction, are functions of endonuclease III in *E. coli, J. Biol. Chem. 259*: 5543.
63. Breimer, L., and Lindahl, T. (1980). A DNA glycosylase from *Escherichia coli* that releases free urea from a polydeoxyribonucleotide containing fragments of base residues, *Nucleic Acids Res. 8*: 6199.
64. Demple, B., and Linn, S. (1980). DNA *N*-glycosylases and UV repair, *Nature (Lond.) 287*: 203.
65. Chetsanga, C. J., and Lindahl, T. (1979). Release of 7-methyl guanine residues whose imidazole rings have been opened from damaged DNA by DNA glycosylases from *Escherichia coli, Nucleic Acids Res. 6*: 3673.
66. Boiteaux, S., O'Connor, T. R., and Laval, J. (1987). Formamidopyrimidine-DNA glycosylase of *Escherichia coli*: cloning and sequencing of the *fpg* structural gene and over expression of the protein, *EMBO J. 6*: 3177.
67. Chetsanga, C. J., and Frenette, G. P. (1983). Excision of aflatoxin B1-imidazole ring opened guanine adducts from DNA by formamidopyrimidine-DNA glycosylase, *Carcinogenesis 4*: 997.
68. Grafstrom, R. H., Park, L., and Grossman, L. (1982). Enzymatic repair of pyrimidine dimer-containing DNA. A 5′ dimer DNA glycosylase: 3′-apyrimidinic endonuclease mechanism from *Micrococcus luteus, J. Biol. Chem. 257*: 13465.
69. Dodson, M. L., and Lloyd, R. S. (1989). Structure-function studies of the T4 endonuclease V repair enzyme, *Mutation Res. 218*: 49.
70. Franklin, W. A., and Lindahl, T. (1988). DNA deoxyribophosphodiesterase, *EMBO J. 7*: 3617.
71. Crosby, B., Prakash, L., Davis, H., and Hinkle, D. C. (1981). Purification and characterization of a uracil-DNA glycosylase from the yeast *Saccharomyces cerevisiae, Nucleic Acids Res. 9*: 5797.
72. Percival, K. J., Klein, M. B., and Burgers, P. M. J. (1989). Molecular cloning and primary structure of the uracil-DNA glycosylase gene from *Saccharomyces cerevisiae, J. Biol. Chem. 264*: 2593.
73. Impellizeri, K. J., Anderson, B., and Burgers, P. M. J. (1991). The spectrum of spontaneous mutations in *Saccharomyces cerevisiae* uracil-DNA glycosylase mutants limits the function of this enzyme to cytosine deamination repair, *J. Bacteriol. 173*: 6807.
74. Berdal, K. G., Bjoras, M., Bjelland, S., and Seeberg, E. (1990). Cloning and expression in *Escherichia coli* of a gene for an alkylbase DNA glycosylase from *Saccharomyces cerevisiae*: a homologue to the bacterial *alkA* gene, *EMBO J. 9*: 4563.
75. Chen, J., Defler, B., Maskati, A., and Samson, L. (1989). Cloning a eukaryotic DNA glycosylase repair gene by the supression of a DNA repair defect in *Escherichia coli, Proc. Natl. Acad. Sci. USA 86*: 7961.

76. Chen, J., Derfler, B., and Samson, L. (1990). *Saccharomyces cerevisiae* 3-methyladenine DNA glycosylase has homology to the AlkA glycosylase of *E. coli* and is induced in response to DNA alkylation damage, *EMBO J. 9*: 4569.
77. Chen, J., and Samson, L. (1991). Induction of *S. cerevisiae* MAG 3-methyladenine DNA glycosylase transcript levels in response to DNA damage, *Nucleic Acids Res. 19*: 6427.
78. Parkash, L., and Prakash, S. (1977). Isolation and characterization of MMS-sensitive mutants of *Saccharomyces cerevisiae, Genetics 86*: 33.
79. Nisson, P. E., and Lawrence, C. W. (1986). The isolation and characterization of an alkylating-agent-sensitive yeast mutant, *nsgl, Mutation Res. 165*: 129.
80. Johnson, A. W., and Demple, B. (1988). Yeast diesterase for 3′ fragments of deoxyribose: purification and physical properties of a repair enzyme for oxidative damage, *J. Biol. Chem. 263*: 18009.
81. Johnson, A. W., and Demple, B. (1988). Yeast DNA repair 3′-diesterase is the major cellular apurinic/apyrimidinic endonuclease: substrate specificity and kinetics, *J. Biol. Chem. 263*: 18017.
82. Popoff, S. C., Spira, A. S., Johnson, A. W., and Demple, B. (1990). The yeast structural gene (APN1) for the major apurinic endonuclease: homology to *E. coli* endonuclease IV, *Proc. Natl. Acad. Sci. USA 87*: 4193.
83. Ramotar, D., Popoff, S. C., and Demple, B. (1991). Complementation of DNA repair deficient *Escherichia coli* by the yeast Apn1 apurinic/apyrimidinic endonuclease gene, *Mol. Microbiol. 5*: 149.
84. Grossman, L., and Yeung, A. T. (1990). The UvrABC endonuclease of *Escherichia coli, Photochem. Photobiol. 51*: 749.
85. Orren, D. K., and Sancar, A. (1989). The (A)BC exinuclease of *Escherichia coli* has only the UvrB and UvrC subunits in the incision complex, *Proc. Natl. Acad. Sci. USA 86*: 5237.
86. Selby, C. P., and Sancar, A. (1990). Structure and function of the (A)BC exinuclease of *Escherichia coli, Mutation Res. 236*: 203.
87. Haynes, R. H., and Kunz, B. A. (1981). DNA repair and mutagenesis in yeast, in *The Molecular Biology of the yeast* Saccharomyces. *Life Cycle and Inheritance* (J. Strathern, E. W. Jones, and J. R. Broach, eds.), Cold Spring Harbor Laboratory Press, Cold Spring Harbor, NY, p. 371.
88. Friedberg, E. C. (1988). Deoxyribonucleic acid repair in the yeast *Saccharomyces cerevisiae, Microbiol. Rev. 52*: 70.
89. Friedberg, E. C. (1991). Eukaryotic DNA repair: glimpses through the yeast *Saccharomyces cerevisiae, Bioessays 13*: 295.
90. Friedberg, E. C. (1991). Yeast genes involved in DNA repair processes: new looks on old faces, *Mol. Microbiol. 5*: 2303.
91. Friedberg, E. C., Siede, W., and Cooper, A. J. (1991). Cellular responses to DNA damage in yeast, in *The Molecular and Cellular Biology of the yeast* Saccharomyces. *Vol. I. Genome Dynamics, Protein Synthesis and Energetics* (J. Broach, E. Jones, and J. Pringle, eds.), Cold Spring Harbor Laboratory Press, Cold Spring Harbor, NY, p. 147.
92. Subramani, S. (1991). Radiation resistance in *Schizosaccharomyces pombe, Mol. Microbiol. 5*: 2311.

93. Dominski, Z., and Jachymczyk, W. J. (1984). Repair of UV-irradiated plasmid DNA in a *Saccharomyces cerevisiae* rad3 mutant deficient in excision repair of pyrimidine dimers, *Mol. Gen. Genet. 193*: 167.
94. Keszenman-Pereyra, D. (1990). Repair of UV-damaged incoming plasmid DNA in *Saccharomyces cerevisiae, Photochem. Photobiol. 51*: 331.
95. Sung, P., Prakash, S., and Prakash, L. (1988). The RAD3 protein of *Saccharomyces cerevisiae* encodes a DNA dependent ATPase, *Proc. Natl. Acad. Sci. USA 84*: 6045.
96. Harosh, I., Naumovski, L., and Friedberg, E. C. (1989). Purification and characterization of RAD3 ATPase/DNA helicase from *Saccharomyces cerevisiae, J. Biol. Chem. 264*: 20532.
97. Schneider, R., and Schweiger, M. (1991). The yeast DNA repair proteins RAD1 and RAD7 share similar putative functional domains, *FEBS Lett. 283*: 201.
98. Schiestl, R., and Prakash, S. (1990). *RAD10*, an excision repair gene of *Saccharomyces cerevisiae*, is involved in the *RAD1* pathway of mitotic recombination, *Mol. Cell. Biol. 10*: 2485.
99. Schiestl, R., and Prakash, S. (1989). Interactions of the *RAD7* and *RAD23* excision repair genes of *Saccharomyces cerevisiae* with DNA repair genes in different epistasis groups, *Current Genet. 16*: 219.
100. Seeberg, E., Nissen-Meyer, J., and Strike, P. (1976). Incision of ultraviolet-irradiated DNA by extracts of *E. coli* requires three different gene products, *Nature* (*Lond.*) *263*: 524.
101. Wood, R. D., Robins, P., and Lindahl, T. (1988). Complementation of the xeroderma pigmentosum DNA repair defect in cell free extracts, *Cell 53*: 97.
102. Schmidt, H., Kapitzka-Fecke, P., Stephen, E. R., and Gutz, H. (1989). Some of the *swi* genes of *Schizosaccharomyces pombe* also have a function in the repair of radiation damage, *Current Genet. 16*: 89.
103. Aboussekhra, A., Chanet, R., Zgaga, Z., Cassier-Chauvat, C., Heude, M., and Fabre, F. (1989). *RADH*, a gene of *Saccharomyces cerevisiae* encoding a putative DNA helicase involved in DNA repair. Characteristics of *radH* mutants and sequence of the gene, *Nucleic Acids Res. 17*: 7211.
104. McKee, R. H., and Lawrence, C. W. (1979). Genetic analysis of gamma-ray mutagenesis in yeast. I. Reversion in radiation sensitive strains, *Genetics 93*: 375.
105. Jentsch, S., McGrath, J. P., and Vershavsky, A. (1987). The yeast DNA repair gene *RAD6* encodes a ubiquitin-conjugating enzyme, *Nature* (*Lond.*) *329*: 131.
106. Sung, P., Berleth, E., Pickart, C., Prakash, S., and Prakash, L. (1991). Yeast *RAD6* encoded ubiquitin conjugating enzyme mediates protein degradation dependent on the *N*-end-recognising E3 enzyme, *EMBO J. 10*: 2187.
107. Larimer, F. W., Perry, J. R., and Hardigree, A. A. (1989). The *REV1* gene of *Saccharomyces cerevisiae*: isolation, sequence and functional analysis, *J. Bacteriol. 171*: 230.
108. Morrison, A., Christensen, R. B., Alley, J., Beck, A. K., Berstine, E. G., Lemontt, J. F., and Lawrence, C. W. (1989). REV3, A *Saccharomyces cerevisiae* gene whose function is required for mutagenesis, is predicted to encode a non-essential DNA polymerase, *J. Bacteriol. 171*: 5659.

109. Game, J. C., and Mortimer, R. K. (1974). A genetic study of X-ray sensitive mutants in yeast, *Mutation Res. 24*: 281.
110. Malone, R. E., Bullard, S., Hermiston, M., Rieger, R., Cool, M., and Galbraith, A. (1991). Isolation of mutants defective in early steps of meiotic recombination in the yeast *Saccharomyces cerevisiae, Genetics 128*: 79.
111. Hoekstra, M., Liskay, R. M., Ou, A. C., DeMaggio, A. J., Burbee, D., and Heffron, F. (1991). HRR25, a putative protein kinase from budding yeast: association with repair of damaged DNA, *Science 253*: 1031.
112. Alani, E., Subbiah, S., and Kleckner, N. (1989). The yeast *RAD50* gene encodes a predicted 153 kD protein containing a purine nucleotide-binding domain and two large heptad-repeat regions, *Genetics 122*: 47.
113. Alani, E., Padmore, R., and Kleckner, N. (1990). Analysis of wild-type *rad50* mutants of yeast suggests an intimate relationship between meiotic chromosome synapsis and recombination, *Cell 61*: 419.
114. Kans, J. A., and Mortimer, R. K. (1991). Nucleotide sequence of the *RAD57* gene of *Saccharomyces cerevisiae, Gene 105*: 139.
115. Broughton, B. C., Barbet, N., Murray, J., Watts, F., Koken, M. H. M., Lehman, A. R., and Carr, A. M. (1991). Assignment of ten DNA repair genes from *Schizosaccharomyces pombe* to chromosomal NotI restriction fragments, *Mol. Gen. Genet. 228*: 470.
116. Guysler-Junker, A., Bodi, Z., and Kohli, J. (1991). Isolation and characterization of *Schizosaccharomyces pombe* mutants affected in mitotic recombination, *Genetics 128*: 495.
117. Ponticelli, A. S., and Smith, G. R. (1989). Meiotic recombination-deficient mutants of *Schizosaccharomyces pombe, Genetics 123*: 45.
118. Jansen, G. J. O. (1969). Location and designation of some *uvs* loci in *Aspergillus nidulans, Aspergillus Newslett. 10*: 23.
119. Jansen, G. J. O. (1970). Survival of *uvsB* and *uvsC* mutants of *Aspergillus nidulans* after UV-irradiation, *Mutation Res. 10*: 21.
120. Jansen, G. J. O. (1970). Abnormal frequencies of spontaneous mitotic recombination in *uvsB* and *uvsC* mutants of *Aspergillus nidulans, Mutation Res. 10*: 33.
121. Shanfield, B., and Kafer, E. (1969). UV sensitive mutants increasing mitotic crossing over in *Aspergillus nidulans, Mutation Res. 7*: 485.
122. Fortuin, J. J. H. (1971). Another two genes controlling mitotic intragenic recombination and recovery from UV damage in *Aspergillus nidulans.* I. UV sensitivity, complementation and location of six mutants, *Mutation Res. 11*: 149.
123. Fortuin, J. J. H. (1971). Another two genes controlling mitotic intragenic recombination and recovery from UV damage in *Aspergillus nidulans.* II. Recombination behaviour and X-ray sensitivity of *uvsD* and *uvsE* mutants, *Mutation Res. 11*: 265.
124. Fortuin, J. J. H. (1971). Another two genes controlling mitotic recombination and recovery from UV damage in *Aspergillus nidulans.* IV. Genetic analysis of mitotic intragenic recombinants from uvs^+/uvs^+, *uvsD/uvsD* and *uvsE/uvsE* diploids, *Mutation Res. 13*: 137.

125. Wright, P. J., and Pateman, J. A. (1970). Ultraviolet-light sensitive mutants of *Aspergillus nidulans, Mutation Res. 9*: 579.
126. Kafer, E., and Mayor, O. (1986). Genetic analysis of DNA repair in *Aspergillus nidulans*: evidence for different types of MMS-sensitive hyperrec mutants, *Mutation Res. 161*: 119.
127. Parag, Y., and Parag, G. (1975). Mutations affecting mitotic recombination frequency in haploids and diploids of the filamentous fungus *Aspergillus nidulans, Mol. Gen. Genet. 137*: 319.
128. Parag, Y. (1977). Genetic analysis of mutations for low (rec) and very high (pop) mitotic recombination frequency in *Aspergillus nidulans, Mol. Gen. Genet. 155*: 319.
129. Oza, K., and Kafer, E. (1990). Cloning of the DNA repair gene, *uvsF*, by transformation of *Aspergillus nidulans, Genetics 125*: 341.
130. Babudri, N., and Politi, M. G. (1989). Different action of MMS and EMS in UV-sensitive strains of *Aspergillus nidulans, Mutation Res. 217*: 211.
131. Babudri, N., and Morpurgo, G. (1988). An uvsB mutant of *Aspergillus* nidulans with high variable spontaneous mutation and intergenic recombination frequencies, *Mutation Res. 199*: 167.
132. Osman, F. (1991). Studies on DNA repair and recombination in the fungus *Aspergillus nidulans,* Ph.D. thesis, University of Liverpool.
133. Osman, F., Cotton, C. M., Tomsett, A. B., and Strike, P. (1991). Isolation and characterization of *nuv11*, a mutation affecting meiotic and mitotic recombination in *Aspergillus nidulans, Biochimie 73*: 321.
134. Osman, F., Tomsett, A. B., and Strike, P. (1992). The effect on spontaneous mitotic intragenic recombination in *Aspergillus nidulans* of mutations isolated as hypersensitive to DNA damaging agents (in preparation).
135. Schroeder, A. L. (1975). Genetic control of radiation sensitivity and DNA repair in *Neurospora*, in *Molecular Mechanisms for Repair of DNA, Part B* (P. C. Hanawalt and R. B. Setlow, eds.), Plenum Press, New York, p. 567.
136. Delange, A. M., and Mishra, N. C. (1981). Characterization of MMS-sensitive mutants of *Neurospora crassa, Genetics 97*: 167.
137. Kafer, E. (1981). Mutagen sensitivities and mutator effects of MMS-sensitive mutants of *Neurospora crassa, Mutation Res. 80*: 43.
138. Kafer, E., and Luk, D. (1988). Properties and strains of additional DNA repair-defective mutants in known and new genes of *Neurospora crassa, Mutation Res. 217*: 75.
139. Kafer, E., and Luk, D. (1989). Sensitivity to bleomycin and hydrogen peroxide of DNA repair-defective mutants in *Neurospora crassa, Mutation Res. 217*: 75.
140. Inoue, H., and Schroeder, A. L. (1988). A new mutagen-sensitive mutant in *Neurospora, mus-16, Mutation Res. 194*: 9.
141. Ishikawa, T., Toh-e, A., Uno, I., and Hasunuma, K. (1969). Isolation and characterization of nuclease mutants in *Neurospora crassa, Genetics 63*: 75.
142. Kafer, E., and Fraser, M. J. (1979). Isolation and genetic analysis of nuclease halo mutants (*nuh*) of *Neurospora crassa, Mol. Gen. Genet. 169*: 117.

143. Newmayer, D., and Galeazzi, D. R. (1978). A meiotic UV-sensitive mutant which causes deletion of duplications in *Neurospora, Genetics 89*: 245.
144. Schroeder, A. L., and Olsen, L. D. (1983). Mutagen sensitivity of *Neurospora* mitotic mutants, *Can. J. Genet. Cytol. 25*: 16.
145. Newmayer, D., and Galeazzi, D. R. (1977). The instability of *Neurospora* duplication Dp(Il-Ir)H4250 and its genetic control, *Genetics 85*: 461.
146. Schroeder, A. L. (1986). Chromosome instability in mutagen sensitive mutants of *Neurospora, Current Genet. 10*: 381.
147. Tuveson, R. W. (1972). Interaction of genes controlling ultraviolet sensitivity in *Neurospora crassa, J. Bacteriol. 112*: 632.
148. Kafer, E., and Perlmutter, E. (1980). Isolation and genetic analysis of MMS-sensitive mutants of *Neurospora, Can. J. Genet. Cytol. 22*: 535.
149. Inoue, H., Harvey, R. C., Callen, D. F., and de Serres, F. J. (1981). Mutagenesis at the *ad-3A* and *ad-3B* loci in haploid UV sensitive strains of *Neurospora crassa*. V. Comparison of dose response curves of single and double mutant strains with wild-type, *Mutation Res. 84*: 49.
150. Kafer, E. (1983). Epistatic grouping of repair deficient mutants in *Neurospora*: comparative analysis of two *uvs-3* alleles, *uvs-6*, and their *mus* double mutant strains, *Genetics 105*: 19.
151. Ishii, C., and Inoue, H. (1989). Epistasis, photoreactivation and mutagen sensitivity of DNA repair mutants *upr-1* and *mus-26* in *Neurospora crassa, Mutation Res. 218*: 95.
152. Worthy, T. E., and Epler, J. L. (1973). Biochemical basis of radiation sensitivity in mutants of *Neurospora crassa, Mutation Res. 19*: 167.
153. MacLeod, H., and Stadler, D. (1986). Excision of pyrimidine dimers from the DNA of *Neurospora, Mol. Gen. Genet. 202*: 321.
154. Baker, T. I., Cords, C. E., Howard, C. A., and Radloff, R. J. (1990). The nucleotide excision repair epistasis group in *Neurospora crassa, Current Genet. 18*: 207.
155. Brockman, H. E., de Serres, F. J., Hung, C. Y., and Ong, T. (1989). Effect of the *uvs*-2 allele of *Neurospora crassa* on the mutagenic potency of 2-*N*-hydroxylaminopurines and 2-aminopurine in the *ad-3* forward mutation test, *Mutation Res. 218*: 1.
156. Calza, R. E., and Schroeder, A. L. (1982). Postreplication repair in *Neurospora crassa, Mol. Gen. Genet. 185*: 111.
157. Resnick, M. A., Boyce, J., and Cox, B. (1981). Postreplication repair in *Saccharomyces cerevisiae, J. Bacteriol. 146*: 285.
158. Kafer, E. (1984). UV-induced recessive lethals in uvs strains of *Neurospora* which are deficient in UV mutagenesis, *Mutation Res. 128*: 137.
159. Tanaka, S., Ishii, C., and Inoue, H. (1989). Effects of heat shock on the induction of mutations by chemical mutagens in *Neurospora crassa, Mutation Res. 223*: 233.
160. Howard, C. A., and Baker, T. I. (1986). Identification of DNA repair and damage induced proteins from *Neurospora crassa, Mol. Gen. Genet. 203*: 462.
161. Howard, C. A., and Baker, T. I. (1989). Inducible responses to DNA damaging or stress inducing agents in *Neurospora crassa, Current Genet. 15*: 47.

162. Holliday, R. (1965). Radiation sensitive mutants of *Ustilago maydis, Mutation Res. 2*: 557.
163. Holliday, R. (1967). Altered recombination frequencies in radiation-sensitive strains of *Ustilago maydis, Mutation Res. 4*: 275.
164. Holliday, R., Halliwell, R. E., Evans, M. W., and Rowell, V. (1976). Genetic characterization of *rec-1*, a mutant of *Ustilago maydis* defective in repair and recombination, *Genet. Res. Camb. 27*: 413.
165. Holliday, R., Taylor, S. Y., Kmiec, E. B., and Holloman, W. K. (1984). Biochemical characterization of *rec-1* mutants and genetic control of recombination in *Ustilago maydis, Cold Spring Harbor Symp. Quant. Biol. 49*: 669.
166. Holden, D. W., Spanos, A., and Banks, G. R. (1989). Nucleotide sequence of the *REC1* gene of *Ustilago maydis, Nucleic Acids Res. 17*: 10489.
167. Holden, D. W., Spanos, A., Kanuga, N., and Banks, G. R. (1991). Cloning the *REC1* gene of *Ustilago maydis, Current Genet. 20*: 145.
168. Kmiec, E., and Holloman, W. K. (1982). Homologous pairing of DNA molecules promoted by a protein from *Ustilago, Cell 29*: 367.
169. Kmiec, E., and Holloman, W. K. (1983). Heteroduplex formation and polarity during strand transfer promoted by *Ustilago rec-1* protein, *Cell 33*: 857.
170. Bauchwitz, R., and Holloman, W. K. (1990). Isolation of the *REC2* gene controlling recombination in *Ustilago maydis, Gene 96*: 285.
171. Katz, K. S., and Ratner, D. I. (1988). Homologous recombination and the repair of double strand breaks during co-transformation of *Dictyostelium discoideum, Mol. Cell. Biol. 8*: 2779.
172. Hanawalt, P. C. (1991). Heterogeneity of DNA repair at the gene level, *Mutation Res. 247*: 203.
173. Bohr, V. A., Phillips, D. H., and Hanawalt, P. C. (1987). Heterogenous DNA damage and repair in the mammalian genome, *Cancer Res. 47*: 6426.
174. Mellon, I., Spivak, G., and Hanawalt, P. C. (1987). Selective removal of transcription-blocking DNA damage from the transcribed strand of the mammalian DHFR gene, *Cell 51*: 241.
175. Terleth, C., van Sluis, C. A., and van de Putte, P. (1989). Differential repair of UV damage in *Saccharomyces cerevisiae, Nucleic Acids Res. 17*: 4433.
176. Terleth, C., Waters, R., Brouwer, J., and van de Putte, P. (1990). Differential repair of UV damage in *Saccharomyces cerevisiae* is cell cycle dependent, *EMBO J. 9*: 2899.
177. Terleth, C., Schenk, P., Poot, R., Brouwer, J., and van de Putte, P. (1990). Differential repair of UV damage in rad mutants of *Saccharomyces cerevisiae*: a possible function of G2 arrest upon UV irradiation, *Mol. Cell. Biol. 10*: 4678.
178. Weinert, T. A., and Hartwell, L. H. (1988). The RAD9 gene controls the cell cycle response to DNA damage in *Saccharomyces cerevisiae, Science 241*: 317.
179. Smerdon, M. J., and Thoma, F. (1990). Site specific DNA repair at the nucleosome level in a yeast minichromosome, *Cell 61*: 675.
180. Peterson, T. A., Prakash, L., Prakash, S., Osley, M. A., and Reed, S. I. (1985). Regulation of *CDC9*, the *Saccharomyces cerevisiae* gene that encodes DNA ligase, *Mol. Cell. Biol. 5*: 226.

181. Siede, W., Robinson, G. W., Kalainov, D., Malley, T., and Friedberg, E. C. (1989). Regulation of the *RAD2* gene of *Saccharomyces cerevisiae, Mol. Microbiol. 3*: 1696.
182. Madura, K., Prakash, S., and Prakash, L. (1990). Expression of the *Saccharomyces cerevisiae* DNA repair gene *RAD6* that encodes a ubiquitin conjugating enzyme, increases in response to DNA damage and in meiosis, but remains constant during the mitotic cell cycle, *Nucleic Acids Res. 18*: 771.
183. Jones, J. S., Prakash, L., and Prakash, S. (1990). Regulated expression of the *Saccharomyces cerevisiae* DNA repair gene *RAD7* in response to DNA damage and during sporulation, *Nucleic Acids Res. 18*: 3281.
184. Madura, K., and Prakash, S. (1990). Transcript levels of the *Saccharomyces cerevisiae* DNA repair gene *RAD23* increase in response to UV light and in meiosis, but remain constant in the mitotic cell cycle, *Nucleic Acids Res. 18*: 4737.
185. Madura, K., and Prakash, S. (1990). The *Saccharomyces cerevisiae* DNA repair gene *RAD2* is regulated in meiosis but not during the mitotic cell cycle, *Mol. Cell. Biol. 10*: 3256.
186. Cole, G. M., Schild, D., and Mortimer, R. K. (1989). Two DNA repair and recombination genes in *Saccharomyces cervisiae, RAD52* and *RAD54*, are induced during meiosis, *Mol. Cell. Biol. 9*: 3101.
187. Jones, J. S., and Prakash, L. (1991). Transcript levels of the *Saccharomyces cerevisiae* DNA repair gene RAD18 increase in UV irradiated cells and during meiosis but not during the mitotic cell cycle, *Nucleic Acids Res. 19*: 893.
188. Walker, G. C. (1984). Mutagenesis and inducible responses to deoxyribonucleic acid damage in *Escherichia coli, Microbiol. Rev. 48*: 60.
189. Angulo, J. F., Schwencke, J., Moreau, P. L., Moustachi, E., and Devoret, R. (1985). A yeast protein analogous to *Escherichia coli* RecA protein whose cellular level is enhanced after UV-irradiation, *Mol. Gen. Genet. 201*: 20.
190. Madura, K., and Prakash, S. (1986). Nucleotide sequence, transcript mapping and regulation of the RAD2 gene of *Saccharomyces cerevisiae, J. Bacteriol. 166*: 914.
191. Robinson, G. W., Nicolet, C. M., Kalainov, D., and Frieberg, E. C. (1986). A yeast excision repair gene is inducible by DNA damaging agents, *Proc. Natl. Acad. Sci. USA 83*: 1842.
192. Cole, G. M., Schild, D., Lovett, S. T., and Mortimer, R. K. (1987). Regulation of *RAD54*- and *RAD52-lacZ* fusions in *Saccharomyces cerevisiae* in response to DNA damage, *Mol. Cell. Biol. 7*: 1078.
193. Ruby, S. W., and Szostak, J. W. (1985). Specific *Saccharomyces cerevisiae* genes are expressed in response to DNA damaging agents, *Mol. Cell. Biol. 5*: 75.
194. McClanahan, T., and McEntee, K. (1984). Specific transcripts are elevated in *Saccharomyces cerevisiae* in response to DNA damage, *Mol. Cell. Biol. 4*: 2356.
195. Johnston, L. H., White, J. H. M., Johnson, A. L., Lucchini, G., and Plevani, P. (1987). The yeast DNA polymerase I transcript is regulated in both the mitotic cell cycle and in meiosis, and is also induced after DNA damage, *Nucleic Acids Res. 15*: 5017.

196. Treger, J. M., Heichman, K. A., and McEntee, K. (1988). Expression of the yeast *UBI4* gene increases in response to DNA-damaging agents and in meiosis, *Mol. Cell. Biol. 8*: 1132.
197. Elledge, S. J., and Davis, R. W. (1987). Identification and isolation of the gene encoding the small sub-unit of ribonucleotide reductase from *Saccharomyces cerevisiae*: DNA damage inducible gene required for mitotic viability, *Mol. Cell. Biol. 7*: 2783.
198. Bardwell, L., Burtscher, H., Weiss, W. A., Nicolet, C. M., and Friedberg, E. C. (1990). Characterization of the *RAD10* gene of *Saccharomyces cerevisiae* and purification of RAD10 protein, *Biochemistry 29*: 3119.
199. Naumovski, L., Chu, G., Berg, P., and Friedberg, E. C. (1985). RAD3 gene of *Saccharomyces cerevisiae*: nucleotide sequence of wild-type and mutant alleles, transcript mapping and aspects of gene regulation, *Mol. Cell. Biol. 5*: 17.
200. Lee, M. G., and Nurse, P. (1987). Complementation used to clone a human analogue of the fission yeast cell cycle control gene *cdc2, Nature* (*Lond.*) *327*: 31.
201. Lambert, C., Couto, L. B., Weiss, W. A., Schultz, R. A., Thompson, L. H., and Friedberg, E. C. (1988). A yeast DNA repair gene partially complements defective excision repair in mammalian cells, *EMBO J. 7*: 3245.
202. Brozmanova, J., Kleibl, K., Vickova, V., Skorvaga, M., Cernakova, L., and Margison, G. P. (1990). Expression of the *E. coli ada* gene in yeast protects against the toxic and mutagenic effects of *N*-methyl-*N'*-nitro-*N*-nitrosoguanidine, *Nucleic Acids Res. 18*: 331.
203. Brozmanova, J., Cernakova, L., Vickova, V., Duraj, J., and Fridrichova, I. (1991). The *Escherichia coli recA* gene increases resistance of the yeast *Saccharomyces cerevisiae* to ionizing and ultraviolet radiation, *Mol. Gen. Genet. 227*: 473.
204. Hoeijmakers, J. H. J., and Bootsma, D. (1990). Molecular genetics of eukaryotic DNA excision repair, *Cancer Cells 2*: 311.
205. Hamilton, K. K., Kim, P. M. H., and Doetsch, P. W. (1992). A eukaryotic DNA glycosylase/lyase recognizing ultraviolet light-induced pyrimidine dimers, *Nature* (*Lond.*) *356*: 725.
206. Chang, D-Y., and Lu, A-L. (1991). Base mismatch-specific endonuclease activity in extracts from *Saccharomyces cerevisiae, Nucleic Acids Res. 19*: 4761.

Index